PHYSICS WITH TRAPPED CHARGED PARTICLES

Lectures from the Les Houches Winter School

PHYSICS WITH TRAPPED CHARGED PARTICLES

Lectures from the Les Houches Winter School

Editors

Martina Knoop
CNRS & Université d'Aix-Marseille, France

Niels Madsen
Swansea University, UK

Richard C Thompson
Imperial College London, UK

ICP

Imperial College Press

Published by

Imperial College Press
57 Shelton Street
Covent Garden
London WC2H 9HE

Distributed by

World Scientific Publishing Co. Pte. Ltd.
5 Toh Tuck Link, Singapore 596224
USA office: 27 Warren Street, Suite 401-402, Hackensack, NJ 07601
UK office: 57 Shelton Street, Covent Garden, London WC2H 9HE

British Library Cataloguing-in-Publication Data
A catalogue record for this book is available from the British Library.

PHYSICS WITH TRAPPED CHARGED PARTICLES
Lectures from the Les Houches Winter School

ISBN 978-1-78326-404-9
ISBN 978-1-78326-405-6 (pbk)

Printed in Singapore by Mainland Press Pte Ltd.

Preface

The content of this book is based on the Winter School on Physics with Trapped Charged Particles hosted by the École de Physique des Houches in January 2012. The Winter School was inspired by the presence of numerous regular conferences on subjects including various aspects of physics with trapped charged particles. In spite of the high level of activity within the field, no appropriate training ground seemed to exist for physics students and early career researchers that would cover the various aspects of the physics of this versatile area. Hence the idea for the Winter School arose.

Many schools of physics consist solely of a (large) number of lectures covering various aspects of the chosen field. Having ourselves experienced such schools or had students attend such schools we felt it necessary to modify this approach for the school in hand. Therefore, rather than model the school on other schools in directly related fields, we modelled it on the long lasting and very successful series of CERN accelerator schools, that mix lectures and tutorial sessions for an improved dissemination of the subject matter. The feedback from the students at the end of the school supports the success of this approach, as the students in fact requested even more "proper" physics lectures as well as more, or at least more elaborate, tutorial sessions. The students clearly recognized and appreciated the lecturers who made an effort to teach, rather than lecture, and this book is meant to follow up on their effort by letting the lecturers put into writing what they went through with the students at the school. The chapters in this book are thus not meant to be new material, but to be reviews targeted at people new to the fields covered.

We would like to acknowledge the international advisory team with Professor Klaus Blaum, Dr. Fred Curell, Professor Michael Drewsen and Professor Piet O. Schmidt, without whom the school would not have had the same complete program, as well as our co-organiser Professor Joel Fajans, who secured a solid team of experts to give the students a good foundation in the physics of trapped charged particles. Finally a big thanks goes to the team in Les Houches for their impeccable organization, and to our secretary Audrey Deidda who helped manage the many payments and other exchanges with the participants.

Niels Madsen, Martina Knoop and Richard Thompson

Group photo of the 2012 Winter School on Physics with Trapped Charged Particles in Les Houches, France.

Program : 4.1.2012

Physics with Trapped Charged Particles

Les Houches, France, 9 January - 20 January, 2012

Time	Monday 9 January	Tuesday 10 January	Wednesday 11 January	Thursday 12 January	Friday 13 January	Saturday 14 January	Sunday 15 January	Monday 16 January	Tuesday 17 January	Wednesday 18 January	Thursday 19 January	Friday 20 January
08:00		BREAKFAST										
08:45 – 09:45		Single particle. dynamics Basic Equilib. D. DUBIN	Modes in Non-neutral plasmas F. ANDEREGG	Dynamics in Paul Traps B. ODOM	Strong Magnetization and Plasma... D. DUBIN	Cooling Techniques D. SEGAL	D	Coulomb Crystals DREWSEN	Storage Rings A. PAPASH	QIP in ion traps C. ROOS	Quantom logic with trapped ions P. SCHMIDT	Multipolar RF Traps R. WESTER
		COFFEE						COFFEE				
10:15 – 11:05		Penning Traps. Strong correlations J. BOLLINGER	Rotating Wall Cent. Separt. J. BOLLINGER	Internal Transport F. ANDEREGG	Recombination in traps ROBICHEAUX	QED Tests with highly charged ions K. BLAUM	A Y	Cooling Techniques II WUNDERLICH	EBIS F. CURELL	Beam Dynamics A. PAPASH	Trapped Molecules R. WESTER	Novel Traps J. GALIANA
11:15 – 12:05		Non-Destructive Diagnostics J. CRESPO	Positron sources, accumulators and plasmas C. SURKO	Tailoring plasmas and beams C. SURKO	Laser based diagnostics M. KNOOP	2D Fluid motion J. FAJANS	O F	Axialisation D. SEGAL	Toroids T. PEDERSEN	Microtraps and QIP MEHLSTAUBLER	Circuit QED J. GALIANA	Multifaceted entanglement C. ROOS
										LUNCH (early)		
12:30		LUNCH					F	LUNCH			LUNCH	
16:00 – 16:50	A R			Tutorial 3 Paul Traps B. ODOM					Quantum Simulations WUNDERLICH	V I S		D E
17:00 – 17:50	R I	Antihydrogen Trapping N. MADSEN	Autoresonance J. FAJANS	Highly charged ions in traps J. CRESPO	Combined Traps (Neutral+Charged) ROBICHEAUX	Tutorial 4 DUBIN ROBICHEAUX FAJANS		Magnetically confined charged particles T. PEDERSEN	EBIT F. CURELL	I T C	Thorium Spectroscopy MEHLSTAUBLER	P A
	V	TEA						TEA		E	TEA	R
18:10 – 19:00	A L	The g-2 exp. on free electron B. ODOM	Destructive Diagnostics M. KNOOP	Precision Mass Measurements K. BLAUM	Traps for radioactive ions F. HERFURTH	Ion Bunching F. HERFURTH		Atomic Clocks in Ion Traps H. MARGOLIS	Quantum optics with ion crystals M. DREWSEN	R N	Fundamental Constants P. SCHMIDT	T U
19:00		Welcome Drink										
19:30		DINNER										R
20:40 – 21:30		Tutorial 1 Penning Traps DUBIN J. BOLLINGER	Poster Session 1	Tutorial 2 NNP ANDEREGG/ DUBIN				Tutorial 5 Cooling Tech. SEGAL / WUNDERLICH	Poster Session 2		Tutorial 7 QIP in ion traps ROOS	E

Timekeepers:

	Mon 9	Tue 10	Wed 11	Thu 12	Fri 13	Sat 14	Sun 15	Mon 16	Tue 17	Wed 18	Thu 19	Fri 20
Morning :		MADSEN	DUBIN	KNOOP	BLAUM	ROBICHEAUX		KNOOP	THOMPSON	WUNDERLICH	MEHLSTAUBLER	THOMPSON
Evening :		FAJANS	SURKO	ANDEREGG	BOLLINGER	SEGAL		DREWSEN	MAGOLIS		WESTER	

Timetable of the Winter School on Physics with Trapped Charged Particles in Les Houches, France.

Contents

Chapter 1

Physics with Trapped Charged Particles

Martina Knoop[1], Niels Madsen[2] and Richard C. Thompson[3]

[1] *CNRS and Université d'Aix-Marseille,*
Centre de Saint Jérôme, Case C21,
13397 Marseille Cedex 20, France
Martina.Knoop@univ-amu.fr

[2] *Department of Physics, College of Science, Swansea University,*
Swansea SA2 8PP, United Kingdom
n.madsen@swansea.ac.uk

[3] *Department of Physics, Imperial College London,*
London SW7 2AZ, United Kingdom
r.thompson@imperial.ac.uk

Ion traps, which were first introduced in the late 1950s and early 1960s, have established themselves as indispensable tools in many areas of physics, chemistry and technology. This chapter gives a brief survey of the operating principles and development of ion traps, together with a short description of how ions are loaded and detected. This is followed by a brief account of some of the current applications of ion traps.

1.1. Introduction

When Wolfgang Paul and Hans Dehmelt were developing the first ion traps in the late 1950s and early 1960s, it is unlikely that they expected to receive Nobel Prizes for this work in 1989. In 2012 another Nobel Prize was awarded in the area of ion traps, to David Wineland. These awards demonstrate how important ion traps have become in the 50 years since their introduction.

The Winter School on "Physics with Trapped Charged Particles" was held at the Les Houches centre in France from 9–20 January 2012. More than 20 speakers gave lectures on a wide variety of topics including the basic principles of ion traps (including storage rings); techniques for cooling, manipulating and detecting the ions; and highly specialized applications such as precision mass measurements and quantum information processing.

Students at the School, including PhD students and postdoctoral researchers, benefited from tutorials with the lecturers as well as the more formal lectures, and were able to have extended discussions with the lecturers outside the timetabled sessions.

This book includes lecture notes from many of the lecturers at the School, though not all were able to contribute. The organizers would like to express their thanks to all the lecturers who have contributed to this review volume, which we hope will be a useful resource for those who attended the School, as well as to newcomers starting out in this exciting field.

This chapter does not attempt to give a thorough review of ion traps; rather, it is to be seen as a short introduction to the subject, giving a flavour of the physics of ion traps and their applications. No attempt is made to give full references, especially as the other chapters in the book give much more detail than can be covered here. For details of the background to ion traps and general techniques that are used for trapped ions, the reader should refer to the books by Werth and collaborators[1,2] and Ghosh.[3]

1.2. History of Ion Traps

The two main types of ion trap (the Paul, or radiofrequency, trap and the Penning trap) were both introduced at about the same time. Wolfgang Paul and his group in Bonn developed the three-dimensional Paul trap from a linear quadrupole mass filter, which can be regarded as a two-dimensional ion trap.[4] The Paul trap established the standard three-electrode structure for ion traps, consisting of a ring electrode and two endcap electrodes which generate the required electric field for trapping (see Section 1.3). Early work with the Paul trap included studies of the lifetime of ions held in the trap and spectroscopic measurements of hyperfine splittings in simple ions. A different research group developed a trap for charged macroscopic dust particles that operated on exactly the same principles as Paul's trap, but in a completely different parameter regime.[5] In this trap it was possible to observe the trajectories of the charged particles directly by using a camera.

The Penning trap, which makes use of a magnetic field for trapping, was reported a little later by a number of groups working independently, including Hans Dehmelt at the University of Washington,[6] and groups in Edinburgh,[7] Bonn[8] and Moscow.[9] It was named the Penning trap by Dehmelt in recognition of the fact that Frans Penning had (in 1936) reported the application of a magnetic field to an electrical discharge to prolong the lifetime of electrons due to the confining effect of the field.[10] The Penning trap was also used for a number of different types of studies, but in particular it was used for precision measurements of hyperfine splittings using techniques such as microwave-optical double resonance, and for measurements of the g-factor of the electron.

1.3. Principles of Ion Traps

All ion traps work by confining the motion of charged particles using electric and magnetic fields. The creation of a static three-dimensional potential well, which would be ideal for trapping, is forbidden by Earnshaw's theorem. Using static electric fields, it is only possible to create a saddle point in the potential. In the simplest case, a quadratic potential is created using three electrodes, as shown in Figure 1.1.

1.3.1. The Penning trap

The Penning trap[6] makes use of a static quadratic potential well along the axis joining the two endcap electrodes (the z-axis). This is created by putting a positive potential (V) on the endcaps relative to the ring (for positively charged particles). The resulting simple harmonic motion in the axial direction has an angular frequency of $\omega_z = (4qV/md^2)^{1/2}$, where q and m are the charge and mass of the ion respectively and d is a distance related to the separation of the endcaps ($2z_0$) and the diameter of the ring

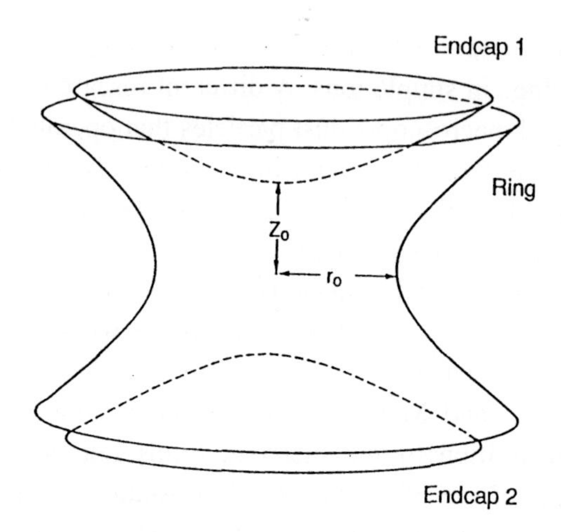

Fig. 1.1. Electrodes of an ion trap.

electrode ($2r_0$) by the equation $d^2 = r_0^2 + 2z_0^2$. Typical values of d range from a few mm to a few cm, depending on the application.

The potential in the radial plane also varies quadratically as a function of x and y, but the potential has a maximum at $x=y=0$, making the radial motion unstable. This is overcome by applying a strong magnetic field along z. The magnetic field gives rise to a cyclotron-type motion of the ions at an angular frequency close to $\omega_c = qB/m$, which is the normal cyclotron frequency for a particle moving in a magnetic field of magnitude B. Its value is shifted down slightly by the presence of the radial electric field, which also gives rise to a second, slower, circular motion in the radial plane called the magnetron motion (at an angular frequency ω_m). The modified cyclotron frequency and the magnetron frequency are given by

$$\omega_c' = \frac{\omega_c}{2} + \sqrt{\frac{\omega_c^2}{4} - \frac{\omega_z^2}{2}} \tag{1.1}$$

$$\omega_m = \frac{\omega_c}{2} - \sqrt{\frac{\omega_c^2}{4} - \frac{\omega_z^2}{2}}. \tag{1.2}$$

The total energy associated with the magnetron motion is negative, due to its low kinetic energy, combined with the negative radial potential energy, and this gives rise to several problems associated with the stability of ions in the trap. It also makes simultaneous cooling of all three degrees of freedom difficult.

Penning traps have been used for many different types of experiment, often using high magnetic fields in the range of 1–10 tesla from superconducting magnets. For trapped electrons this gives cyclotron frequencies up to tens of GHz. For protons the cyclotron frequencies are in the range of tens of MHz and for singly charged atomic ions they are typically of the order of 1 MHz. The axial frequency is restricted to values lower than $\omega_c/\sqrt{2}$ as otherwise the argument of the square root in Eq. 1.1 and Eq. 1.2 becomes negative.

Space charge limits the maximum number density of particles in the trap to values of the order of $10^8 - 10^9$ cm^{-3} for atomic ions.

Penning traps also come in so-called open versions, referred to as cylindrical Penning traps or Penning–Malmberg traps, where there are no endcaps as such, but rather a number of co-axial cylindrical electrodes to generate the axially confining electric fields. These traps, while less suitable for precision measurements, offer increased versatility and facility of loading and unloading particles and are used in particular for studies of non-neutral plasmas, and for antimatter experiments.

1.3.2. The radiofrequency (RF) Paul trap

The RF trap[4] makes use of a dynamic trapping mechanism to overcome the instability associated with a static electric field. As was pointed out above, the saddle point from a positive potential applied to the endcaps gives stable motion in the axial direction and unstable motion in the radial direction. However, if the potential is reversed the radial motion is now stable and the axial motion unstable. The equations of motion for an oscillating potential (at angular frequency Ω) are examples of the well-known Mathieu equation. There are solutions that are stable in both the axial and radial directions, for a particular range of combinations of voltages and oscillation frequencies (at a given value of q/m). The final motion is then an oscillation in an effective potential that arises as a result of the driven motion. This so-called *pseudopotential* has a three-dimensional quadratic minimum at the centre of the trap and gives rise to simple harmonic motion in both the axial and radial directions. If the applied potential is $V_{dc} + V_{ac}\cos(\Omega t)$ then we find (in a given parameter range[3]) the approximate oscillation frequencies

$$\omega_z = \frac{\Omega}{2}\sqrt{a_z + \frac{1}{2}q_z^2} \tag{1.3}$$

$$\omega_r = \frac{\Omega}{2}\sqrt{a_r + \frac{1}{2}q_r^2}, \tag{1.4}$$

where the stability parameters a_z and q_z are given by

$$a_z = -2a_r = -16qV_{dc}/m\Omega^2 d^2 \tag{1.5}$$

$$q_z = 2q_r = 8qV_{ac}/m\Omega^2 d^2. \tag{1.6}$$

Typically, the parameters are chosen such that the value of q_z is in the region of 0.4 and a_z is much less than q_z. Under these conditions the radial and axial frequencies are typically one-tenth of the applied frequency. The typical size of a trap designed for use with large clouds of ions is a few mm to a few cm.

Some experiments with small numbers of trapped atomic ions are carried out in miniature RF traps with electrode separations in the region of 1 mm. These traps are operated with applied frequencies of the order of a few MHz and a few hundred volts amplitude.

One consequence of the trapping mechanism in the RF trap is that there is always a small-amplitude driven motion at the applied frequency Ω. This is referred to as the *micromotion.* (The only exception to this is if an ion is located exactly at the centre of the trap where the fields are zero.) The displacement from the centre can give rise to heating effects in some situations and also the micromotion can give rise to sidebands or broadening in the spectra of trapped ions.

One particularly important feature of the RF trap is its ability to reach the Lamb–Dicke regime.[11] This arises in optical spectroscopy, when the ion is better localized than a fraction of the wavelength of the interrogating laser light. As a consequence, the first order Doppler effect is eliminated and is replaced by a small number of sidebands separated by the oscillation frequency of the ion. This allows ultrahigh-resolution optical spectroscopy of forbidden transitions to be carried out. For optical transitions, this parameter regime can only be reached in a micromotion-free zone and is the indispensable condition for many experiments in frequency metrology or quantum information.

1.3.3. The linear RF trap

A variant of the three-dimensional RF trap is the linear trap, where electrodes are again used to create an oscillating quadratic electrical potential. However, in this case the electrodes are four parallel rods and the oscillating potential is applied between the rods, with opposite pairs being connected together. If we take the z-axis as the direction parallel to the rods, then the RF potential gives confinement in the x and y directions. In order to provide axial confinement as well, a direct current (DC) potential well is created, typically by applying positive voltages to two endcap electrodes at the ends of the rods.

The radial confinement strength is similar to that obtained in a conventional Paul trap, except that this type of trap is easier to make much smaller. The oscillation frequencies can therefore be much higher, around 10 MHz in some cases. The axial confinement depends only on the effective potential along the z-axis and is usually weaker than the radial one.

Linear RF traps have the advantage that they have a line where the micromotion is eliminated, rather than just a single point. They are therefore often used in experiments where a small number of ions are required to be free of micromotion, for example in quantum information processing applications. For this type of application they are often manufactured using microfabrication techniques, giving typical electrode separations in the region of a few hundred μm.

1.3.4. Low-energy storage rings

An alternative way to store charged particles is to let them circulate in a storage ring. While storage rings are usually large machines intended for high-energy physics experiments, the last two decades or so have seen many machines built for low-energy applications.[12]

In a storage ring, or circular accelerator, particles are confined radially (i.e. perpendicular to their direction of motion) by alternating quadrupole fields that, depending on energy, may be either electrical (low energy) or magnetic (high energy) in origin. As a given quadrupole field will focus only in one plane, alternating focusing ensures confinement of the particles due to their forward motion. The similarity with the Paul trap is striking: in a storage ring it is the forward motion of the particles that ensures that the average transverse potential is confining, and in the Paul trap it is the oscillating electric field that does the same. This setup with alternating focusing and defocusing elements is also referred to as an alternating gradient design.

Storage rings are beneficial for studies where, for example, a charge change is to be studied, allowing the charge-changed state to easily escape and be detected. By merging different beams they may also allow studies of very low-energy collisions, and with the somewhat recent reintroduction of electrostatic storage rings (over the last decade or so), storage and study of biomolecules has become possible.

For more details about the operation and applications of electrostatic storage rings, see the chapters by Papash and co-workers in this volume.[13,14]

1.4. Creation, Cooling and Detection of Ions

Here we give a very short summary of important issues concerning the use of ion traps: how are ions loaded into the trap, how can they be cooled down for experiments and how are they detected?

1.4.1. Creation of ions

There are two main approaches to the creation and loading of ions into an ion trap: either the ions need to be created inside the trap, or, if they are created elsewhere, they need to be captured by the trap.

In the first instance, ions can be created in two different ways. Initially, many experiments used electron bombardment to ionize atoms inside the trap volume (from a small atomic beam oven, for example). This is a very simple technique but is not at all selective, so any species that is present in the vacuum system may be ionized and trapped. However, the technique is very easy to implement.

More recently, photo-ionization has been used, particularly in experiments with single ions or small numbers of ions. This requires typically one laser source, tuned to a resonance transition in the neutral atom, and a second light source (also usually a laser) to excite the atom into the continuum by a further one or two steps, one of which may be resonant. Not only is this technique element-selective, it is also isotope-selective if the first laser has a narrow linewidth and can be tuned accurately to the resonance transition. In this way it has been possible to load even very low-abundance isotopes with high efficiency. The availability of a wide range of wavelengths from tunable diode lasers has enabled many research groups to move to photo-ionization as their preferred method of creating ions.

When ions are created inside the trap, they will automatically remain trapped (so long as the stability conditions are fulfilled). On the other hand, ions that have been produced externally (for example, from an ion source or accelerator) cannot simply be loaded into the trap because, although they will be attracted towards the centre of the trap, they will also have enough energy to exit the trap again. It is therefore usually necessary to find some way of turning the trap off, to allow the ions into the centre of the trap, and then turning the trap on again. So long as the ions do not have enough energy to escape, they will then remain trapped.

Alternatively, if there is a strong cooling mechanism present (e.g. buffer gas in the case of RF traps), ions that fly through the trap may be cooled enough while inside the trapping region to force them to remain trapped. In this case it is possible to accumulate ions over a period of time in order to fill the trap.

1.4.2. Cooling of ions

When ions are first loaded into a trap, it is likely that they will have large energies, comparable to the depth of the trap (that is, the energy required for an ion to escape). For example, if the ions are created throughout the trapping region, they will have a range of energies up to that which is required to exit the trap. Some of the higher energy particles will escape, and the rest will thermalize to a lower average energy, but this will still be higher than required for many experiments, especially where any spectroscopic technique is employed that is limited by the Doppler effect.

It is therefore frequently necessary to slow the ions down, i.e. to cool them. There are several useful outcomes from cooling ions: the energy of each particle is reduced, the ions are better localized at the centre of the trap, the ion density increases and the lifetime of the ions in the trap is increased. An increase in the number density may also allow plasma effects to be observed, in particular if the Debye length becomes much smaller than the dimensions of the ion cloud. Under these circumstances the ion cloud has a uniform density with a sharp boundary and is referred to as a non-neutral plasma.

Another effect that can arise at very low temperatures is crystallization of the ion cloud. This occurs if the thermal energy is much less than the Coulomb interaction energy between adjacent ions. The ions then take up fixed positions relative to each other and are said to form an *ion Coulomb crystal*. These crystals (the simplest of which is a string of ions along the axis of a trap) can be observed in images of cold ions in traps.

A number of techniques exist for reducing the energy of trapped ions. They have different advantages and are applicable to different types of experiments and ion species. More details can be found in the books referred to earlier[1–3] and in the chapter by Segal and Wunderlich in this volume.[15]

1.4.2.1. Buffer-gas cooling

The simplest technique is buffer-gas cooling. This allows the exchange of energy between the trapped particles and atoms (or molecules) of a low-pressure buffer gas, generally chosen to be light and inert (e.g. helium). The energy exchange takes place through thermal collisions and the final temperature achieved is limited to that of the apparatus, which is usually at room temperature. If the pressure is low enough, the stability characteristics of the trap are unaffected. The technique is not element-specific and is often used with clouds of ions in RF traps. The choice of the buffer-gas species is important in order to ensure that collisions are elastic. The collisions with some molecular gases can be inelastic and change the internal energy state of the trapped ions, provoking quenching and state-mixing.

Buffer-gas cooling is complicated in Penning traps by the fact that the total energy of the magnetron motion is negative, so the thermalization process tends to lead to an increase in the magnetron radius rather than a decrease. However, it is used for cooling high-energy positrons as they are accumulated in a Penning trap. Buffer-gas cooling of both ions and positrons may be used in conjunction with other techniques that prevent the magnetron orbit size increasing (e.g. the rotating-wall technique, which uses an additional rotating electric field to prevent radial expansion of the plasma).

1.4.2.2. Resistive cooling

A technique that can be applied in the Penning trap is resistive cooling. In this arrangement a resistor is connected between the two endcap electrodes. As the ions move in the trap they induce image charges in the electrodes. These charges are continuously changing due to the motion of the ions, leading to a movement of real charges between the electrodes through the resistor. Energy is therefore dissipated in the resistor as heat, and this energy comes from the motional energy of the ions. It can be shown that this leads to an exponential decay of the energy of the axial motion. However, it is only the centre-of-mass motion that is cooled efficiently if a cloud consists of only one ion species. This is because the centre of charge then coincides with the centre of mass and this is stationary for all modes of oscillation except the centre-of-mass mode. Any internal energy of the ion cloud therefore does not generally couple strongly to the resistor. This can be improved, for example, by putting the resistor between one endcap and the ring rather than between the two endcaps so that the coupling is asymmetric and therefore becomes sensitive to internal modes of oscillation of the ion cloud. The coupling is also increased if there are ions of different species present, as the centres of charge and mass are now separate.

Again, the cooling can only reduce the ion temperature down to the temperature of the apparatus. This is because it is limited by Johnson noise in the resistor, which depends on temperature. One solution is to reduce the temperature of the apparatus (including the resistor) to around 4 K using liquid helium. In order to achieve a high rate of cooling, the effective value of the resistor can be increased by using a high quality factor (high-Q) tuned circuit.

Resistive cooling is effective for axial and cyclotron motions for atomic ions in the Penning trap but again is unsuitable for the magnetron motion. However, the magnetron motion can be coupled to either the cyclotron or axial motion by use of an additional oscillating potential, which is applied between appropriate electrodes at the sum frequency (the electrodes may need to be segmented for this purpose). If this is done, then energy is exchanged between the coupled motions and the resistive cooling becomes effective for all three motions. This is referred to as *sideband cooling* or *axialization* and is often used in experiments designed to measure the mass of trapped particles.

Note that for electrons in a trap the cyclotron motion automatically cools to the ground state in a short time via synchrotron radiation. This is only significant in the case of electrons and positrons because the cyclotron frequency is so high (up to tens of GHz at a field of a few tesla).

1.4.2.3. Laser cooling

The cooling technique that is able to reach the lowest temperatures is laser cooling. The simplest version, Doppler cooling, is very efficient and can typically take the temperature down to 1 mK. It involves irradiating the ion with laser light tuned close to a strong transition from the ground state of the ion. However, laser cooling is only applicable to a small number of species of ion: it has only been used for approximately ten different species, all of which are singly charged atomic ions (many of them alkali-like ions). Other species either have an energy level structure that is too complicated or do not have wavelengths that are accessible with tunable continuous-wave lasers.

Laser cooling works by transferring momentum from the laser beam to the ion and using the Doppler effect to ensure that the ion only absorbs light when it is moving towards the laser. The ion therefore slows down by a small amount each time a photon is absorbed from the beam. In principle this requires a two-level system so that after ions are excited by the laser they return by spontaneous emission to the initial state. However, in many species of interest there is a metastable state into which the ion can also decay. A second laser therefore needs to be provided to repump ions back out of this state into the cooling cycle, as otherwise no more cooling can take place. This is a disadvantage, but on the other hand the metastable states are often useful, for instance to use as a quantum bit ("qubit") for coherent processes in experiments in the area of quantum information processing, or for a clock transition for frequency standards.

Doppler laser cooling can be used in any sort of trap, but it is generally easier to apply in RF traps, as the strong magnetic field of the Penning trap leads to large Zeeman splittings and a requirement for many more laser frequencies. However, in some species without a metastable state, a single laser can still be used for Doppler cooling, taking advantage of optical pumping techniques.

In some experiments it is necessary to cool beyond the Doppler limit, even down to the ground state of the vibrational motion in the trap. This motion corresponds to a quantum mechanical simple harmonic oscillator and has states labeled by the vibrational quantum number n. In order to cool below the Doppler limit (which corresponds to the value of n being typically of the order of 10) it is necessary to employ other techniques, the most common of which is sideband cooling. (This is not to be confused with the technique described above in the context of resistive cooling.) In this case the ion is cooled initially using Doppler cooling and then it is cooled further by exciting it on a narrow optical transition (generally a forbidden transition to a metastable state). If the transition is narrow enough, and the laser linewidth is small enough, the spectrum becomes a carrier with sidebands that arise from the vibrational motion in the trap. The sidebands are spaced at the vibrational frequency and their amplitudes are dependent on the degree of excitation of the motion. For appropriate trapping conditions, the ion will be close to the Lamb–Dicke regime and therefore there will only be a small number of sidebands present.

Assume that the initial vibrational quantum number is n. If the laser is tuned to the first sideband below the carrier frequency (the so-called red sideband) then when the ion is excited there is a loss of one quantum of vibrational excitation, so after excitation the ion is in the state $n - 1$. Spontaneous decay back to the ground state generally does not change the value of n. If this process of excitation and decay is repeated, eventually the ion will reach the ground state of the motion ($n = 0$), at which point the red sideband amplitude becomes zero as there is no state $n = -1$ available to the ion to be excited into.

Sideband cooling is used routinely in several laboratories for preparing individual ions (or several ions in a string) in the ground state of motion as a preparation for coherent operations such as quantum gates. These techniques are also used for trapped-ion atomic clocks (frequency standards).

1.4.2.4. Sympathetic cooling

Laser-cooling techniques are limited to a small number of ion species, as mentioned above. However, in some cases it is possible to use one laser-cooled species to cool a different species (present in the same trap) by a process of *sympathetic cooling*, mediated by the exchange of energy in thermal collisions between the ions. Sympathetic cooling has been used in both Penning and RF traps. It has been shown that in some cases a small number of laser-cooled ions can cool a much larger number of simultaneously trapped ions of other species by this process. Sometimes it results in a radial separation of the different species. Sympathetic cooling is also used for cooling of antiprotons by electrons held in the same trap. It has frequently been used with large clouds of ions but it also works when there are just two ions in the trap: this is the *quantum logic* approach, used in some recent frequency standards work. Here a laser-cooled ion is used to cool and also probe the electronic state of a second ion which has a highly stable and narrow optical transition, used as the clock transition.

1.4.3. Detection of ions

Since atomic ions at low temperatures have very low energy, and often there are not many ions in a trap, the detection of trapped ions is often difficult. Here we list the commonly used methods for detecting the presence of ions in a trap. These methods fall into two groups: destructive detection and non-destructive detection. For more details on all aspects of the detection of ions in traps, see the chapter by Knoop in this volume.[16]

One method for detecting ions in a destructive manner is simply to release the ions from the trap and accelerate them towards a detector such as microchannel plate (MCP). This is a reliable and sensitive method of detection but, because it is destructive, it is necessary to reload the trap for each run of the experiment. This technique is often used in mass measurements for unstable isotopes created at accelerators.

One non-destructive technique is to use the image charges induced in the trap electrodes by the motion of ions in the trap – the same physical process used for resistive cooling. In this case the voltage across the resistor is monitored. When ions are present in the trap the induced current in the resistor gives rise to a voltage across it, which can be measured with a sensitive voltmeter. However, this is often too small to be seen directly, so an alternative is to make the electrodes part of a high-Q resonant circuit and to drive the motion using a weak excitation of this circuit. If ions are present, energy will be absorbed from the circuit, lowering its Q-value. This can be detected as a drop in voltage across the electrodes.

The most sensitive technique is to detect the laser fluorescence. This requires much the same energy level structure as laser cooling; indeed it is often the fluorescence from lasers used for laser cooling that is also employed to observe the ions. On resonance an ion may absorb and re-emit up to several million photons per second. With a solid angle for detection of 1% of 4π and an overall quantum efficiency of detection of up to 10%, this means that over 10 000 counts per second may be observed from a single laser-cooled ion. Indeed it is just possible to see an ion with the naked eye if the fluorescence is in the visible region of the spectrum (e.g. for Ba^+).

This signal may be recorded with a photon-counting photomultiplier. However, many experiments have used highly sensitive CCD cameras to record images of single ions or strings and crystals of laser-cooled ions and to study the properties of the structures that the ions form.

Laser fluorescence detection is also able in some cases to determine the electronic state of an ion. This needs an ion with a metastable state that is not addressed by the laser-cooling wavelengths. When the ion occupies this state, it is not able to absorb light, so no fluorescence is seen (until it decays from that state). Since otherwise thousands of photons would have been detected, this allows extremely efficient and sensitive detection of the electronic state of the ion. This was named the *electron shelving technique* by Dehmelt, who first introduced it, and it is now widely used in experiments involving coherent interactions between lasers and ions. It is routinely used to observe *quantum jumps* as an ion makes transitions into and out of a metastable state.

1.5. Applications of Ion Traps

Ion traps are now used in a large variety of applications throughout physics. Here we list some of the more important applications in order to give a flavour of the versatility of these devices.

Precision measurements of fundamental quantities in atomic physics

Due to the fact that ions in traps are so well isolated from the environment (e.g. no collisions) and that they generally have low velocities, they are suitable for precision measurements of various sorts. An excellent example is the measurement of g-factors of electrons, positrons and atomic ions. In these experiments it is necessary to measure both the cyclotron frequency of the particle and the spin-flip frequency. The ratio of the two can be used to measure the g-factor to a high level of precision, without the need to know the strength of the magnetic field. These measurements make use of the fact that frequency is the physical quantity that can straightforwardly be measured to the highest accuracy (compared, for example, to length). The determination of the g-factor of the electron in a Penning trap has resulted in the most accurate value of a fundamental constant to date (with a precision of 3×10^{-13}).[17] Similar measurements with bound electrons in hydrogen-like ions have been used for tests of quantum electrodynamics (QED), as the g-factor is modified by the presence of the nucleus.

Mass spectrometry

Since the cyclotron frequency of an ion is inversely proportional to its mass, a measurement of the ratio of the cyclotron frequencies of two ions yields their mass ratio to a high level of precision. This approach has been used not only for the determination of the mass ratios of fundamental particles to the highest precision, but also for the measurements of long strings of stable and unstable isotopes at accelerator facilities such as CERN. These studies give vital information on nuclear binding energies used in the development of models of nuclear structure.[18]

Storage of antimatter

The extreme isolation of particles in a trap allows even exotic particles such as positrons and antiprotons to be stored for long periods of time. This has enabled studies of the creation of antihydrogen atoms to be carried out at CERN. For more details on the trapping of positrons in traps, see the chapters by Surko in this volume.[19,20] The production of antihydrogen is discussed in the chapter by Madsen in this volume.[21]

Non-neutral plasma studies

A large cloud of charged particles in an ion trap constitutes a non-neutral plasma. So long as the Debye length is much less than the dimensions of the plasma, it will have a uniform density when it is in equilibrium. The confining potential of the trap is equivalent physically to a sea of particles of the opposite charge to those that are trapped. Non-neutral plasmas have been studied extensively, mainly in Penning traps, and techniques have been developed for manipulating the plasma density and shape as well as probing its modes of oscillation. Such experiments have been carried out with electrons, positrons and atomic ions. For more details of experiments on non-neutral plasmas in traps, see the chapters by Anderegg in this volume.[22,23]

Optical and microwave spectroscopy

Since the early days of work with ion traps, they have been used as a way of confining a sample for study by spectroscopy. In this sense the detailed properties of the trap are not of interest as its main function is to keep the sample in the right place. One difficulty with this is that the maximum density of ions that can be held in a trap is limited by the Coulomb repulsion to typically 10^8 ions per cm^3, which is similar to the number density of molecules in an ultrahigh vacuum system at a pressure of 10^{-9} mbar. This may result in signal levels in conventional spectroscopic investigations that are small compared to those carried out in low-pressure gases, for example. However, the extended interaction time available in an ion trap compensates for this disadvantage and so many important spectroscopic investigations have been carried out in the RF, microwave and optical regimes.

Laser cooling

As alluded to earlier, laser cooling enabled a wide range of experiments to be carried out in ion traps that were previously impossible, such as the observation of a single atomic particle at rest. Early experiments with laser cooling showed that single ions or small clouds of ions could be cooled to a temperature close to the Doppler limit fairly easily. This allowed the types of spectroscopic measurements referred to above to be taken to new levels of sensitivity and accuracy due in particular to the absence of the Doppler effect. Furthermore, developments in the theory of laser cooling showed that even lower temperatures could be reached using sideband cooling. In this way ions could be prepared in the ground state of the potential with high probability, opening the door to many new experiments. For more details on laser cooling, see the chapter by Segal and Wunderlich in this volume.[15]

Microwave frequency standards

One major application of laser-cooled ions in traps is the development of new frequency standards, both in the microwave region and the optical region of the spectrum. Penning traps are particularly well suited to microwave frequency standards as they can hold large numbers of laser-cooled ions in a single trap. The clock transition in the ions is driven by radiation from a stable oscillator, and is close to resonance with a microwave transition in the ions (which may be a hyperfine or Zeeman transition). The response of the ions to this radiation is detected by observation of the fluorescence from the ions, which depends on which state the ions are in. Using feedback to the oscillator that depends on the observed fluorescence, the oscillator can be stabilized to the ionic transition and can then be used as a highly stable reference, i.e. a frequency standard. Although some ion trap-based microwave clocks have been developed, they did not offer large benefits compared to well-developed atomic beam frequency standards such as the caesium clock.

Optical frequency standards

In the optical domain, ion traps have had a huge impact on frequency standards and are now performing at a much higher level of stability than the caesium clock. Most ion trap-based optical clocks use single ions that can be laser cooled on a strongly allowed optical transition. The ion also needs a narrow optical transition to a long-lived excited state that can be used both for sideband cooling and also as the clock transition. Much of the effort that has gone into ion trap research has been driven by the desire to develop new optical clocks. For this to succeed, it is necessary to be able to prepare a single ion in a trap, to cool it to the ground state of motion, to probe it with ultra-stable laser radiation, to detect its electronic state with high efficiency and to feed back to the laser in order to stabilize its frequency. These efforts have been so successful that a new definition of the second is likely to be based on an ion trap clock in due course. For more details on the development of ion trap-based atomic clocks, see the chapter by Margolis in this volume.[24]

Ion Coulomb crystals

When ions in traps are subject to strong laser cooling, eventually the thermal energy becomes much less than the energy of interaction of neighbouring ions via the Coulomb interaction. At some point the Coulomb interaction dominates so much that the ions fall into a crystal-like structure called an ion Coulomb crystal (ICC). As in a conventional crystal, this means that the ions take up fixed positions relative to each other. In very small crystals, and at the lowest temperatures, the ions do not exchange places. The simplest such structure is a linear string of ions, but it is also possible to see two-dimensional planar crystals and three-dimensional solid crystals. They have been observed in both Penning and RF traps with sizes ranging from a few ions up to many thousands of ions. This is a rich field of research in itself but also has applications, for example, in the study of chemical reaction rates between ions and neutral gases, where reaction rates may be measured by counting the number of ions in a crystal as a function of time.

Quantum effects

It is perhaps not surprising that the preparation of a single isolated atomic particle in the ground state of its motion allows interesting and novel quantum mechanical effects to be observed. Indeed trapped ions have been used in many such experiments, including the observation of quantum jumps in a single ion, the preparation and analysis of Schrödinger cat states, and the study of the quantum Zeno effect.

Quantum information processing

Perhaps the most exciting application of trapped ions has been in the field of quantum information processing. This field was initiated by Cirac and Zoller in 1995, who first proposed that a pair of laser-cooled trapped ions could be used to create a *quantum gate* – the quantum mechanical equivalent of a classical computer gate.[25] Practical realizations of their ideas followed and the field has grown rapidly ever since. Ion traps remain the system in which the most promising demonstrations of quantum information processing have been carried out. A large fraction of the current activity with trapped ions constitutes research in this area. It now includes quantum simulation as well as quantum computing applications. This work builds on many of the areas mentioned above, including laser cooling, spectroscopy and ICCs. For more details on all aspects of quantum information studies using ions in traps, see the chapter by Roos in this volume.[26] An alternative implementation of quantum information processing using ion traps is with trapped electrons as qubits. This is discussed in detail in the chapters by Verdú in this volume.[27, 28]

Quantum logic

A recent development is the application of techniques from quantum information processing to frequency standards. One problem for frequency standards is to find an ion that has a suitable atomic structure for laser cooling but that also incorporates a narrow clock transition. A new solution is to separate these two functions so that one ion can be laser cooled but a different ion has the clock transition. When both are trapped at the same time, they form a Coulomb crystal of two ions and so the laser cooling is effective for both ions through the Coulomb interaction. Then a transition that takes place in the clock ion can be detected via a measurement technique based on the principles used in ion quantum gates, also mediated by the Coulomb interaction between the ions. This powerful technique has already been used to develop the most accurate ion-based optical clock to date, and another important application will be the measurement of possible variations of the fundamental constants with time.[24]

Molecular Physics

Electrostatic storage rings allow for the storage of a much larger mass range than conventional magnetic ones, and therefore several rings around the world are under construction or under commission for molecular physics experiments. Further, they are often small, and can therefore be cooled to cryogenic temperatures, an important feature for avoiding or studying blackbody radiation effects. An example from the negative ion world is the study of the lifetime of the negative SF_6 ion in the ELISA storage ring.[29] For more applications see the chapters by Papash and co-workers in this volume.[13,14]

1.6. Conclusions and Outlook

Much has been achieved in the area of trapped ion research since ion traps were first invented in the late 1950s and early 1960s. These remarkable devices have found applications in many different areas of science and technology and have already led to the awarding of three Nobel Prizes. There is every sign that they will continue to find new applications in the coming years and will have even more impact in science and technology.

Acknowledgments

This work was supported in part by the COST ACTION MP1001 "Ion Traps for Tomorrow's Applications", the CNRS, EPSRC, CERN and École de Physique des Houches.

References

1. F. G. Major, V. N. Gheorghe and G. Werth, *Charged Particle Traps* (Springer, Berlin, Heidelberg, New York, 2004).
2. G. Werth, V. N. Gheorghe and F. J. Major, *Charged Particle Traps II: Applications* (Springer, Berlin, Heidelberg, New York, 2009).
3. P. K. Ghosh, *Ion Traps* (Oxford University Press, Oxford, 1995).
4. E. Fischer, Die dreidimensionale Stabilisierung von Ladungsträgern in einer Vierpolfeld, *Z. Phys.* **156**, 1–26 (1959).
5. R. F. Wuerker, H. Shelton and R. V. Langmuir, Electrodynamic containment of charged particles, *J. Appl. Phys.* **30**, 342–349 (1959).
6. H. G. Dehmelt, Radiofrequency spectroscopy of stored ions. I. Storage, *Adv. At.. Mol. Phys.* **3**, 53–72 (1967); H. G. Dehmelt, Radiofrequency spectroscopy of stored ions. II. Spectroscopy, *Adv. At.. Mol. Phys.* **5**, 109–154 (1969).
7. J. Byrne and P. S. Farago, On the production of polarized electrons by spin-exchange collisions, *Proc. Phys. Soc.* **86**, 801–815 (1965).
8. G. Gräff and E. Klempt, *Z. Naturforschung* **22a**, 1960–1962 (1967).

9. A. A. Sokolov and Yu. G. Pavlenko, *Optics and Spectroscopy* **22**, 1–3 (1967).
10. F. M. Penning, Introduction of an axial magnetic field in the discharge between two coaxial cylinders, *Physica* **3**, 873–894 (1936).
11. R. H. Dicke, The effect of collisions upon the Doppler width of spectral lines, *Phys. Rev.* **89**, 472–473 (1953).
12. L. H. Andersen, O. Heber and D. Zajfman, Physics with electrostatic storage rings and traps, *J. Phys. B* **37**, R57–R88 (2004).
13. A. I. Papash and C. P. Welsch, "Basics of beam dynamics and applications to electrostatic storage rings", in Knoop, M., Madsen, N. and Thompson, R. C. (eds), *Physics with Trapped Charged Particles* (Imperial College Press, London, 2013) pp. 305–326.
14. A. I. Papash, A. V. Smirnov and C. P. Welsch, "Electrostatic storage rings at ultra-low energy range" in Knoop, M., Madsen, N. and Thompson, R. C. (eds), *Physics with Trapped Charged Particles* (Imperial College Press, London, 2013) pp. 327–358.
15. D. M. Segal and Ch. Wunderlich, "Cooling techniques for trapped ions" in Knoop, M., Madsen, N. and Thompson, R. C. (eds), *Physics with Trapped Charged Particles* (Imperial College Press, London, 2013) pp. 43–81.
16. M. Knoop, "Detection techniques for trapped ions" in Knoop, M., Madsen, N. and Thompson, R. C. (eds), *Physics with Trapped Charged Particles* (Imperial College Press, London, 2013) pp. 25–42.
17. D. Hanneke, S. Fogwell and G. Gabrielse, New measurement of the electron magnetic dipole moment and the fine structure constant, *Phys. Rev. Lett.* **100**, 120801 (2008).
18. K. Blaum, Yu. N. Novikov and G. Werth, Penning traps as a versatile tool for precise experiments in fundamental physics, *Contemp. Phys.* **51**, 149–175 (2010).
19. C. M. Surko, "Accumulation, storage and manipulation of large numbers of positrons in traps I. – the basics" in Knoop, M., Madsen, N. and Thompson, R. C. (eds), *Physics with Trapped Charged Particles* (Imperial College Press, London, 2013) pp. 83–128.
20. C. M. Surko, J. R. Danielson and T. R. Weber, "Accumulation, storage and manipulation of large numbers of positrons in traps II. – selected topics" in Knoop, M., Madsen, N. and Thompson, R. C. (eds), *Physics with Trapped Charged Particles* (Imperial College Press, London, 2013) pp. 129–172.
21. N. Madsen, "Antihydrogen formation and trapping" in Knoop, M., Madsen, N. and Thompson, R. C. (eds), *Physics with Trapped Charged Particles* (Imperial College Press, London, 2013) pp. 219–238.
22. F. Anderegg, "Waves in non-neutral plasma" in Knoop, M., Madsen, N. and Thompson, R. C. (eds), *Physics with Trapped Charged Particles* (Imperial College Press, London, 2013) pp. 173–193.

23. F. Anderegg, “Internal transport in non-neutral plasmas” in Knoop, M., Madsen, N. and Thompson, R. C. (eds), *Physics with Trapped Charged Particles* (Imperial College Press, London, 2013) pp. 195–218.
24. H. S. Margolis, “Optical atomic clocks in ion traps” in Knoop, M., Madsen, N. and Thompson, R. C. (eds), *Physics with Trapped Charged Particles* (Imperial College Press, London, 2013) pp. 261–274.
25. J. I. Cirac and P. Zoller, Quantum computations with cold trapped ions, *Phys. Rev. Lett.* **74**, 4091–4094 (1995).
26. C. F. Roos, “Quantum information processing with trapped ions” Knoop, M., Madsen, N. and Thompson, R. C. (eds), *Physics with Trapped Charged Particles* (Imperial College Press, London, 2013) pp. 239–260.
27. J. Verdú, “Trapped electrons as electrical (quantum) circuits” in Knoop, M., Madsen, N. and Thompson, R. C. (eds), *Physics with Trapped Charged Particles* (Imperial College Press, London, 2013) pp. 289–303.
28. J. Verdú, “Novel traps” in Knoop, M., Madsen, N. and Thompson, R. C. (eds), *Physics with Trapped Charged Particles* (Imperial College Press, London, 2013) pp. 275–287.
29. J. Rajput, Measured lifetime of SF_6^-, *Phys. Rev. Lett.* **100**, 153001 (2008).

Chapter 2

Detection Techniques for Trapped Ions

Martina Knoop

CNRS and Université d'Aix-Marseille,
Centre de Saint Jérôme, Case C21,
13397 Marseille Cedex 20, France
Martina.Knoop@univ-amu.fr

Various techniques are used to detect the presence of charged particles stored in electromagnetic traps, their energy, their mass, or their internal states. Detection methods can rely on the variation of the number of trapped particles (destructive methods) or the use of the ion's interaction with electromagnetic radiation as a non-destructive tool to probe the trapped particles. This chapter gives an introduction into various methods, discussing the basic mode of operation completed by the description of recent realizations.

The direct observation of a trapped sample can be easily made for macroscopic particles contained in a Paul or Penning trap. Today, the main interest of the use of ion traps lies in the manipulation and interrogation of atomic or molecular ions. Detection of trapped ions needs amplification and imaging techniques which can rely either on the light emitted by the ions or on their physical presence in the trapping electromagnetic field. Most of today's techniques have been developed not only to make sure that a sample is trapped but moreover to obtain additional information: the size and shape of the ion sample, the number of stored particles, their temperature, the quantum state of the ions, etc.

Two main classes of detection methods exist: destructive techniques which will destroy the trapped ion sample at least partially, and non-destructive techniques, which detect the presence of ions in the trap, keeping their number constant. This latter tool often relies on a perturbing effect to retrieve additional information about the sample. In this chapter, the distinction has been made between techniques which employ the monitoring of photons emitted by the trapped ions (fluorescence techniques), and those which electronically detect the presence of ions in the trap.

The ultimate detection method chosen for a given experiment is of course determined by the species of the trapped ion, the size of the sample, and the dimensions of the trap. Actually, the choice of the detector itself is guided by a large number of technical criteria: the spectral response of the device, its temporal and/or spatial resolution, the gain curve, the dark noise of the device, the environment in which the detector will be installed (vacuum conditions, contaminants, ..), as well as the electric and magnetic environment. This chapter gives a review of the major techniques and instruments used, without being exhaustive. In every paragraph, additional references will give details on the described technique or present major results of its implementation.

2.1. Electronic Techniques

Many trapped ions cannot be made visible by resonant excitation, either because they do not dispose of a closed fluorescence cycle giving a large enough number of scattered photons, or because the wavelength corresponding to the absorption and emission of photons is difficult to generate with a sufficiently powerful source. Due to experimental and geometric constraints, it is much easier to obtain a reasonable signal-to-noise ratio of the sample of trapped ions if the excitation process is performed by a laser rather than a lamp. However, the generation of a resonant laser wavelength is technically more demanding than its non-coherent counterpart.

Over the years, different methods have been developed in order to detect the presence of ions in the trap, even if they do not emit fluorescence. After a review of the different detectors employed, the main techniques for electronic detection of the trapped sample will be presented.

2.1.1. *Instruments*

As mentioned in the introduction, technical specifications, such as the spectral response or the temporal resolution, are decisive in the selection of the best detector. Three main devices exist and are in use today.[1]

2.1.1.1. *Faraday Cup*

The Faraday cup is a metallic recipient open on one side in order to absorb a beam of charged particles. The cup is grounded via an electrometer capable of measuring currents in the pico Ampere range. The detection limit is of course due to the sensitivity of the electrometer. The device is primarily destined to measure large particle numbers and can reach a lower detection threshold of a couple of hundred ions in the best case scenario.

2.1.1.2. *Electron multiplier*

Electron multipliers (EMs) are versatile tools for ion or electron detection. They are composed of a large number of stacked polarized dynodes which will produce an amplified cascade of secondary electrons. In general, the first dynode is polarized around -3 keV for positive ion detection. It is possible to detect single ions with this device as the gain can reach 10^8 and the dark count under high vacuum conditions is below one count per second. The effective detector surface is very similar to that of a photomultiplier, typically around 100 mm^2. EMs exist in two versions: the discrete-dynode and the continuous-dynode (or channel) EM. EMs must work under reduced pressure conditions. Due to the electronic conversion of analog pulse signals into transistor–transistor logic (TTL)-pulses, an EM always presents a dead time in its operation process. Durations below 10 ns have been achieved.[2] EMs are often coupled to a pre-amplifier and a discriminator generating a direct TTL output.

2.1.1.3. *Microchannel plate*

Microchannel plates (MCPs) rely on the same principle of operation as EMs: a cascade of secondary electrons is generated in an array of very small continuous-dynode amplifiers.[3] The device is composed of a million channel-EMs each having a diameter of about 10 μm. The gain of the MCP is proportional to the length to diameter ratio, L/d, of the individual channels following the relation $g = \exp(G \times (L/d))$, where G is the secondary emission characteristic of the channel called gain factor. L/d can be as high as several hundred, for a total absolute gain value above 10^4. Higher gain is achieved by MCPs configured with multiple sets of channels mounted in line. Depending on their respective geometry this can be in the "chevron" mode (two stacks), or in the three-stack "Z" configuration, see Figure 2.1.

An MCP has an entry converter in order to convert the impinging particles into photons or electrons. Furthermore, the MCP stack is always followed by a read-out device to detect the produced electron avalanche. Depending on the application this can be a simple anode giving a one-dimensional output (or a more complex anode array for a two-dimensional

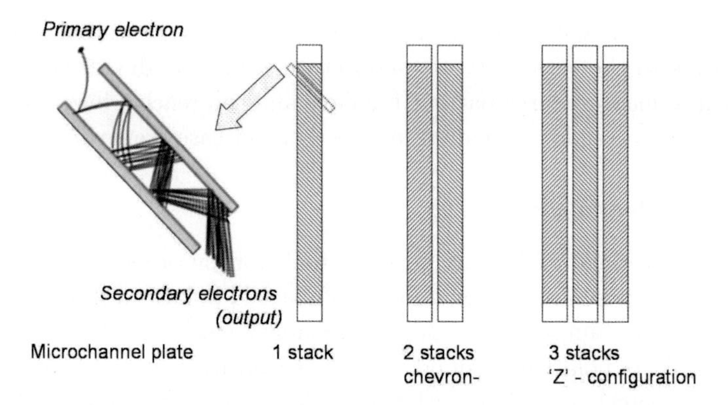

Fig. 2.1. Scheme of an MCP. Left: the principle of operation, right: stacking multiple MCPs increases the gain of the detector.

signal), or a phosphor screen followed by a charged-coupled device (CCD) in order to keep the spatial information of the detector.

The temporal resolution of the MCP is limited by the electronics and read-out device, but is typically inferior to a nanosecond.

2.1.2. *Techniques*

Many of the electronic detection techniques rely on the ejection of the trapped ion cloud in order to be monitored by a detector external to the storage device. However, as the contained sample is always charged, the image current which the sample induces on the electrodes is a non-destructive means for detection. A short description of some techniques is given in Werth's book on charged particle traps.[4]

2.1.2.1. *Ion loss*

The most simple detection method of a trapped ion sample is to switch off the trapping field and record a synchronized signal on an ion detector. The fine-tuning of this measurement will have to take into account the solid angle under which the ions leave the trapping region, their flight path to the detector, and the phase of the trapping field at the moment it is turned off.

More sophisticated versions of detection will therefore make use of ion optics in order to collect as many charged particles as possible and guide them to the detection region. As "turning off" the trapping field is often not an instantaneous action (in particular if a magnetic field or a radiofrequency field is present), a technically simpler solution is the ejection of the ions with an acceleration pulse without modification of the trapping field. Ions can be ejected along all axes of the trapping field.

A first application of this technique is to measure the trapping time constant of the trap. Recording the relative ion signal for different durations since the first filling of the trap gives the loss rate of the trap. Due to space-charge effects, it can be shown that a completely filled trap will lose ions more rapidly than a trap which contains only a small ion cloud. The observed exponential decay of the signal is often ruled by two time constants: one which is due to the trap and the depth of its potential, and one which is determined by the collisions with the residual gas of the vacuum vessel. This latter influence can also be beneficial for the ion cloud, as the addition of a light atomic gas in the vacuum vessel can cool the ion cloud as a consequence of elastic collisions.

2.1.2.2. *Depletion techniques*

These detection methods record the variation of the trapped ion number as a function of additional interaction processes. The technique is often employed in chemistry or physical chemistry experiments in order to investigate the reaction of a trapped species with an introduced reactant gas or an applied source of electromagnetic radiation. The variation of the number of trapped ions gives an insight into the rate of the process, under the condition that pressure and temperature parameters are extremely well controlled.

In order to *actively* produce a spectrum, the main idea is to modify the m/Z ratio of the trapped molecular species by absorption of one or several photons. Depending on the energy of the photon, various reactions can be triggered. Photofragmentation of a chemical bond requires an ultraviolet (UV) photon or multiple infrared (IR) photons, whereas photodissociation can be made by one high-frequency IR or multiple IR photons. Even lower energies are required for the photodissocaition of a weak bond or the photodetachment of an electron. For a detailed review see Rizzo *et al.*[5]

A recent example is the monitoring of the resonance enhanced multiphoton dissociation in H_2^+-ions, where the study of the photodissociation process has been made by monitoring the number of ions still present in the trap as a function of the number of laser shots and of the energy of the dissociation laser.[6]

2.1.2.3. *Ejection with or without additional perturbation*

Active ejection of the ions from the trap is a way to measure relative ion numbers. As mentioned above, an optimized experiment will take into account the phase of the radiofrequency trapping field for the timing of the ejection pulse. Higher sensitivity in the detection rate can be achieved by collecting and guiding the ejected ions by a set of (electrostatic) Einzel lenses. Optimization of experimental features can be made numerically with good success.[7]

A major experiment in ion trapping using this technique is the work carried out in the Mainz group on the trapping performances of a radiofrequency trap. Alheit *et al.*[8] recorded the number of trapped ions for the complete first stability range using samples of a few thousand H_2^+-ions in a trap with a radius of 20 mm. With this experiment, they were able to characterize the ion trap and demonstrate that the trapping behavior in a large interval of storage parameters is not homogeneous. Actually, at a given resonance condition,

$$n_r\omega_r + n_z\omega_z = \kappa\Omega, \tag{2.1}$$

with κ an integer value, and $|n_r| + |n_z| = N$, resonances of order N appear in the trapping zone. If this condition is fulfilled, ions can take up energy from the trapping field, they are heated and then lost from the trap. Strong absences in the trapping capacity appear if these nonlinear resonance conditions are fulfilled. The phenomenon is, of course, amplified for larger clouds, as space charge becomes an issue for all ion numbers larger than one. The same feature can be used for isotope separation,[9] as the trap operation is different for varying m/Z-ratios.

Rather than scanning the whole stability range, insight into the actual trapping parameters can also be gained by using resonant excitation of the stored ions. The addition of a small oscillating voltage on the trap electrodes can excite motional frequencies if the frequency of the oscillating voltage coincides with the a macromotion component of the cloud. It is thus possible to record the spectrum of secular frequencies ω_i, by repeating a protocol of application of the described additional "tickle" frequency, and a subsequent ejection of the ion cloud, measuring *de facto* the residual ion number as a function of the applied excitation frequency[10] (see Figure 2.2).

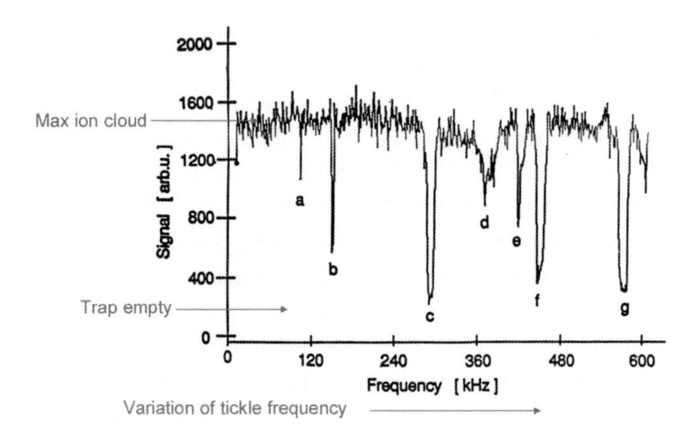

Fig. 2.2. Motional frequency spectrum of a trapped cloud of calcium ions at trapping frequency $\Omega/2\pi = 1$ MHz. Different frequencies and their coupling terms can be identified : $\mathbf{a} : \omega_r/2$, $\mathbf{b} : \omega_r$, $\mathbf{c} : \omega_z$, $\mathbf{d} : \omega_z + \omega_r/2$, $\mathbf{e} : 2\omega_z - \omega_r$, $\mathbf{f} : \omega_r + \omega_z$, $\mathbf{g} : 2\omega_z$.

2.1.2.4. *Time-of-flight profiles*

An even more refined technique of the ejection method is the recording of the time-of-flight (TOF) profile of an ejected ion cloud on a detector situated at a fixed distance. The distribution of arrival times of the ejected ions will reflect their mass distribution, and it can be described by a set of experimental parameters as there are the m/Z ratio of the species, the velocity distribution of the ions in the trap, and the distance, d, of the detector. The longer the flight path the better the temporal resolution of ions arriving at the detector. A typical experiment is composed of an extraction zone next to the trap, followed by a drift zone before the detector. An energy ramp might eventually be added in order to modify the potential energy of the ions with respect to the detector.[11] For a single species in a Paul trap, the recorded profiles of an ejected ion cloud show a double bump as a result of the velocity distribution of the ions in the trap. Moreover radiofrequency residuals might also distort the profile. The extension of the flight path by means of ion mirrors to almost infinite length in a multi-turn device allows the increase of the mass resolution $m/\Delta m$ beyond 30 000.[12]

An extension to the time-of-flight method coupled with the excitation of an ion cyclotron resonance (ICR) makes it possible to reach unprecedented precision in the mass resolution.[13] The precise determination of the cyclotron frequency $\omega_c = qB/(2\pi m)$ of a single cooled ion is the basis for this advanced technique of mass measurement in Penning traps.

The ion with negligible radial kinetic energy is excited to a well-defined magnetron radius by external dipole excitation. Application of an additional oscillating quadrupole field on the segmented trap converts the magnetron into cyclotron motion. This manipulation drastically increases the radial energy, as the frequency values for the various motional states differ by several orders of magnitude. Once the ion is ejected from the trap, its radial energy is transformed into axial energy by coupling the magnetic moment of the orbital motion to a magnetic field gradient. It is then possible to record the time-of-flight of the ejected ions as a function of the exciting frequency, the shortest time-of-flight value corresponding to the cyclotron frequency of the ions. This technique has been realized with mass resolving powers of more than 10^7.

Even higher resolutions can be achieved by applying the exciting pulse not in a continuous way, but using a temporal Ramsey method to achieve still narrower resonance lines. Like many other methods, absolute values can only be determined by calibration with a well-known reference mass. Best results (i.e., a good signal-to-noise ratio) are obtained with a single ion in the trap and a few hundred repetitions.

2.1.2.5. *Image currents*

The recording of the image current which is induced by a trapped ion cloud onto the trap's electrodes allows non-destructive ion detection.[14] The oscillation of the trapped ions leads to an oscillating current between the trap electrodes. The sensitivity of this method can reach the single-ion level. The signal-to-noise ratio for this detection method is given by

$$S/N = \frac{\sqrt{\pi}}{2}\frac{r_{ion}}{D}Z\sqrt{\frac{\nu}{\Delta\nu}}\sqrt{\frac{Q}{kTC}}, \tag{2.2}$$

with r_{ion} the ion's motional radius and Z its charge. D is the effective electrode distance, and T, C, and Q the temperature, capacity, and quality factor of the detection system, respectively. $\frac{\nu}{\Delta\nu}$ is the ratio of the ion's motional frequency to its width. In the case of an individual singly charged ion, the induced image current is only about a few hundred femtoampere (fA). For multiply charged ions, this current increases proportionally to the charge. Thermal noise is one of the main limiting factors in these experiments, and better results can be obtained by cooling the detection electronics and the traps either with liquid nitrogen or with liquid helium.

Detection and resistive cooling of a single trapped ion was demonstrated as early as 1998 in a cryogenic Penning trap.[15] In this experiment, the narrow-band detection circuit consisted of a superconducting coil which formed a parallel LC resonant circuit with the parasitic capacities of the device. The signal-to-noise ratio was determined by the quality factor Q of the circuit, which had to reach a value of a few thousand.[16] The image current was detected by the proportional voltage drop across the trap, which was amplified, and then Fourier transformed in order to determine the corresponding frequency spectrum. Necessary calibrations were made by trapping reference ions in the setup.

2.2. Fluorescence Techniques

The interaction of atoms with light is one of the most powerful tools developed in ion-trapping experiments. Laser light is often used to create the ions, to prepare them in different quantum states, to cool them, or to heat them. Using the photons emitted upon (resonant) excitation of an atomic transition is the most reliable way to detect and identify the trapped particles.

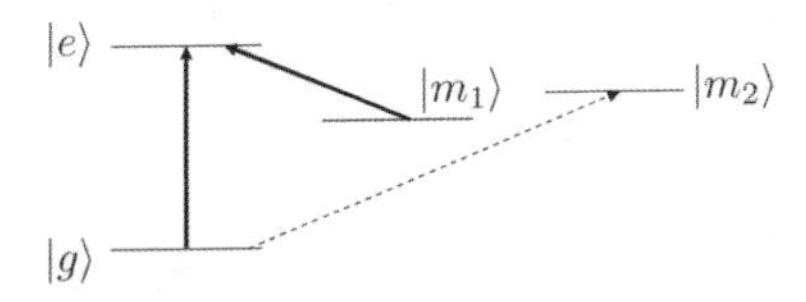

Fig. 2.3. First energy levels of a typical three-level ion. $|g\rangle$, $|e\rangle$, and $|m\rangle$ denote the ground, excited, and metastable states, respectively.

The direct detection of the ion's fluorescence is typically made on a strong transition, most often the resonance line. Many ions trapped today have energy schemes very similar to that pictured in Figure 2.3. The strongest transition (the resonance line) is the one connecting the ground state $|g\rangle$ to the excited state $|e\rangle$. The lifetime of the excited state $|e\rangle$ is of the order of some tens of ns, giving rise to a natural linewidth of the corresponding atomic transition of the order of some tens of MHz. The metastable states $|m_i\rangle$ have a lifetime of the order of a second, the transition connecting them to the ground state are forbidden in the dipole approximation. The branching ratio from the $|e\rangle$-state towards the metastable states is typically several percent compared to the de-excitation towards the ground state. Given the large difference in lifetimes, a repumper laser from $|m_1\rangle$ to $|e\rangle$ might be necessary in order to close the cycle and obtain an observable fluorescence signal. Quenching with a buffer gas is a less efficient but simple alternative method; inelastic collisions with a heavy or molecular buffer gas reduce the lifetime of the metastable state.[17]

If the resonance line is excited by a laser reaching the saturation limit, a few million photons can be scattered per second per atom. Taking into account the finite collection angle, the losses in the optical system, and the quantum efficiency of the detector, a typical setup has a detection efficiency of about 10^{-4}. A single-ion signal is then about 10 000 photons per second and can easily be recorded with a photomultiplier in photon-counting mode. A good imaging optics is, however, mandatory in order to spatially separate the ion's fluorescence signal from the background noise due to scattered light by the exciting laser beams (see Section 2.2.8 for details).

The most direct example of fluorescence detection is the observation of a single Ba^+-ion by the human eye in an experiment performed at the University of Hamburg.[18] In fact, this experiment had a microscope objective mounted on one of its viewports, and the trapped ion could be directly visualized during its life in the trap. The necessary condition for this experiment is, of course, to have a strong transition in the visible, preferably in the center of the eye's spectral sensitivity.

Absorption imaging is a standard tool in ultracold matter experiments with neutral atoms; it is not commonly used in ion-trapping experiments due to the very low density of ion clouds. Only very recently, a group at Griffith University has recorded the absorption image of a single Yb^+-ion.[19] The observed image contrast of 3.1(3)% is the maximum theoretically allowed for the imaging resolution of the employed setup.

2.2.1. *Lineshape*

The observation of the fluorescence of an atomic ensemble can also be made while varying the frequency of the exciting laser beam. The spectral response of the ions will reflect the excitation probability as a function of wavelength and will provide a multitude of additional information. The recorded linewidth is, in most cases, larger than the theoretical natural linewidth of the transition. The main broadening factor is the Doppler effect, which can generate linewidths of the order of several GHz for an uncooled ion cloud. Many experiments use laser-cooling techniques, which make the recording of a Doppler profile slightly more complex, as probing and cooling are not uncorrelated. A good estimation of the temperature of a laser-cooled target is a profile of the repumping transition keeping the laser on the resonance line fixed.

Scanning the cooling laser frequency may provoke a structural transition from a cloud to a Coulomb crystal configuration where the ions are better localized. Structural phase transitions of the atomic ensemble can be indicated by a change in the line profile, as they are marked by a change in the recorded atomic lineshape.[20]

Under the influence of laser cooling and in a trap where micromotion perturbations are compensated, the ion may reach the strong coupling limit where its motional amplitude is smaller than a fraction of the excited wavelength.[21] The spectral response of the ion will then split up into a central carrier and sidebands, which can serve to determine the vibrational state that the ions are in.[22,23] For an ion cooled to the lowest vibrational levels, the number of sidebands is reduced to very few, the ion is said to be in the Lamb–Dicke regime.

The monitored lineshape can also give information on the environment of the trapped ions. It can be used to measure the influence (or absence) of collisions and of magnetic or electric fields. Depending on the geometric configuration of laser beams, dark resonances appear in a three-level system when a coherent superposition of the involved states is created. These dark states depend on the involved laser linewidths and powers, but also on the micromotion amplitude of the probed ions.[24]

2.2.2. *Single-ion detection*

Counting the number of trapped ions is a perfect way to control the experimental conditions, to calibrate the necessary parameters, and to obtain a complete insight into the underlying physics process. However, the determination of a precise ion number can only be made after having identified the net fluorescence signal for a single ion. Quantum jumps can be observed in a three-level atom, when one of the excited states is a metastable state (Figure 2.3). In this configuration, the cooling laser beams are applied on the cycling transitions $|g\rangle - |e\rangle - |m_1\rangle$, and the fluorescence is observed on the resonance line. Applying an additional excitation to the $|g\rangle - |m_2\rangle$ transition will provide optical pumping to the $|m_2\rangle$- state which has a very long lifetime.[25] With a single trapped ion, the observed fluorescence signal will then become binary: "ON" when the ion is scattering photons on the cooling cycle transitions, and "OFF" when the ion is pumped to the metastable state.[26–28] The duration that the ion will remain in the "shelved" dark state corresponds to the lifetime of the level.[29]

A single trapped ion is a well-localized point source; storing two ions in a trap allows the investigation of the interaction between two extremely well-defined sources. A spectacular realization of this is Young's double-slit experiment which recorded the interferences of two laser-cooled ions trapped within the same device.[30]

2.2.3. *Motional frequencies*

Secular motion Measuring the secular motional frequencies (macromotion) of the trapped ions provides information about the stiffness of the trap, eventual screening effects, and possible asymmetries. Many electronic methods have been developed in order to precisely measure frequencies since this is the basic tool for all mass spectrometry experiments (see Section 2.1). Motional frequencies can also be measured via the fluorescence emitted by the ions, the easiest application using a cloud of laser-cooled ions. If an additional small oscillating voltage (a "tickle") is applied to the trap, the ions will absorb energy and be slightly heated, if they are in resonance with the tickle frequency. This response can be made visible on the fluorescence signal by a drop in the maximum amplitude (see Figure 2.4). Scanning the tickle frequency will thus reveal all motional frequencies of the trapped ions if the applied tickle amplitude is not too strong and the scanning frequency not too fast, as the cloud has to be re-cooled after each excitation.[31]

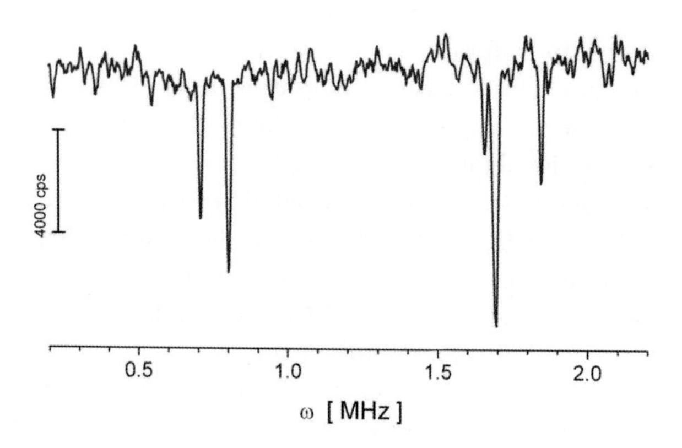

Fig. 2.4. Motional frequency spectrum of a laser-cooled cloud of calcium ions. On this graph the ion's fluorescence is recorded while varying the excitation tickle frequency.

Micromotion The micromotion amplitude of a trapped ion increases with its distance from the trap center. Moving ions in the trap is a way of obtaining information about the trapping field. Manipulation of the probe can be made by applying additional static voltages. Traps which are destined to trap a single ion are, in general, laid out to reach the Lamb–Dicke regime. In order to reach this, and to correct for eventual asymmetries of the trap (due to mounting or coating by the atomic beam during ion creation), these traps are mounted with a set of compensation electrodes, which allow – via the application of direct current (DC) voltages – the movement of the ion in the trap potential, such that it reaches the minimum of the trapping field.

Different techniques are used in order to compensate for excess micromotion which is induced by a defect or asymmetry in the potential. Large asymmetries can be observed on a camera, as the ion (or ion cloud) will move if the potential well is lowered. The fine tuning of the micromotion compensation is often carried out by recording the correlation signal of the photon count with the trapping radiofrequency.[32] The more micromotion the ions experience the larger the amplitude of this correlation signal, and this therefore constitutes an extremely precise tool for canceling defects of the trapping potential.

2.2.4. *Ions as a spatial probe*

In a seminal experiment, Guthöhrlein *et al.* used a trapped ion to explore the mode structure of the cavity which has been superimposed onto the trap.[33] They probed different Hermite–Gauss modes with an axial resolution of 170 nm, where the excitation rate of the trapped ion served as a sensitive detector.

2.2.5. *Temporal Ramsey*

The first atomic frequency standard based on the interrogation of laser-cooled ions was reported in 1985 by the National Institute of Standards and Technology (NIST) Ion Storage Group.[34] The probed clock transition was a hyperfine transition of the Be^{+}-ion in the microwave region. Ramsey's method of separated oscillatory fields has been implemented in a temporal regime. Actually, a radiofrequency pulse of duration t was applied, followed by a free precession interval of variable duration T (which could be as large as 19 s) and a second radiofrequency pulse of duration t coherent with the first pulse. This technique permitted the resolution of the atomic transition to better than 25 mHz, corresponding to a residual systematic uncertainty of 9.4×10^{-14}.

2.2.6. *Quantum logic spectroscopy*

Recently, the development of an innovative probing method has allowed the application of precision spectroscopy to ions that do not have a suitable atomic transition for laser cooling or atomic state preparation.[35] In a quantum logic spectroscopy experiment, two ions of different species are trapped. The first, "logic", ion is laser cooled, its atomic states are prepared via optical pumping and the read-out is made via the recording of its fluorescence. This ion provides sympathetic cooling to the second, "spectroscopy", ion. Entanglement of both ions is made via the common motional modes, and this allows a coherent transfer of the spectroscopy ion's internal state onto the logic ion, where it can then be measured with high efficiency. This technique allows the probing of ions which are not directly "visible" in the experiment, and is the basis for the most accurate optical atomic clocks.[36]

2.2.7. *Imaging techniques*

The visual observation of a single Ba^+-ion by Toschek's group in Heidelberg[37] was the first in a long line of spatial resolution images that have been taken of trapped ions. A couple of years later the first images of crystallized structures were published by Walther's group in Munich.[23] Today, images of ion clouds and crystals with single-ion resolution are a standard tool in Paul trap experiments.[38] The aspect ratio of a laser-cooled cloud gives insight into the stiffness of axial and radial potential in a linear trap. Dark sites in the crystal (due to other isotopes or other ions), allow the monitoring of the rearrangement of the ions on the crystal sites. Reaction rates of trapped ions can be determined by recording cloud sizes, shapes, and evolutions in the controlled atmosphere of a reactant.[39,40] In a perfect potential the observed crystal structures should rotate around the symmetry axis. Simulations have shown that very small defects in the potential introduce sufficient asymmetry for this not to happen in Paul traps.[41] In Penning traps, however, laser cooling will generate a rotating plasma of stable structure. Active stroboscopic techniques have to be implemented in order to synchronize the observation to the plasma rotating frequency.[42] This latter technique has allowed the direct observation of the structural phases of a crystallized ion plasma.[43]

2.2.8. *Optical setup and instrumentation*

Different detectors can be used in order to record the fluorescence of the trapped atoms. Traditionally, a photomultiplier (most often in the photon-counting mode) will collect the emitted photons. In order to gain information about the spatial distribution, CCD cameras are often used. These devices have to be coupled to image intensifiers in order to amplify the signal emitted by individual ions. Actually, intensifiers consist of a (wavelength sensitive) photocathode, an MCP, and a phosphor screen; these can be integrated to the camera or purchased externally. One drawback of the use of a camera is the slow response rate, rarely more than about a hundred images per second. Gateable devices can be synchronized onto well-defined trigger events and can overcome this limitation. Electron multiplier CCDs (EMCCDs) add supplementary gain to the detection process by impact ionization, although they are not gateable.

For the design of the optical setup, various external parameters have to be taken into account. The numerical aperture of the imaging objective will determine the solid angle for photon collection, which is one of the main factors in the determination of the detection efficiency. The working distance of the objective will depend on the position of the objective. If it is inside the vacuum vessel, it can be placed close to the trap, a perturbation of the trapping potential by the isolating (glass) surfaces must however be avoided. A recent idea[44] is the collection of atomic fluorescence by an optical fiber which is easy to implement in the trapping device.

In any configuration, the objective will be aligned in order to give an intermediate image. Only in this image plane can spatial filtering be made without losses of the signal amplitude; a diaphragm or pinhole can separate the atomic fluorescence signal from the scattered light of the exciting laser, which is in many cases at the same wavelength. The choice of the optics is vast, it can range from multi-surface systems to very simple lenses,[45] depending on the ultimate goal of the experiment. A very clever use of the individual ion detection is made by the Innsbruck group, where the high-resolution optical setup also serves to individually address single ions in a trapped chain.[46]

References

1. G. Knoll, *Radiation Detection and Measurement* (Wiley, Oxford 2010).
2. Y. A. Liu and R. H. Fleming, Reduced electron multiplier dead time in ion counting mass spectrometry, *Rev. Sci. Instrum.* **64**(6), 1661–1662 (1993).

3. J. L. Wiza, Microchannel Plate Detectors, *Nucl. Instrum. Methods* **162** (13), 587–601 (1979).
4. F. Major, V. Gheorge and G. Werth, *Charged Particle Traps*, Vol. 37, *Springer Series on Atomic, Optical, and Plasma Physics* (Springer, Heidelberg, 2005).
5. T. R. Rizzo, J. A. Stearns and O. V. Boyarkin, Spectroscopic studies of cold, gas-phase biomolecular ions, *Int. Rev. Phys. Chem.* **28**(3), 481 (2009).
6. J. P. Karr, A. Douillet and L. Hilico, Photodissociation of trapped H_2^+ ions for REMPD spectroscopy, *Appl. Phys. B* **107**, 1043–1052 (2012).
7. http://www.simion.com. (2011).
8. R. Alheit, C. Henning, R. Morgenstern, F. Vedel and G. Werth, Observation of instabilities in a Paul trap with higher-order anharmonicities, *Appl. Phys. B* **61**, 277–283 (1995).
9. R. Alheit, K. Enders and G. Werth, Isotope separation by nonlinear resonances in a Paul trap, *Appl. Phys. B* **62**, 511–513 (1996).
10. F. Vedel, M. Vedel and R. E. March, New schemes for resonant ejection in r.f. quadrupolar ion traps, *Int. J. Mass Spectrom.* **99**(12), 125–138 (1990).
11. M. Lunney, F. Buchinger and R. Moore, The temperature of buffer-gas cooled ions in a Paul trap, *J. Mod. Opt.* **39**(2), 349–360 (1992).
12. S. Shimma, H. Nagao, J. Aoki, K. Takahashi, S. Miki and M. Toyoda, Miniaturized high-resolution time-of-flight mass spectrometer Multum- S II with an infinite flight path, *Anal. Chem.* **82**(20), 8456–8463 (2010).
13. K. Blaum, High-accuracy mass spectrometry with stored ions, *Phys. Rep.* **425**(1), 1–78 (2006).
14. D. J. Wineland and H. G. Dehmelt, Principles of the stored ion calorimeter, *J. Appl. Phys.* **46**(2), 919–930 (1975).
15. M. Diederich, H. Häffner, N. Hermanspahn, M. Immel, H. J. Kluge, R. Ley, R. Mann, W. Quint, S. Stahl and G. Werth, Observing a single hydrogen-like ion in a Penning trap at T = 4 K, *Hyperfine Interact.* **115**, 185–192 (1998).
16. H. Häffner, T. Beier, S. Djeki, N. Hermanspahn, H. J. Kluge, W. Quint, S. Stahl, J. Verd, T. Valenzuela and G. Werth, Double Penning trap technique for precise g-factor determinations in highly charged ions, *Eur. Phys. J. D* **22**, 163–182 (2003).
17. M. Knoop, M. Vedel and F. Vedel, Investigations of a rf stored Ca+ ion cloud and observation of the S - D forbidden transitions, *J. Phys. II* **4**, 1639 (1994).
18. W. Neuhauser, M. Hohenstatt, H. Dehmelt and P. Toschek, Visual observation and optical cooling of electrodynamically contained ions, *Appl. Phys.* **17**, 123–129 (1978).
19. E. W. Streed, A. Jechow, B. G. Norton and D. Kielpinski, Absorption imaging of a single atom, *Nature Communications* **3**, 933 (2012).
20. F. Diedrich, E. Peik, J. Chen, W. Quint and H. Walther, Observation of a Phase Transition of Stored Laser-Cooled Ions, *Phys. Rev. Lett.* **59**, 2931–2934 (1987).
21. R. H. Dicke, The effect of collisions upon the Doppler width of spectral lines, *Phys. Rev.* **89**, 472 (1953).

22. D. J. Wineland and W. M. Itano, Laser cooling of atoms, *Phys.Rev.A* **20**, 1521 (1979).
23. F. Diedrich, J. Bergquist, W. Itano and D. Wineland, Laser cooling to the zero-point energy of motion, *Phys. Rev. Lett.* **62**, 403 (1989).
24. C. Lisowski, M. Knoop, C. Champenois, G. Hagel, M. Vedel and F. Vedel, Dark resonances as a probe for the motional state of a single ion, *Appl. Phys. B* **81**, 5–12 (2005).
25. H. Dehmelt, Proposed 10^{14} $\Delta\nu < \nu$ Laser Fluorescence Spectroscopy on Tl^+ Mono-Ion Oscillator II (spontaneous quantum jumps), *Bull. Am. Phys. Soc.* **20**, 60 (1975).
26. W. Nagourney, J. Sandberg and H. Dehmelt, Shelved optical electron amplifier: Observation of quantum jumps, *Phys. Rev. Lett.* **56**, 2797–2799 (1986).
27. T. Sauter, R. Blatt, W. Neuhauser and P. Toschek, Observation of quantum jumps, *Phys. Rev. Lett.* **57**, 1696–1698 (1986).
28. J. Bergquist, R. Hulet, W. Itano and D. Wineland, Observation of quantum jumps in a single atom, *Phys. Rev. Lett.* **57**, 1699–1702 (1986).
29. M. Knoop, C. Champenois, G. Hagel, M. Houssin, C. Lisowski, M. Vedel and F. Vedel, Metastable level lifetimes from electron-shelving measurements with ion clouds and single ions, *Eur. Phys. J. D* **29**, 163–171 (2004).
30. U. Eichmann, J. Bergquist, J. Bollinger, J. Gilligan, W. Itano, D. Wineland and M. Raizen, Young's Interference Experiment with Light Scattered from Two Atoms, *Phys. Rev. Lett.* **70**, 2359–2362 (1993).
31. C. Champenois, M. Knoop, M. Herbane, M. Houssin, T. Kaing, M. Vedel and F. Vedel, Characterization of a miniature Paul-Straubel trap, *Eur. Phys. J. D* **15**, 105 (2001).
32. D. Berkeland, J. Miller, J. Bergquist, W. Itano and D. Wineland, Minimization of ion micromotion in a Paul trap, *J. Appl. Phys.* **83**, 5025 (1998).
33. G. R. Guthöhrlein, M. Keller, K. Hayasaka, W. Lange and H. Walther, A single ion as a nanoscopic probe of an optical field, *Nature* **414**, 49 (2001).
34. J. J. Bollinger, J. D. Prestage, W. M. Itano and D. J. Wineland, Laser-cooled-atomic frequency standard, *Phys. Rev. Lett.* **54**, 1000–1003 (1985).
35. P. O. Schmidt, T. Rosenband, C. Langer, W. M. Itano, J. C. Bergquist and D. J. Wineland, Spectroscopy using quantum logic, *Science* **309**(5735), 749–752 (2005).
36. C. W. Chou, D. B. Hume, J. C. J. Koelemeij, D. J. Wineland and T. Rosenband, Frequency comparison of two high-accuracy Al^+ optical clocks, *Phys. Rev. Lett.* **104**(7), 070802 (2010).
37. W. Neuhauser, M. Hohenstatt, P. Toschek and H. Dehmelt, Localized Visible Ba^+ Mono-Ion Oscillator, *Phys. Rev. A* **22**, 1137 (1980).
38. L. Hornekær and M. Drewsen, Formation process of large ion Coulomb crystals in linear Paul traps, *Phys. Rev. A* **66**(1), 013412 (2002).
39. S. Willitsch, M. T. Bell, A. D. Gingell, S. R. Procter and T. P. Softley, Cold reactive collisions between laser-cooled ions and velocity-selected neutral molecules, *Phys. Rev. Lett.* **100**, 043203 (2008).
40. F. H. J. Hall, M. Aymar, N. Bouloufa-Maafa, O. Dulieu and S. Willitsch, Light-assisted ion-neutral reactive processes in the cold regime: Radiative

molecule formation versus charge exchange, *Phys. Rev. Lett.* **107**, 243202 (2011).
41. M. Marciante, C. Champenois, A. Calisti, J. Pedregosa-Gutierrez and M. Knoop, Ion dynamics in a linear radio-frequency trap with a single cooling laser, *Phys. Rev. A* **82**(3), 033406 (2010).
42. X.-P. Huang, J. J. Bollinger, T. B. Mitchell and W. M. Itano, Phase-locked rotation of crystallized non-neutral plasmas by rotating electric fields, *Phys. Rev. Lett.* **80**, 73–76 (1998).
43. T. B. Mitchell, J. J. Bollinger, D. H. E. Dubin, X.-P. Huang, W. M. Itano and R. H. Baughman, Direct observations of structural phase transitions in planar crystallized ion plasmas, *Science* **282**(5392), 1290–1293 (1998).
44. W. Lange, Private communication, 2009.
45. D. J. Berkeland, Linear Paul trap for strontium ions, *Rev. Sci. Instrum.* **73** (8), 2856–2860 (2002).
46. H. C. Nägerl, D. Leibfried, H. Rohde, G. Thalhammer, J. Eschner, F. Schmidt-Kaler and R. Blatt, Laser addressing of individual ions in a linear ion trap, *Phys. Rev. A* **60**, 145 (1999).

Chapter 3

Cooling Techniques for Trapped Ions

Daniel M. Segal
The Blackett Laboratory, Imperial College London, London SW7 2AZ, United Kingdom
d.segal@imperial.ac.uk

Christof Wunderlich
Department Physik, Universität Siegen, 57068 Siegen, Germany
wunderlich@physik.uni-siegen.de

This chapter gives an introduction to, and an overview of, methods for cooling trapped ions. The main addressees are researchers entering the field. It is not intended as a comprehensive survey and historical account of the extensive literature on this topic. We present the physical ideas behind several cooling schemes, outline their mathematical description, and point to relevant literature useful for a more in-depth study of this topic.

3.1. Introduction

Trapped ions are used in a wide variety of areas of experimental physics from high-precision measurements and metrology through to applications in quantum information processing (QIP) and quantum simulation (QS). The key to the success of many of these applications is the ability to cool the trapped ions to low temperatures (or, for single ions, more correctly, to reduce their kinetic energies to very low values). This article is based on a series of lectures given at the Winter School on Trapped Ions held in January 2012 at Les Houches in France. The aim of the lectures was to give an up-to-date account of the cooling techniques that have been employed for trapped ions in the past and to give an overview of new techniques currently being developed. Though other techniques are discussed, the article concentrates on the extremely successful approach of laser cooling (Section 3.3).

It is worth first considering how hot ions might typically be when they are first loaded into an ion trap. A typical trap has a depth somewhere in the region of 0.1–100 eV. Ions will remain trapped if they have this sort of kinetic energy when they are near the centre of the trap. Using $E \sim k_B T$ we find that ions with a temperature of $T \sim 1 \times 10^3$K can easily be trapped. In comparison to the kinds of traps available for neutral atoms these traps are extraordinarily deep. Furthermore traps can be made to be very small so that oscillation frequencies in ion traps can be very high, easily in the region of tens of MHz.

How hot the ions initially are depends critically on the loading technique. Ions may be either injected into a trap from outside or can be created *in situ* from neutral atoms passing through the trap. In the former technique ions are typically loaded in bunches. A typical trap has a direct current (DC) voltage applied to some sort of endcaps (which may be hollow to allow the passage of ions). The voltage of one endcap is temporarily brought to a low value while a bunch of ions enters the trap. The endcap voltage is then ramped up quickly so that the ions that bounce off the potential created by the far endcap find themselves trapped. The initial temperature of ions loaded in this way depends on the internal temperature of the initial ion bunch and upon the details of how the electrode pulsing is implemented, but typically initial temperatures are high and similar to the actual trap depth. In order to conduct useful experiments some form of cooling is often required so that the ions become well localized at the centre of the trap rather than filling its entire volume. Of course the density of ions near the centre of the trap is limited by their mutual Coulomb repulsion so that final densities are many orders of magnitude lower than they can be for trapped neutral atoms.

As an example of the injection loading technique, the HITRAP facility at the Gesellschaft für Schwerionenforschung (GSI) in Darmstadt[1] will load a variety of different experimental Penning traps with highly charged ions generated by smashing an energetic beam of heavy ions into a stripping target. Exotic ions up to and including hydrogen-like uranium (a uranium nucleus with only a single orbiting electron remaining) can be created in this way. After many stages of slowing and pre-cooling, bunches of highly charged ions at a temperature of around 4 K will be directed into an experimental trap. One such trap is operated by the SPECTRAP collaboration whose aim is to use cold trapped highly charged ions to make optical spectroscopic measurements of hyperfine transitions that, in ordinary ions, would be in the radio or microwave region of the spectrum. These experiments will allow very sensitive tests of quantum electrodynamics (QED) in situations where electric and magnetic fields are huge, so that perturbative approaches to QED are stretched to their limits. Despite the initial internal temperature of the ion bunch being quite low it is expected that imperfect control of the transport of these ions into the final trap and "shutting the door" on them will lead to relatively high initial temperatures in the trap in the region of an eV or more. A variety of in-trap cooling techniques will therefore be brought to bear once the bunch of highly charged ions has been captured.

3.2. Non-laser Cooling Techniques

3.2.1. *Electron cooling*

The pre-cooling of ions in the HITRAP facility will be accomplished using electron cooling. This is a technique often used in so-called "nested" Penning traps. A Penning trap can be comprised of a series of hollow cylindrical electrodes. By making a structure with a large number of such electrodes along a line, an array of traps for positively and negatively charged particles can be formed. By generating a W-shaped potential as shown in Figure 3.1, electrons and positive ions can be trapped in the same region of space wherein they can interact (the electrons are confined to the inverted well owing to their negative charge). An interesting example of this sort of cooling is in the field of antihydrogen production. In these experiments negatively charged antiprotons are cooled through collisions with positrons (see Madsen's chapter in this volume and the web pages of the ALPHA, Atomic Spectroscopy And Collisions Using Slow Antiprotons (ASACUSA), and Antihydrogen trap (ATRAP) collaborations at CERN).

As a precursor to these experiments, Hall *et al.*[3] performed a pilot experiment in order to develop the technique, in which protons were cooled through collisions with electrons. In their experiment the cylindrical electrodes could have voltages up to 150 V, their magnetic field was B=6 T and the trap was held in a cryostat at T=4.2 K. A 40 nA, 1 kV electron beam from a field emission point was fired along the B field towards a metal plate at the end of the trap. Hydrogen ejected from the plate was then ionized by the electrons. The protons generated were initially trapped in a shallow harmonic well (to the right of the W-shaped potential shown in Figure 3.1) formed by applying appropriate potentials to the trap electrodes.

The protons could then be loaded into the W-shaped potential well with a well-known initial energy. The protons could then be ejected from the trap and detected. By measuring the profile of their time-of-flight a temperature for the protons could be inferred. This temperature was found to be commensurate with the energy they had when they were injected into the trap. It was also possible to use the electron beam to pre-load some electrons into their trap. The electrons thermalize with the 4.2 K cryogenic background in about 0.1 s by emitting synchrotron radiation. Elastic collisions with the protons cool the protons and when they were subsequently ejected they were found to emerge at a significantly reduced temperature.

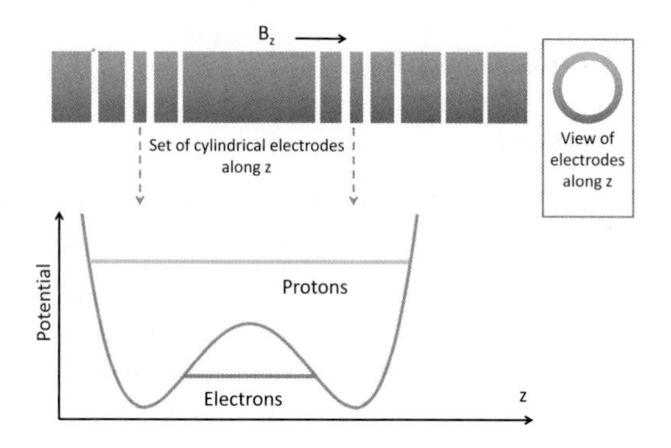

Fig. 3.1. Multiple electrodes along the trap axis allow complicated potentials to be created. A W-shaped potential can hold a small cloud of electrons in the middle of a larger cloud of protons. Recombination (to form neutral H which would then be lost from the trap) is surprisingly rare so that elastic collisions dominate leading to thermalization between the electrons and the protons. The electrons in their cyclotron orbits radiate rapidly removing kinetic energy from the system leading to "sympathetic" cooling of the protons. Note - for the antihydrogen work recombination is the goal so strategies for enhancing this process are necessary (adapted from reference[3]).

3.2.2. *Resistive cooling*

The idea here is to couple a trapped ion to an external circuit in such a way that the energy is dissipated in the circuit, cooling the ion. Following the approach outlined by Holzscheiter[4] we imagine a charged particle (mass m, charge q) oscillating between a pair of infinite parallel plates separated by $2z_0$ connected by a resistor R. The oscillating particle induces an image charge given by[4]

$$q' = (z_0 \pm z)q/2z_0. \tag{3.1}$$

This induces a current in the resistor

$$i = qv_z/2z_0. \tag{3.2}$$

As a result energy is dissipated at a rate

$$dE/dt = -i^2R = q^2RE/4mz_0^2. \tag{3.3}$$

This in turn leads to exponential cooling of the charged particle with a time constant given by[4]

$$t_{\mathrm{nat}} = 4mz_0^2/q^2R. \tag{3.4}$$

If the resistor is cooled, for instance by placing it in a cryogenic environment, then eventually the charged particle will come into thermal equilibrium with the resistor. Thus, this technique can, in principle, result in temperatures in the region of 4 K (liquid He), and, since initial ion temperatures can by as high as the trap depth, resistive cooling is often well worth implementing. In practice the technique is usually operated in a resonant fashion whereby the circuit connected to the ion is not simply resistive but has some reactance (in this case the external circuit is known as a "tank circuit"). While in principle this makes the technique more effective it adds a complication for clouds of ions where many resonant frequencies may be apparent due to the many modes of oscillation of the ion cloud. Resistive cooling also requires cryogenic operation which has a clear experimental overhead. As a result, unless cryogenic operation is required for some other reason, it is worth considering other cooling methods.

3.2.3. *Buffer-gas cooling*

Again, bearing in mind the very high temperatures that newly loaded ions may have in traps, even cooling the ions down to room temperature could be extremely beneficial for some experiments. In radiofrequency (RF) traps this can be accomplished by allowing a small amount of neutral "buffer gas" into the trap along with the trapped ions. The buffer-gas particles, which are in thermal equilibrium with the trap container through frequent collisions with the walls, collide with the ion, removing kinetic energy and transporting it to the walls of the chamber as heat. To avoid chemical reactions occurring between the neutral particles and the ion, the buffer-gas is chosen to be inert and typically the lightest noble gas available, helium, is used. A pressure of He in the region of 10^{-6} mbar suffices. This technique is used extensively in linear ion trap mass spectrometry to cool samples and increase sensitivity.

Buffer-gas cooling in a Penning trap is possible but requires an extra technique to be employed. In a Penning trap a static electric quadrupole potential is used to confine the ions in the axial direction. This static potential actually repels ions away from the centre of the trap towards the walls. The radial confinement in this trap comes from the imposition of a strong axial magnetic field. As a result the ions move around the top of a radial electrostatic potential hill in circular "magnetron" orbits. On a smaller scale the ions perform cyclotron orbits at a frequency that is reduced by the presence of the electric potential – the "modified cyclotron frequency". If energy is taken away from the ions by collisions with the buffer gas then they will tend to move down the electric potential and thus towards the walls of the trap.

The magnetron motion is therefore unusual insofar as it is unstable. The modified cyclotron motion is "normal" in the sense that removing energy from it results in smaller cyclotron loops. Buffer-gas cooling can be made to work in the Penning trap by a trick that couples the "normal" modified cyclotron motion to the unstable magnetron motion. In this way energy is transferred periodically between the two motions. While the ion's motion is predominately cyclotron-like the ion is cooled strongly by the buffer-gas collisions. For the right parameters this strong cooling can be made to dominate over the tendency of the ions to move outwards in the trap when the ion is moving in a more purely magnetron-like orbit.[5]

3.3. Laser Cooling

The idea of using lasers to cool samples of atomic particles was independently suggested for ions[6] and for neutral atoms[7] in 1975. The first experimental demonstrations of laser cooling of trapped atomic ions came a few years later in 1978,[8,9] soon followed by a thorough theoretical analysis specifically for trapped ions.[10] Laser cooling for a single ion was demonstrated in 1980.[11] Since these early days laser cooling has become a ubiquitous technique used extensively to produce extremely cold samples of both ions and neutral atoms. Some of the main ideas have been covered extensively elsewhere in the literature (e.g.,[12–15,17]), and we will limit ourselves to giving a basic description of the process here. However, there have been a number of recent developments specific to ion traps and we will focus more closely on these.

When an atom (or ion) absorbs a photon from a laser beam the momentum of the photon $\hbar k$ is imparted to the atom (k is the wavenumber). For visible photons $\hbar k$ is small, in the region of 1×10^{-27}kgm/s. Though this is indeed small, it is not entirely negligible compared to the relatively small momentum of a room temperature atom. If the laser beam is directed against the direction of motion of the atom this momentum kick slows the atom down. Laser cooling harnesses this interaction to bring gaseous samples of atoms or ions down to very low temperatures. However, there are some difficulties that need to be overcome. Once the atom has absorbed a photon it is in an excited state. After a short time (typically a few ns) the atom re-emits the photon spontaneously, receiving another momentum kick in the process. The atom is then ready to go around the cycle of absorption and emission again. At first sight it would seem that nothing has been gained, but closer consideration shows that the momentum kicks from the absorption events are additive, since all the photons involved come from the laser beam, whilst the spontaneously emitted photons have a nearly isotropic distribution, so that the recoil momentum kicks tend to cancel each other out over many cycles of the process. The fact that atoms can absorb and emit photons so rapidly means that scattering can provide an enormous deceleration to a moving atom – in the region of $10^5 g$ (g is the earth's gravitational acceleration). However, in order to realise the potential of this force, the Doppler effect must be taken into account.

Imagine a beam of atoms moving towards an oncoming laser beam. The problem is that atomic transitions are relatively narrow in frequency. In order for laser cooling to work the laser must be deliberately detuned to the red side of the natural transition frequency of the atom. In this way the moving atom is shifted into resonance with the laser and can scatter a photon. The next problem is that as an atom slows down, the Doppler effect takes it out of tune with the incoming radiation, potentially switching the process off after a few scattering events. It is easy to show that of the order of 10^4 scattering events are required to bring a room temperature atom to rest. There has been an enormous amount of activity in the area of laser cooling of neutral atoms and a variety of techniques have been developed that overcome the problems described above, allowing large numbers of scattering events to take place and eventually leading to ultra cold samples of neutral atoms which can be stored in a variety of neutral atom traps. Most of these techniques are not relevant to trapped ions. However it is important to understand that this type of cooling – called Doppler cooling – whilst being extremely effective, does have its limitations. In particular the description above assumes that atoms only have two energy levels. This is a gross simplification and in real atoms multiple laser frequencies are often required in order to pump atoms out of metastable energy levels where they could otherwise become stranded for long periods of time.

Although very low temperatures can be achieved using Doppler cooling there is a limit – the Doppler limit – and this limit *is* relevant to trapped ions. The argument above is rather too simplistic to capture the physical process that leads to this limit. Firstly, even a stationary atom does not absorb light at a single well-defined frequency – the atomic transition has a "lineshape" due to natural broadening that is Lorentzian. Secondly, the atoms or ions one wishes to cool do not typically all move in a set direction but instead move about randomly undergoing collisions with each other. At any given time some atoms in the gas will be moving towards the red-detuned laser and will thus be more likely to scatter a photon, however even atoms that are moving away from the laser beam will have some probability of absorbing a photon in the wing of the absorption line. For neutral atoms a strategy often employed is to use pairs of counter-propagating laser beams so that whatever direction the atom may be moving in, it will find itself Doppler shifted into resonance with a laser beam that tends to slow it down. By considering the scattering of photons from atoms moving in one dimension in counter-propagating red-detuned laser beams at this level of complexity it can be shown that the radiation pressure or "scattering force" F is given by

$$F \sim \hbar k^2 v, \tag{3.5}$$

where $\sim$ indicates that the right-hand side (r.h.s) of Eq. 3.5 is actually preceded by a numerical factor. If the laser power is low so that the transition is not overly saturated and the detuning is of the order of the natural linewidth of the transition γ then this numerical factor will be of order unity. This is a velocity dependent force and can therefore be viewed as a form of friction. This friction removes energy from the atom at the rate

$$-F.v \sim \hbar k^2 v^2. \tag{3.6}$$

In order to understand the Doppler limit it is necessary to revisit the above argument concerning the near-isotropic emission events causing momentum kicks that tend to cancel each other out over many absorption–emission cycles. Although this is broadly true, the recoil momentum kicks during spontaneous emission cannot be ignored completely. If we imagine an initially stationary atom undergoing this process in isolation we see it making a random walk in momentum space. If the atom can scatter photons at a rate R photons per second, then its average momentum will grow as

$$\Delta p = \sqrt{n}\hbar k = \sqrt{Rt}\hbar k. \tag{3.7}$$

Again, for well-chosen values of the detuning and the saturation parameter the scattering rate can be of order γ. We will henceforth replace R with γ but bear in mind that the equations that follow are only approximate. Over a time t, n photons will be scattered and the average increase in kinetic energy due to the random walk is

$$\frac{\Delta p^2}{2m} = \frac{\hbar^2 k^2 \gamma t}{2m}. \tag{3.8}$$

This means that the random spontaneous scattering of photons leads to a heating rate given by $\hbar^2 k^2 \gamma / 2m$. The limit to the achievable temperature can be found by equating this heating rate to the cooling rate and finding out what temperature results from this equilibrium:

$$\frac{\hbar^2 k^2 \gamma}{2m} = \hbar k^2 v^2 , \tag{3.9}$$

so that

$$v^2 = \frac{\hbar \gamma}{2m} . \tag{3.10}$$

The average kinetic energy in one dimension realised in this way can be related to temperature through

$$\langle \frac{mv^2}{2} \rangle = \frac{\hbar \gamma}{4} = \frac{1}{2} k_B T. \tag{3.11}$$

giving the famous Doppler limit[a]

$$\frac{\hbar\gamma}{2} = k_B T \ . \tag{3.12}$$

Amongst the first neutral atoms to be cooled in this way were Rb and Na for which the Doppler limits are 144 μK and 240 μK respectively (using the D line for laser cooling); astonishingly low temperatures by any reckoning. The ingenuity of workers in the field of laser cooling has led to a number of strategies for circumventing the Doppler Limit and reaching even lower temperatures. Most of these techniques are of limited relevance to trapped ions for which alternative "sub-Doppler" cooling techniques have been developed. These techniques will be the focus of the remainder of this chapter. To understand these techniques it is important to ask the question: what is fundamentally the lowest temperature that a collection of trapped ions can reach? It is important to remember that the ions are held in a harmonic potential well. At some level the quantization of the motion of the ion in this potential well must become manifest. Indeed the lowest temperature that the ions can reach is set by the zero-point or motional ground state energy of an ion held in this potential well. Since ion traps can be made so tightly confining, the ultimate temperature that trapped ions can reach is higher than the lowest temperatures achievable for trapped neutral atoms where the traps are in effect comparatively weakly confining. Nonetheless, as we will see, it is possible to cool trapped ions to the extent that they spend the vast majority of their time in the motional ground state and in this state the ions are as cold as they physically can be.

Trapping neutral atoms is often achieved using a so-called magneto-optical trap (MOT). In this device a number of laser beams derived from the same laser intersect over a small volume in a vacuum chamber where there is a spatially varying magnetic field. A combination of the laser detuning and the Zeeman effect ensure that an atom feels a scattering force directed towards the centre of the volume. This combination of a static magnetic field landscape and the light field lead to a velocity and position dependent force, and thus forms a "trap" for atoms.

[a]This simple approach captures the basic physics and produces the Doppler limit formula. That it includes the well-known factor of two in the denominator is fortuitous since it leaves out a number of important subtleties. It is based on a more rigorous treatment outlined in a series of masters level lectures on laser cooling of neutral atoms given by Professor E.A. Hinds. For a more comprehensive treatment aimed specifically at the special case of trapped ions see reference[10]

By contrast ions, being charged, can be trapped even without the imposition of a light field, and even when they are very hot. Laser cooling can then be applied as a separate consideration. This means that a trapped ion can in principle be laser cooled using a single laser beam. Indeed some species of ion (notably Be^+ and Mg^+) really do require only a single laser frequency and can be cooled with a single beam. These ions are unusual in that they do not have metastable levels below the first electronically excited state used for laser cooling. When such metastable levels are present the ion can be lost to them for long periods of time. The requirement for three-dimensional cooling is fulfilled by ensuring that the single laser beam is directed into the ion trap making an appreciable angle to all the principal axes of the trap.

It is important to realise that a single trapped ion does not move randomly in an ion trap – it oscillates harmonically at a set of well-defined motional frequencies. For Doppler cooling the normal situation is that these motional frequencies are much lower than the photon scattering rate. The strategy for laser cooling is therefore straightforward. A laser beam is sent into the trap and detuned to the red of the natural frequency of the ion. Most of the time the ion does not interact with the laser, but the ion's oscillations in the trap periodically bring it into resonance with the laser due to the Doppler effect. During brief intervals within each oscillation period the ion scatters some photons. Because of the judicious red detuning, this only occurs when the ion happens to be moving towards the laser beam. The ion thus slows down and its oscillatory amplitude is reduced. Even if only a few photons are scattered per oscillation cycle the ion gradually loses momentum in the trap and its motion is reduced. For a cloud of ions this constitutes cooling. Collisions between the ions complicate the overall motion but don't change the essential outcome.[13]

Despite the fact that, for ions, trapping and laser cooling are really separable techniques, and despite the significant differences in the details of how laser cooling is applied for ions, the physics behind the Doppler limit is unchanged and therefore this limit also applies to trapped ions. The rest of this chapter considers a variety of techniques that have been developed, or are currently under development, that allow cooling well below the Doppler limit to the quantum mechanical ground state of the motion in the trap potential.

So far we have considered an idealized atom or ion with only two energy levels: the ground state and a single excited state. Most trapped ions are significantly more complicated than this. The singly charged positive ions of the alkaline earths are the most commonly used ions for laser cooling. This is because the neutral atoms have two electrons outside full shells, which means that with one electron removed their spectra become particularly simple due to them having a single electron in the outer shell. They thus have spectra that resemble those of the neutral alkalis (which are the most popular neutral atom species for laser cooling). $^{40}Ca^+$ is one of the most commonly used ion species, largely because the laser frequencies required are all obtainable using solid state lasers. A simplified energy level diagram for Ca^+ is given in Figure 3.2. Laser cooling is performed on the dipole allowed $S_{1/2} - P_{1/2}$ transition at 397 nm. The ion can spontaneously emit to the $D_{3/2}$ level which is metastable, having a lifetime of $\sim$ 1 s. After scattering about 16 photons the ion will end up in the $D_{3/2}$ level, switching off the laser cooling for around a second. The result is that the ion would appear dark most of the time and cooling would be very ineffective. To avoid this a "repumper" laser resonant with the $D_{3/2} - P_{1/2}$ transition at 866 nm is shone into the trap so that the ion is rapidly put back into the S to P cooling cycle as soon as it drops into the $D_{3/2}$ level. The 397 nm laser is red-detuned by around a half a natural linewidth to provide laser cooling while the 866 nm laser is tuned to the centre of the $D_{3/2} - P_{1/2}$ transition.

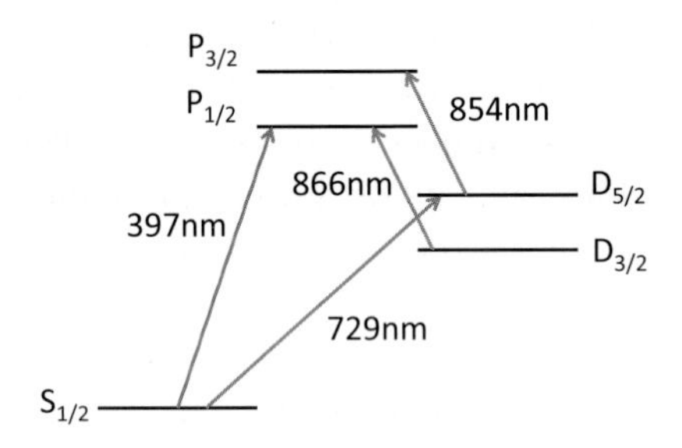

Fig. 3.2. Ca^+ level scheme.

One of the limitations of laser cooling is the large experimental overhead involved in setting up the laser systems required to cool a new species of ion. This typically requires a suite of lasers, all of which need to be carefully controlled. Fortunately, the long range of the Coulomb interaction between ions furnishes us with a very effective means of cooling a very wide range of ions that cannot be directly cooled with lasers. The approach is to co-trap laser-cooled ions with other ions which may be of interest for other purposes. These other ions undergo long-range Coulomb collisions with the laser-cooled ions and are thus "sympathetically" cooled in the process. These other ions may be different atomic ions or even molecular ions. Provided the mass difference between the laser-cooled species and the species to be sympathetically cooled is not too large, the cooling can be very efficient. If the mass difference is very large then the efficiency of the process is lost because the different species of ions settle into positions in the trap that are rather well separated, so that the interactions between the directly cooled ions and the sympathetically cooled ones become weak. In an extreme example of this process a small cloud of around 200 organic molecules of mass 410 Da were sympathetically cooled to a 115 mK by a cloud of Ba^+ ions.[18]

For a typical miniature RF ion trap, Doppler cooling leads to sufficiently low temperatures for the ion's motion in the trap to be manifestly quantum mechanical. In general this motion is a mixed state of the harmonic oscillator states of the ion in the trap potential. These states are characterized by a principal quantum number n and values of $\overline{n} \sim 10$ are quite normal, with lower values down to ~ 1 having been achieved for very small, steep traps.[19] So, although Doppler cooling is extremely effective, it is not normally capable of cooling ions into the quantum mechanical ground state for which $n = 0$. This is because of a "catch 22" situation: the Doppler limit is higher for transitions with high scattering rates. So one strategy might be to simply laser cool on a narrower transition with a lower scattering rate. However for such a transition the cooling rate is also very low. If there are other heating effects present, then achieving very low temperatures will be difficult, if the cooling rate is too low. In fact at this level there are indeed a number of heating effects in ion traps (some of which are not well understood) which are very difficult to control. For this reason a different strategy is needed to go beyond the Doppler limit. However it is important to realise that even though strategies do exist to achieve this, Doppler cooling is always used to pre-cool ions near to the Doppler limit before engaging these other approaches. Since the motion of a trapped ion at low temperatures is manifestly quantum mechanical, the starting point for understanding sub-Doppler cooling in ion traps must be to write down the Hamiltonian for the system of a trapped ion interacting with a laser field.

For reasons hinted at above, a narrow transition will be required in order to perform sub-Doppler cooling of the trapped ion. Fortunately, a very narrow transition is available, for example, in Ca^+ which is the $S_{1/2} - D_{5/2}$ transition shown in Figure 3.2. This transition is dipole forbidden but allowed as an electric quadrupole transition. The $D_{5/2}$ state decays to the ground state with a natural lifetime of around 1 s. It can be driven using a laser at 729 nm and a few microwatts at 397 nm is all that is required to saturate the normal Doppler cooling transition. In order to achieve excitation within a few ms on the 729 nm transition a great deal more laser power is required to oversaturate this weak transition. Typically a few milliwatts are required, however, this is easily within the reach of diode laser-based systems and Ti:Sapphire laser systems. The basic strategy for ground state cooling is to pre-cool the ion to the Doppler limit using the 397 nm and 866 nm lasers and then to switch to a different cooling scheme involving the narrow transition at 729 nm to remove the last few motional quanta from the system, leaving it in the ground state of its motion in the trap.

3.3.1. *Ion–laser interaction Hamiltonian*

Although the ion motion must be treated quantum mechanically, for most purposes we will find that a classical treatment of the radiation field will suffice. We will assume the ion is a two-level atom (TLA) with a ground state $|g\rangle$ and an excited state $|e\rangle$. It should be borne in mind that the ion will be cooled to the Doppler limit before beginning the sub-Doppler cooling process we are about to discuss. This process involves the narrow $S_{1/2} - D_{5/2}$ transition so that we may identify $|g\rangle$ with $S_{1/2}$ and $|e\rangle$ with $D_{5/2}$. In the following, we consider coherent deterministic (de-)excitation of the ion by the laser field only. For laser cooling, as discussed here, in addition a dissipative process is required in the form of spontaneous decay. This ensures that entropy is removed and the ion is cooled. An adequate mathematical treatment that includes spontaneous emission would require solving a master equation (e.g.,[13]). However, the treatment given below is sufficient to gain physical insight into different mechanisms that lead to cooling in the remainder of this chapter. More detailed developments of the material presented in this section are available in a number of places[b] (e.g.,[20–22]).

We assume the atomic wavefunctions are known so we can write the

[b]Many excellent PhD theses in Ion Trap Quantum Information Science are available online that present aspects of this material in greater detail than the published literature. In particular the authors have found a number of theses from the group of Rainer Blatt in Innsbruck to be an excellent resource and we have mostly followed the notation used by this group. Of particular interest in the context of this chapter is the thesis by C. Roos.

time independent Schrödinger equation for an isolated atom (ion) as

$$H_A|i\rangle = \hbar\omega_i|i\rangle\ ,\ i = g, e. \tag{3.13}$$

Using the closure theorem and orthonormality of $|i\rangle$ allows us to write the atomic part of the Hamiltonian as

$$H_A = \hbar\omega_g|g\rangle\langle g| + \hbar\omega_e|e\rangle\langle e|. \tag{3.14}$$

We define the zero of energy as being midway between $\hbar\omega_g$ and $\hbar\omega_e$ such that $\omega_e = \omega_a/2$ and $\omega_g = -\omega_a/2$ giving

$$H_A = \frac{\hbar\omega_a}{2}\{|e\rangle\langle e| - |g\rangle\langle g|\} = \frac{\hbar\omega_a}{2}\sigma_z, \tag{3.15}$$

where we have implicitly defined σ_z which is one of the Pauli matrices. The full Hamiltonian we are interested in must take into account both the ion's motion in the trap and its interaction with an applied laser field linking $|g\rangle$ and $|e\rangle$. We therefore write

$$H = H_A + H_m + H^{(i)}, \tag{3.16}$$

where H_m relates to the motion of ion of mass m in a one-dimensional harmonic trap and $H^{(i)}$ relates to the ion's interaction with the laser field. H_m can be written

$$H_m = \frac{p^2}{2m} + \frac{1}{2}m\nu^2 z^2 = \hbar\nu(a^\dagger a + 1/2), \tag{3.17}$$

where a and $a^\dagger$ are related to z and p through the usual relationships for a harmonic oscillator

$$z = \sqrt{\frac{\hbar}{2m\nu}}(a + a^\dagger), \qquad p = \sqrt{\frac{\hbar m\nu}{2}}(a^\dagger - a), \tag{3.18}$$

and ν is the angular frequency of the harmonic oscillation. We have in mind a single ion or a small chain of ions in a linear trap for which the harmonic motion along the trap axis z has a lower motional frequency (weaker binding) than the other two axes of the trap. We will only consider the motion in this one dimension. The Hamiltonian for the interaction between the ion and the applied laser field is given by

$$\begin{aligned} H^{(i)} &= -\mathbf{E}\cdot\mathbf{D} = E\hat{\mathbf{x}}\cos(\omega_L t - kz)\cdot\mathbf{D} \\ &= -\frac{E}{2}\{e^{i(\omega_L t - kz)} + e^{-i(\omega_L t - kz)}\}\hat{\mathbf{x}}\cdot\mathbf{D} \end{aligned} \tag{3.19}$$

which describes the laser propagating along the z direction polarized along x and $\mathbf{D}$ is the atomic dipole operator.

Now an operator O can be written

$$O = \sum_{i,j} O_{ij} |i\rangle\langle j|, \tag{3.20}$$

where $O_{ij} = \langle i|O|j\rangle$.
So the operator $\hat{\mathbf{x}} \cdot \mathbf{D}$ can be written

$$\hat{\mathbf{x}} \cdot \mathbf{D} = \sum_{i,j} \langle i|\hat{\mathbf{x}} \cdot \mathbf{D}|j\rangle |i\rangle\langle j| = \langle e|\hat{\mathbf{x}} \cdot \mathbf{D}|g\rangle \{|e\rangle\langle g| + |g\rangle\langle e|\}. \tag{3.21}$$

This gives

$$H^{(i)} = \frac{\hbar\Omega}{2}(\sigma_+ + \sigma_-)\{e^{i(\omega_L t - kz)} + e^{-i(\omega_L t - kz)}\}, \tag{3.22}$$

where we have introduced the definitions $\sigma_+ = |e\rangle\langle g|$ and $\sigma_- = |g\rangle\langle e|$ and we define the Rabi frequency Ω

$$\Omega = -\frac{E}{\hbar}\langle e|\hat{\mathbf{x}} \cdot \mathbf{D}|g\rangle. \tag{3.23}$$

Now $z = z_0(a + a^\dagger)$ ($z_0 = \sqrt{\hbar/2m\nu}$ is the root mean square extension of the ground state wavefunction), and we define the Lamb–Dicke parameter (LDP) $\eta = kz_0$ giving

$$H^{(i)} = \frac{\hbar\Omega}{2}(\sigma_+ + \sigma_-)\{e^{i\eta(a+a^\dagger)}e^{-i\omega_L t} + e^{-i\eta(a+a^\dagger)}e^{i\omega_L t}\}. \tag{3.24}$$

We will return later to an interpretation of the important LDP. For now we consider the physical interpretation of this Hamiltonian. The laser field can have two notable effects: it can (i) affect the motion of the active electron in the ion (σ_+, σ_-) and it can (ii) affect the motion of the ion in the trap ($a, a^\dagger$). The first of these actions is fairly obvious since the laser radiation can be in resonance with the atomic transition. The second action is less obvious since the motional frequency is very low and so direct resonant coupling of this motion to the laser field is out of the question. Critically, it transpires that (i) and (ii) go hand in hand and the key to understanding the interaction is to see how this works.

This is best done by making a transformation to the "interaction picture". This is a useful step when a Hamiltonian can be separated into time independent parts and a part which contains all the explicit time dependence. Looking back at Eq. 3.16 we note that H_A and H_m are themselves time independent and so lead only to the usual exponential time dependence associated with their respective eigenfunctions. On the other hand $H^{(i)}$ is itself a time dependent operator. The benefit of the "interaction picture" is that it allows us to home in on the effects caused by the time dependent interaction without the clutter of the normal time dependence associated with H_A and H_m.

In general if we can write a Hamiltonian as $H = H_0 + V(t)$ then H obeys the Schrödinger equation:

$$H\Psi = i\hbar\frac{\partial\Psi}{\partial t}. \tag{3.25}$$

Now, if we define a new wavefunction $\Psi' = e^{iH_0t/\hbar}\Psi$, then we also have $\Psi = e^{-iH_0t/\hbar}\Psi'$. It can be shown that the full Hamiltonian in the interaction picture, denoted $\bar{H}$, is given by

$$\bar{H} = e^{iH_0t/\hbar}Ve^{-iH_0t/\hbar} \tag{3.26}$$

where this Hamiltonian obeys the Schrödinger equation

$$i\hbar\frac{\partial\Psi'}{\partial t} = \bar{H}\Psi'. \tag{3.27}$$

Substituting $H_0 = H_A + H_m$ into this equation leads (after some algebra involving Taylor expansions of exponentiated operators) and after making the rotating wave approximation (neglecting terms that evolve with frequency $\omega_L + \omega_a$) to an expression for the full Hamiltonian for our system in the interaction picture . To simplify the notation we define $\tilde{a} = ae^{-i\nu t}$, $\tilde{a}^\dagger = a^\dagger e^{i\nu t}$ and $\Delta = \omega_L - \omega_a$ giving

$$\bar{H} = \frac{\hbar\Omega}{2}\{e^{i\eta(\tilde{a}+\tilde{a}^\dagger)}\sigma_+e^{-i\Delta t}\} + H.c. \tag{3.28}$$

Below we will see how the experimentally adjustable detuning Δ plays a decisive role in determining the response of the trapped ion's motion and internal state to the applied radiation field. In the following it is shown that choosing the right detuning picks out specific changes to the motional state to go hand in hand with the excitation or de-excitation of the internal state.

3.3.2. *Lamb–Dicke regime*

The motion of the ion in the trap means that, in its rest frame, it sees the monochromatic laser as being comprised of a range of frequencies. The fact that the motion is simple harmonic means that only certain discrete frequencies are present in the spectrum seen by the ion in its rest frame. For large amplitude oscillations the spectrum consists of an unshifted (carrier) frequency plus an array of positive (blue) and negative (red) sidebands, separated by the motional frequency ν (the amplitude of the motion acts as a "modulation index" in frequency modulation theory). For small oscillations the spectrum becomes simpler with just the carrier and a single pair of sidebands (one red and one blue) having an appreciable amplitude. We will put these notions on a firm mathematical footing later. For now we note that we are free to choose the laser frequency we apply. If we choose to apply the laser with $\omega_L = \omega_a$ then the ion in its rest frame will interact with the carrier. On the other hand if we apply radiation that is detuned by the motional frequency $\omega_L = \omega_a \pm \omega$ then the ion, in its rest frame, will interact with one of the sidebands.

Returning to the lab frame we should ask how it is that a *detuned* photon can be absorbed by the ion, changing its internal state by more (or less) than the photon energy, and yet conserving energy. The answer is that the ion's motion can be de-excited (or excited) in the process restoring the energy balance. On the other hand, if we tune the laser to resonance then there is no change to the ion's motional energy as a result of the photon absorption.

We will restrict ourselves to the situation where the motion of the ion is small enough that a single pair of sidebands is present. This is known as the Lamb–Dicke regime. For typical miniature RF traps, Doppler cooling will put an ion into the Lamb–Dicke regime. In the Lamb–Dicke regime the laser can cause a limited range of interactions: (i) it couples $|g,n\rangle$ to $|e,n\rangle$ by choosing it to be resonant, (ii) it couples $|g,n\rangle$ to $|e,n+1\rangle$ by detuning it to the blue sideband or (iii) it couples $|g,n\rangle$ to $|e,n-1\rangle$ by detuning it to the red sideband. We are therefore choosing $\Delta = \omega_L - \omega_a = \delta \pm \nu$ where δ allows for the un-ideal situation where the laser isn't exactly tuned to a sideband. We will show below how, mathematically, the choice of the detuning picks out the relevant interaction.

How small should the oscillations be in order for the spectrum to simplify to a carrier plus two sidebands? Classically the amplitude of the oscillation needs to be much smaller than $\lambda/2\pi$. The quantum mechanical analogue of the amplitude of the oscillation is given by the spread of the wavefunction $\hat{z}_n$, which is given by $(\langle z^2\rangle)^{1/2}$ where

$$\langle z^2\rangle = (2\langle n\rangle + 1)\hbar/2m\nu, \tag{3.29}$$

so that

$$\eta\sqrt{2\langle n\rangle + 1} = \hat{z}_n \frac{2\pi}{\lambda}. \tag{3.30}$$

The classical Lamb–Dicke criterion says that the r.h.s. must be much smaller than 1. Starting with the Hamiltonian of Eq. 3.28, below we will expand the operator with a and $a^\dagger$ in the exponent. Such an expansion to lowest order in η adequately describes the ion's dynamics only if

$$\eta\sqrt{2\langle n\rangle + 1} \ll 1, \tag{3.31}$$

which defines the quantum mechanical Lamb–Dicke regime.

At this stage it is useful to consider a Taylor expansion of the operator $e^{i\eta(\tilde{a}+\tilde{a}^\dagger)}$. In the Lamb–Dicke regime we can limit the Taylor expansion to first order

$$e^{i\eta(\tilde{a}+\tilde{a}^\dagger)} = 1 + i\eta(ae^{-i\nu t} + a^\dagger e^{i\nu t}). \tag{3.32}$$

The full Hamiltonian of Eq. 3.28 is completely general, but in the Lamb–Dicke regime, and given a particular laser tuning, it is well approximated by a much simpler Hamiltonian. In the same way as the rotating wave approximation picks out certain terms on account of their relatively slow time dependence ("stationary terms") the combination of Δ and ν in the above equation picks out particular terms when Eq. 3.32 is inserted into the full Hamiltonian of Eq. 3.28.

For $\Delta = 0$ (the carrier) and using Eq. 3.32 we find

$$\bar{H} = \frac{\hbar\Omega}{2}(\sigma_+ + \sigma_-). \tag{3.33}$$

For $\Delta = \nu$ (blue sideband) we have

$$\bar{H} = \frac{\hbar\Omega}{2}\eta(a^\dagger\sigma_+ + a\sigma_-). \tag{3.34}$$

For $\Delta = -\nu$ (red sideband) we have

$$\bar{H} = \frac{\hbar\Omega}{2}\eta(a\sigma_+ + a^\dagger\sigma_-). \tag{3.35}$$

This takes exactly the same form as the Jaynes–Cummings Hamiltonian of quantum optics with the exception that here a and $a^\dagger$ destroy and create quanta of motion in the harmonic well rather than photons of the electromagnetic field (the quanta of the motional state of an ion are often called phonons following the nomenclature used in condensed matter physics). We can read it as meaning that a phonon is destroyed when the ion is promoted from the $|g\rangle$ to $|e\rangle$ and a phonon is created when the ion makes a transition from $|e\rangle$ to $|g\rangle$. For this reason the Hamiltonian for the blue sideband is often referred to as the an anti-Jaynes–Cummings Hamiltonian, which is a new possibility revealed in this context.

3.3.3. *Coupling strength*

The strength of the coupling between an atom and the radiation field is given by the Rabi frequency. Building upon the ideas presented in the last section it is straightforward to calculate the coupling strength for interactions when the laser is detuned to the sidebands. This is taken into account through a modified Rabi frequency $\Omega_{m+n,n}$. In the Lamb–Dicke regime, only three interactions need to be considered, one for the carrier

$$\Omega_{n,n} = \Omega(1 - \eta^2 n) \tag{3.36}$$

and one for each of the two sidebands:

$$\Omega_{n+1,n} = \eta\sqrt{n+1}\,\Omega \quad \text{(blue sideband)}, \tag{3.37}$$

$$\Omega_{n-1,n} = \eta\sqrt{n}\,\Omega \quad \text{(red sideband)}. \tag{3.38}$$

For small η, these relations demonstrate the expected result that interaction on the sidebands is weaker than that on the carrier. In classical frequency modulation theory the red and blue sidebands have equal weight, however, note here that there is a purely quantum mechanical effect. For a given n the coupling on the red sideband is a little weaker than on the blue sideband. This imbalance gets greater as the system pushes deeper into the quantum regime where n is small. When $n = 0$ the coupling on the red sideband is zero. This is exactly what one would expect – with the laser tuned to the position of the red sideband and the ion already in $|n = 0\rangle$ there is no motional state below $|n = 0\rangle$ to which the excitation can take place. By tuning the laser to the red sideband the effect on the motional state is therefore *conditional* upon the initial motional state of the ion. If it is in an excited motional state then an interaction can occur whereas if it is already in the motional ground state nothing happens. This conditionality was exploited by Cirac and Zoller[23] in their original proposal for a two-qubit quantum gate using trapped ions (see Figure 3.3).

It is worth noting that while small η is desirable, since it simplifies the interactions, if η is *too* small then the interactions on the sidebands will be very weak. For applications such as normal sideband cooling described in the following section it is therefore necessary to use visible radiation since microwave transitions lead to very small values of η and, as such, do not allow for the simultaneous changes to the motional state of an ion that accompany changes to the internal state. Another way of viewing this is in terms of the momentum $\hbar k$ of the photons involved. For small k, as found in the microwave region of the electromagnetic spectrum, the photons simply do not carry enough momentum to give the ion the required momentum kick in order to change its motional state.

3.3.4. *Sideband cooling*

Imagine a single ion that has already been cooled on a strongly allowed transition to the Doppler limit in a trap which is sufficiently stiff that the ion is in the Lamb–Dicke regime with small $\eta \lessapprox 1$. To remove the last few quanta of motional energy we switch to "sideband cooling" on a narrow transition. By tuning the laser to the red sideband a photon will be absorbed which excites the ion internally but removes a quantum of motional energy (see Figure 3.3). The fact that the ion is in the Lamb–Dicke regime means that, in emission, the carrier transition is favoured so that when the ion returns to the internal ground state the motional state is on average unaltered. Just a few cycles of this process should leave the ion in the motional ground state. Once the ion is in the motional ground state the system uncouples from the radiation since there is no motional state below the ground state to which a transition might occur (as described above).

As described so far, this scheme has a major drawback. The cycle time is set by the natural lifetime of the upper state which is necessarily very long (since it is associated with a narrow transition). For example with Ca^+, it might take a second or longer for the ion to complete a cycle, if the upper state is metastable. Heating rates in small traps are usually not insignificant at this level and so some method for achieving a modest speed-up of the process is needed. This can be achieved by using another laser to couple the upper state of the narrow transition via an allowed transition to a higher lying state which *does* have a dipole allowed transition to the internal ground state of the ion (for Ca^+ this is the $D_{5/2} \to P_{3/2}$ transition at 854 nm, see Figure 3.2). The effect is to "lend some width" to the narrow transition so the ion recycles to the ground state more rapidly.[24]

An intercombination resonance in In^+ is sufficiently narrow (natural linewidth $\Gamma = 2\pi \times 360$ kHz) to be able to resolve motional sideband in typical ion traps, but still provides fast spontaneous decay to close the sideband cooling cycle. Sideband cooling of In^+ is reported in reference[25] .

There are some practical limits to the efficacy of sideband cooling. One

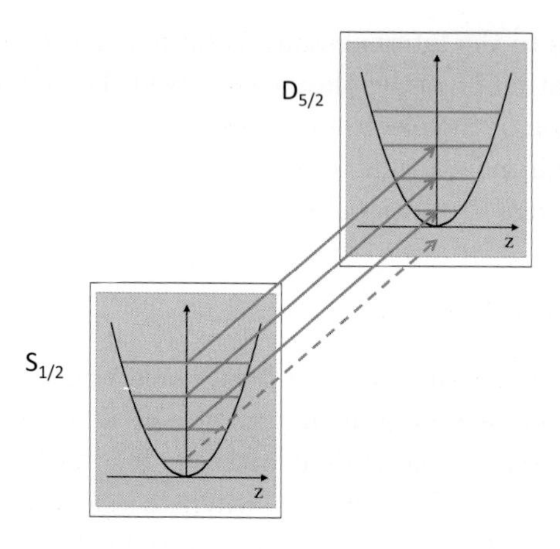

Fig. 3.3. The state of the ion is determined by its internal state ($S_{1/2}$ or $D_{5/2}$ for Ca^+) and its quantum state of motion in the trap potential. By tuning a laser to the red sideband, transitions that result in the loss of one quantum of motional energy are excited. When the ion returns to the lower internal state it does so whilst predominately conserving the motional quantum number. The ion thus steps down the ladder of motional states until it is in the ground state. Once the ground state is reached the interaction with the laser automatically switches off since there is no state below the ground state to which excitation can occur.

of these is associated with an oversimplification we have made in the discussion of the process. If a laser is tuned to the red sideband it is still capable of resulting in "off resonant" excitation on the carrier or, worse, on the blue sideband. Though these latter excitations will be rare, they actually put energy *into* the ion motion. Schemes for overcoming this drawback have been developed (see electromagnetically induced transparency (EIT) cooling below).

We have assumed that one privileged direction in the ion trap has a lower oscillation frequency than the others and have discussed sideband cooling in the context of the motion in this one direction. If cooling to the motional ground state in three dimensions is required, then the process needs to be stepped through the different frequencies of motion, adding a complication. If coupling between the motion in different directions is weak (as is usually the case) it may suffice to perform ground state cooling in only one dimension, leaving the ions relatively hot in the other modes (directions).

Sideband cooling was first performed for a single Hg^+ ion.[26] It has since been performed for a number of other species of ion (e.g.[25,27–29]). It has also been achieved for short strings of ions in RF traps.[30,31] An added complication for strings of ions is the additional motional modes of oscillation. For a single ion there is only one mode of oscillation in one dimension – the centre-of-mass (COM) mode. For two ions, there are two modes – the COM mode in which the two ions move in phase with each other and the "breathing mode" in which two ions move in anti-phase. Each extra ion added to a string of ions in the trap brings an extra mode in one dimension. In reality the spectrum of even a short string of ions in a trap is significantly more complicated than that alluded to in the discussion above of a carrier and a single pair of sidebands.

3.3.4.1. *Raman sideband excitation*

In the sections above it became clear why, for sideband cooling, laser light was used. The Lamb–Dicke parameter η, which is a measure for how well the internal dynamics of trapped ions can be coupled to their vibrational dynamics, should not be too small in order to be able to efficiently excite a motional sideband for cooling. Thus, typical Zeeman and hyperfine resonances in the RF regime (of order 1 MHz – 10 GHz) would not be suitable for sideband cooling. However, if a two-photon Raman transition between two such states is driven, then cooling becomes possible. The physical mechanism behind Raman sideband cooling will be outlined below.

If the two light beams driving a stimulated Raman transition between hyperfine states were parallel, that is, $\hat{k}_1 = \hat{k}_2$, where $\hat{k}_i \equiv \vec{k}_i/|\vec{k}_i|, i = 1, 2$ is a unit vector pointing along the propagation direction of the respective light beam, then the absorption of a photon from one beam and subsequent stimulated emission into the second Raman beam would lead to a negligible net transfer of momentum to the atom. The momentum transferred to the atom, potentially useful for cooling, would amount to $\hbar\vec{k}_1 - \hbar\vec{k}_2$, where the photons have a frequency difference corresponding to a typical Zeeman or hyperfine splitting. This is, in typical traps, not sufficient to efficiently (de-)excite vibrational motion. However, if the two Raman beams are antiparallel (i.e., $\hat{k}_1 = -\hat{k}_2$), then the net momentum transfer is $\hbar\vec{k}_1 + \hbar\vec{k}_2$ making (de-)excitation of vibrational motion efficient.[32,33] The difference vector $\hat{k}_1 - \hat{k}_2$ points along the direction of the vibrational motion which is to be cooled, for example, an axial or radial vibrational mode in a linear ion trap.

3.3.5. *Sideband cooling using RF radiation*

Above, we saw that Zeeman or hyperfine resonances of the electronic ground state can be used for Raman sideband cooling. Driving such a resonance in one-photon absorption and emission using RF radiation[c] results in negligible momentum transfer to the ion and thus does not allow for efficient cooling. For example, the Lamb–Dicke parameter $\eta \approx 10^{-7}$ for the ground state hyperfine resonance in $^{171}\mathrm{Yb}^{+}$. The following paragraphs are devoted to outlining a scheme which allows for controlling the motion of trapped ions using RF radiation, despite η having a negligible magnitude.

An illustration from classical physics may be useful in order to better understand how the ionic oscillator may be excited even though essentially no linear momentum is transferred from the absorbed photon. As a prototype of a harmonic oscillator characterized by angular frequency ν we consider a mass m attached to a massless spring exerting a force $\vec{F_S} = -m\nu^2\vec{z}$ on m, as sketched in Figure 3.4(a). We start with a motionless oscillator. In order to set the mass in motion, we can apply a force F for a time Δt resulting in a momentum kick $\Delta p = F/\Delta t$ and, as a consequence, the mass oscillates around its equilibrium position. In phase space, the impulsive excitation corresponds to a shift along the momentum axis followed by motion on a circle in phase space (Figure 3.4(a)). This motion is associated with cyclic transformation of the initial kinetic energy $(\Delta p)^2/2m$ into potential energy and *vice versa*. This way of exciting an oscillator would correspond to the motional excitation of a trapped ion by the absorption of a photon associated with a momentum kick. In the case of a quantum mechanical oscillator, energy conservation is satisfied by appropriately tuning the light field to a sideband (see Eqs 3.34 and 3.35).

Alternatively, the suspension point of the spring may be (nearly) instantaneously moved to a new equilibrium position (i.e., on a timescale Δt fast compared to the oscillation frequency of the mass), Δz away from the initial position. Due to its inertia, the mass m does not initially move. However, now it is no longer located at the equilibrium position of the harmonic potential (corresponding to the spring in its relaxed state). Instead, it is displaced from the spring's new equilibrium position, has acquired the potential energy $m\nu^2(\Delta z)^2$ and thus will start to oscillate around this new equilibrium position. The corresponding trajectory in phase space is again a circle signifying harmonic motion. A quantum mechanical analogue to

[c]Here we adhere to the definition of RF according to the Oxford English Dictionary: frequencies between about 300 GHz and 3 kHz.

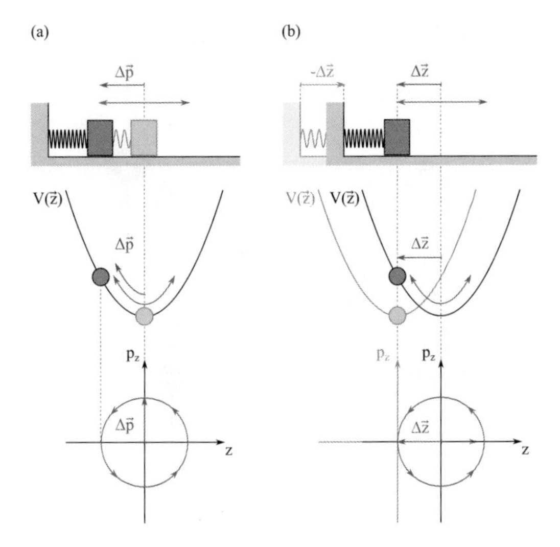

Fig. 3.4. Two ways of exciting a classical harmonic oscillator. a) A momentum kick Δp sets the mass m in motion. The trajectory in phase space is a circle. b) The oscillator's equilibrium position is nearly instantaneously shifted to the right by an amount Δz resulting again in harmonic motion of m around the new (shifted by Δz) equilibrium position.

this type of excitation is discussed in the following paragraphs.

Adding a state dependent force to the effective harmonic force confining a trapped ion has a similar effect. This is illustrated in Figure 3.5. Upon excitation from state $|0\rangle$ into state $|1\rangle$ the oscillator's equilibrium position is shifted by

$$\Delta z = F/(m\nu^2), \tag{3.39}$$

where $F = |\partial_z E_1(z)|$ or, using $E_1 = \hbar\omega$,

$$F = (\hbar/2)|\partial_z \omega(z)|, \tag{3.40}$$

that is, the force F is proportional to the magnitude of the gradient $|\partial_z \omega(z)|$ of the ionic angular resonance frequency $\omega(z)$. Such a gradient can be achieved by applying a spatially varying magnetic field $B(z)$ to the trapped ion.[34] The parameter that replaces the usual LDP and serves as a measure for how likely the motional excitation of an ion is when exciting its internal resonance is

$$\kappa \equiv \Delta z / z_0, \tag{3.41}$$

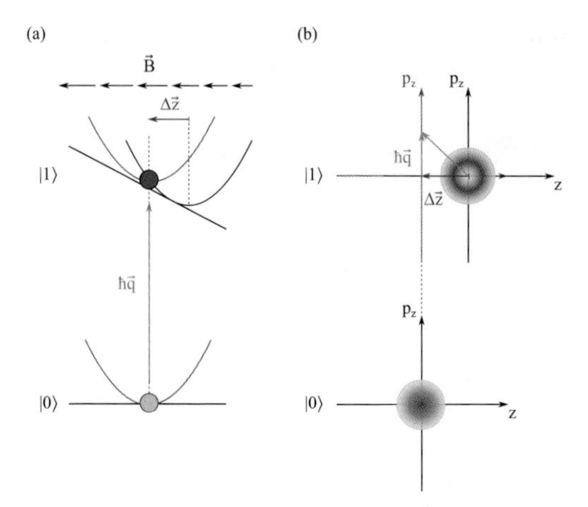

Fig. 3.5. Excitation of a quantum mechanical harmonic oscillator by a state dependently shifting its equilibrium position. (a) Upon excitation to state $|1\rangle$ the equilibrium position of the harmonic oscillator is shifted by an amount Δz. This shift is caused here by an additional linear potential superimposed with the harmonic oscillator potential. This energy shift may be due to a spatially varying magnetic field giving rise to a position dependent Zeeman shift of state $|1\rangle$. (b) Illustration in phase space. Here, the oscillator is excited from its ground state to its first excited state (a Fock state characterized by vibrational quantum number $n = 1$). In addition, if the energy splitting between the atomic states, $\hbar\omega_0$is sufficiently large, the atom may experience an appreciable momentum kick $\hbar\omega_0/c$ associated with the absorption of a photon. This is not the case if the resonance ω_0 lies in the RF regime.

with the root mean square extension of the ground state wavefunction, z_0 of a trapped ion. Using Eqs 3.39 and 3.41, the coupling strength κ can also be expressed as

$$\kappa = \frac{z_0|\partial_z \omega(z)|}{\nu}, \tag{3.42}$$

which indicates that κ measures the change in the ion's resonance frequency when the ion is moved over a distance z_0 in units of the harmonic oscillator frequency. Thus, coupling between internal and motional dynamics becomes efficient, and cooling by scattering of photons is possible, even when RF radiation is used.[34,36,37]

A proper mathematical description of the interaction between a trapped atom and electromagnetic radiation in the RF regime shows that the the usual LDP η is replaced by a new effective LDP,[34]

$$\eta_{\text{eff}} = \eta + i\kappa \equiv \eta' e^{i\theta} \,. \tag{3.43}$$

Even if the motional excitation of the ion relies on the state dependent oscillator potential as described above, there is linear momentum associated with the absorption of a photon (i.e., LDP $\eta \neq 0$). However, this momentum is appreciable only if radiation of high enough frequency is employed to drive the ionic resonance. In Eq. 3.43, η accounts for absorption of linear momentum while κ accounts for the effect due to the gradient in the resonance frequency. This expression reduces to $\eta_{\text{eff}} = \kappa$ (neglecting a global phase factor) when $\eta \approx 0$, which is the case for RF radiation. Then, the Hamiltonian reads[34] (in an interaction picture under the rotating wave approximation, expanded up to first order in η_{eff} [compare Eqs 3.28 and 3.32])

$$\tilde{H}_l = \frac{1}{2}\hbar\Omega_R \left[e^{-i[\Delta t+\phi]}\sigma_+ \left[1 + i\eta_{\text{eff}}\left(a^\dagger e^{i\nu_z t} + ae^{-i\nu_z t}\right)\right] + H.c.\right]. \quad (3.44)$$

If we set the detuning $\Delta = -\nu_z$, then we obtain the Hamiltonian Eq. 3.35 with η replaced by η_{eff}, which is the Hamiltonian that was shown to make sideband cooling possible (the phase ϕ in the above Hamiltonian is not relevant here). Thus, in order to achieve cooling, the red motional sideband can be driven by long-wavelength RF radiation.[35] Such resonances in the electronic ground state are characterized by the absence of spontaneous emission (for all practical purposes). Therefore, after absorption of a photon on the red sideband, and the ensuing loss of one motional quantum, the ion has to be returned to its initial state by a suitable optical pumping process, as was mentioned in Section 3.3.4.

The field that leads to magnetic gradient induced coupling (MAGIC) between internal and motional states of ions, simultaneously, makes ions in an ion Coulomb crystal individually distinguishable by their resonance frequency.[34,37,38] This is the case if the ionic resonance exhibits a Zeeman shift. For this purpose, instead of applying a magnetic gradient one could also use a spatially varying ac Stark shift induced by an additional light beam[39] to individually address ions by selecting their respective resonance frequency. In addition, this Stark shift gradient gives rise to an effective LDP that could potentially be used to drive motional sidebands in the long-wavelength range. The magnitude of this effective LDP induced by an ac Stark shift gradient will depend on atomic properties that, for the typically used ion species, are known or could be measured, and on the intensity profile of the light beam, and its detuning. Another possibility for sideband cooling is to take advantage of the gradient of an RF field driving a hyperfine transition to induce coupling between internal and motional states and thus sideband cool trapped ions.[40]

3.3.5.1. *Simultaneous cooling of many vibrational modes*

Usually, if one wants to sideband cool N modes of an N-ion crystal (considering one spatial direction only), the radiation driving an internal atomic resonance is red-detuned by the frequency of the motional mode to be cooled. Thus, only this one mode loses energy during repeated cooling cycles and is cooled efficiently. If cooling of more than one mode is desired, then the light field has to be tuned sequentially to the resonances corresponding to the respective modes (or, alternatively, light fields at multiple sideband resonances are applied simultaneously). In a Coulomb crystal consisting of N trapped ions, the magnetic gradient that induces η_{eff}, at the same time, shifts the resonance frequency of each ion depending on its position in the crystal. This effect can be put to use for simultaneously cooling a number of vibrational modes[36] as is outlined below.

The magnetic gradient could be designed so that the red sideband resonance, corresponding to the COM mode at frequency ν_1, of ion 1 coincides with the red sideband, corresponding to the mode with the next higher frequency ν_2, of ion 2, and so on for the other ions . In this way, a red sideband resonance of each one of N ions is used for cooling a particular vibrational mode. Since all these resonances coincide, it is sufficient to irradiate all ions with just a single frequency field in order to achieve simultaneous cooling of all N vibrational modes.

3.4. Laser Cooling Using Electromagnetically Induced Transparency

We have seen how the simultaneous cooling of several vibrational modes could be achieved by suitably shaping the atomic absorption spectrum such that several motional red sidebands are simultaneously excited with a fixed-frequency field. A suitably shaped atomic spectrum that allows for the simultaneous cooling of more than one mode may also be achieved by employing electromagnetically induced transparency.

As mentioned in Section 3.3.4, when applying usual sideband cooling, a limit on the achievable cooling rate is imposed by power broadening of the red sideband resonance which leads to non-resonant excitation of the carrier resonance and even of the blue sideband, thus giving rise to transitions that contribute to the heating of the ions. If these spurious transitions could be suppressed, then the cooling limit and the cooling rate would benefit. Therefore, a second effect of a deliberately and suitably modified atomic absorption spectrum (in addition to the cooling of multiple modes) – useful for cooling – can be such that it effectively suppresses unwanted resonances contributing to heating of the ions; for example, the carrier and blue sideband resonance.

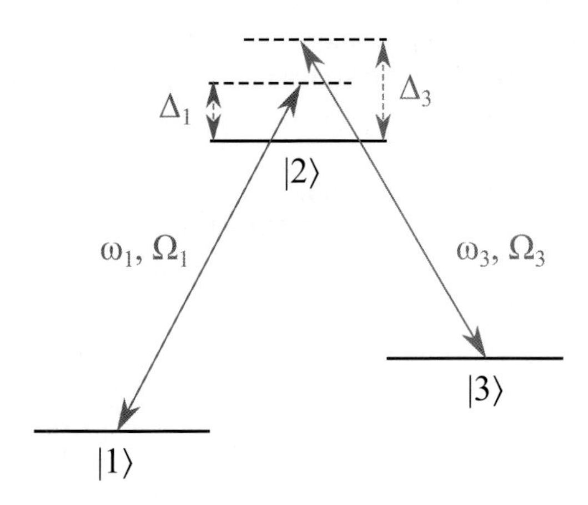

Fig. 3.6. Generic 3-level system that exhibits electromagnetically induced transparency useful for the cooling of trapped ions.

We will now outline how the atomic spectrum can be modified in a desired way using coherent driving fields. For our purposes, the latter typically are in the RF range, or between the infrared and ultraviolet range of the electromagnetic spectrum. For brevity, we will refer to such an electromagnetic field as the "light field" or simply "field". Different theoretical treatments of the light–atom interaction may be used here. A perturbative treatment is valid only under particular conditions imposed on the relative strength of the parameters characterizing the atom interacting with the field(s). The optical Bloch equations are exact but do not always provide a clear insight into the physical processes that take place in the atom–field interaction. The dressed state picture can be useful in identifying the states and scattering amplitudes that play a role for a given set of parameters. Here, we view the atom–field interaction from each one of these three different angles, depending on the context and on the insight we wish to obtain.

For the moment we are interested in coherent dynamics only, and do not take into account spontaneous emission. If we deal with hyperfine and Zeeman states connected via magnetic dipole resonances in the RF regime, then this is a valid approach. However, for laser cooling an additional dissipative process is required. If a three-level system (as shown in Figure 3.6) is implemented with optical resonances, this requirement is realized by dipole allowed spontaneous emission from state $|2\rangle$ into states $|1\rangle$ and $|3\rangle$ with rates $\beta \times \Gamma$ and $(1-\beta) \times \Gamma$, respectively ($\beta \in [0,1]$). Incidentally, this implies that there is no dipole allowed transition between states $|1\rangle$ and $|3\rangle$. Long-lived hyperfine ground states may be coupled to an extra level by laser light to induce an effective spontaneous decay of one of these states.

Now we consider a three-level atom and two light fields as depicted in Figure 3.6. First, it will be shown how the atomic response to a probe field (here, at frequency ω_3) can be modified when applying a second driving field (here, at frequency ω_1). For explaining how the spectral response of an atom can be modified using coherent light, the dressed atom picture is useful[41–43] where the electromagnetic field is quantized.

In order to appreciate how the absorption spectrum is modified, let us first consider the atomic system reduced to two levels $\{|1\rangle, |2\rangle\}$ driven by a field with frequency ω_1. The dipole interaction between the field $\vec{F_1} \cos(\omega_1 t)$ with (electric or magnetic) amplitude $\vec{F_1}$ field and the atomic (electric or magnetic) dipole $\vec{d_1}$ is

$$V = -\vec{d_1} \cdot \vec{F_1}/\hbar \propto (b_1 + b_1^\dagger) \tag{3.45}$$

($b_1^\dagger$ and b_1 are the creation and annihilation operator, respectively for the field mode at frequency ω_1). For a linearly polarized quantized field whose axis of polarization is adopted as the x-axis, this interaction reads (using $\sigma_x = \sigma_+ + \sigma_-$, not considering a time dependence)

$$V = \frac{\hbar\Omega_1}{2}\left[(\sigma_+ b_1 + \sigma_- b_1^\dagger) + (\sigma_+ b_1^\dagger + \sigma_- b_1)\right] . \tag{3.46}$$

The Rabi frequency Ω_1 (here chosen to be a real number) characterizes the strength of the coupling. Below, we will retain only the first two terms on the r.h.s. of Eq. 3.46 which describe resonant coupling between atom and field (corresponding to the rotating wave approximation in a semi-classical description). These terms describe the excitation of the atom while simultaneously taking away one photon from the field, or *vice versa.*

Thus, the Hamiltonian describing the atom–field system is

$$H_1 = \frac{\hbar\omega_a}{2}\sigma_z + \hbar\omega_1 b_1^\dagger b_1 + \frac{\hbar\Omega_1}{2}(\sigma_+ b_1 + \sigma_- b_1^\dagger) , \tag{3.47}$$

where we have included the field itself, not only the dipole interaction with the atom, into the Hamiltonian. The first two terms on the r.h.s. of Eq.

3.47 describe the eigenenergies of the uncoupled atom–field system. In Figure 3.7(a) these energy levels are sketched for the resonant case (i.e., $\Delta_1 = \omega_a - \omega_1 = 0$); the atomic level $|1\rangle$ in the presence of $N+1$ photons is degenerate with atomic level $|2\rangle$ in the presence of N photons.

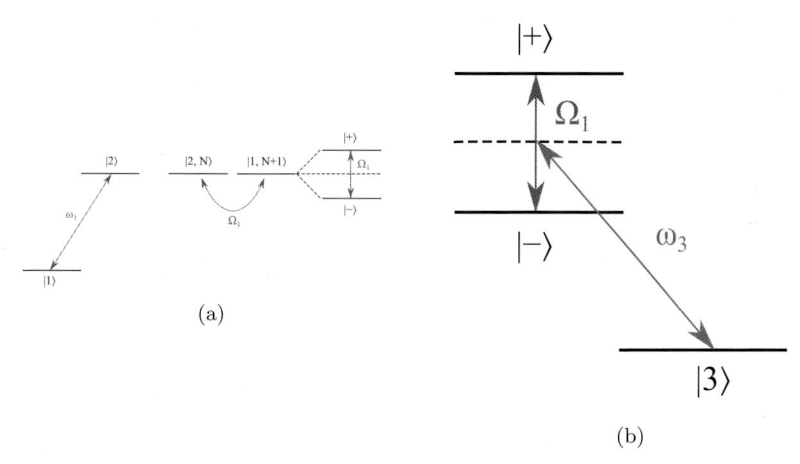

Fig. 3.7. (a) Atomic states $\{|1\rangle, |2\rangle\}$ dressed by the field with frequency ω_1. When the field is included explicitly into the Hamiltonian, then state $|1\rangle$ with $N+1$ photons is degenerate with state $|2\rangle$ in the presence of N photons. The dipole coupling between field and atom characterized by Rabi frequency Ω_1 lifts this degeneracy and gives rise to the eigenstates (dressed states) $|+\rangle$ and $|-\rangle$. (b) The weak probe field at frequency ω_3 is swept across the the level structure formed by the dressed states $|+\rangle$ and $|-\rangle$ (here, the probe field is tuned midway between the two expected resonances).

When the coupling term (the last term in Eq. 3.47) is taken into account, then the field ω_1 gives rise to the dressed states ($\Delta_1 = 0$, Figure 3.7(a))

$$|+, N\rangle = 1/\sqrt{2}(|1, N+1\rangle + |2, N\rangle), \tag{3.48}$$

$$|-, N\rangle = 1/\sqrt{2}(|1, N+1\rangle - |2, N\rangle), \tag{3.49}$$

where N is the number of photons at frequency ω_1 present in the field (Figure 3.7(a)). These are stationary eigenstates of H_1 (eq. 3.47). Although we use the dressed state picture that arises when the field is quantized, here we don't take into account particular effects due to field quantization (the "graininess" of the field). Instead, we consider strong coherent fields with the field amplitude proportional to $\sqrt{\bar{N}}$, and with a large mean photon number $\bar{N}$ (i.e., small relative fluctuations $\Delta N/\bar{N} \ll 1$, but $\Delta N \gg 1$). We are not interested in small changes to the field due to the presence of the atom, and, therefore, omit the label N henceforth.

Obviously, atomic motion is not included in the Hamiltonian Eq. 3.47 (recall that $b^\dagger$ and b refer to the electromagnetic field). At this point we are interested only in understanding how the atomic absorption profile can be modified.

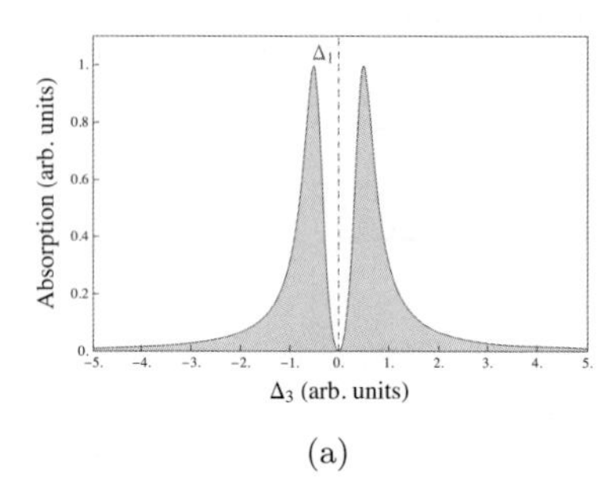

(a)

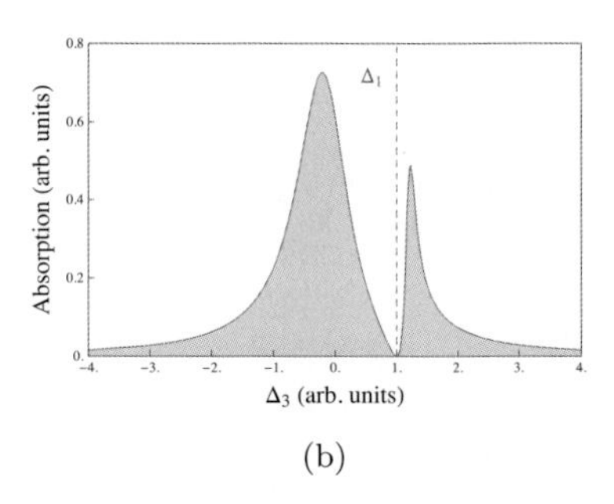

(b)

Fig. 3.8. Calculated absorption spectra of the 3-level system shown in Figures 3.6 and 3.7 as a function of the detuning Δ_3 of the probe field. (a) The field dressing states $|1\rangle$ and $|2\rangle$ are tuned to exact resonance ($\Delta_1 = 0$). Two absorption maxima are observed while zero absorption occurs when the probe field is tuned midway between the two dressed states $|+\rangle$ and $|-\rangle$ (compare Figure 3.7(b)), that is, $\Delta_3 = 0$. Parameters (dimensionless, in units of the decay rate Γ of state $|2\rangle$): $\beta = 0.5$, $\Omega_1 = 1$, $\Omega_3 = 0.01$, $\Delta_1 = 0$. (b) Detuning $\Delta_1 > 0$. The zero of the absorption probability occurs again at the two-photon-resonance condition $\Delta_3 = \Delta_1$. Parameters (dimensionless, in units of Γ): $\beta = 0.5$, $\Omega_1 = 1$, $\Omega_3 = 0.2$, $\Delta_1 = 1$.

The eigenstates of the atom–field system $|+\rangle$ and $|-\rangle$ are split by $\hbar\Omega_1$ (Figure 3.7). Now, we add the weak probe field with frequency ω_3 and Rabi frequency $\Omega_3 = -\vec{d_3} \cdot \vec{F_3}/\hbar$ (see Figure 3.6) scanning the level structure induced by the first field ω_1. For now, we assume $\Omega_3 \ll \Omega_1$ so that the probe field does not appreciably alter the level structure [d].

From the level structure displayed in Figure 3.7 we expect to obtain two resonances, when monitoring the absorption of this probe field as a function of its detuning, Δ_3. This is indeed the case as is shown in Figure 3.8(a). The width of the resonances seen in Figure 3.8 is determined by the spontaneous decay of state $|2\rangle$ with rate $(1-\beta)\Gamma$ into state $|1\rangle$ and with rate $\beta \times \Gamma$ into state $|3\rangle$. The fact that the absorption probability goes to zero for $\Delta_3 = 0 = \Delta_1$ can be understood by considering the amplitudes

[d]This assumption is introduced only for pedagogical reasons. It is not necessary for explaining the absorption spectra and will be dropped later.

for dipole transitions between state $|3\rangle$ and the states dressed by the ω_1-field, $|+\rangle$ and $|-\rangle$, respectively. These transition amplitudes are (in the perturbative limit regarding the interaction between field 3 and the atom) proportional to

$$\frac{\langle +|\vec{d}|3\rangle}{\Omega_1/2 - \Delta_3 + i\Gamma/2} \tag{3.50}$$

and

$$\frac{\langle -|\vec{d}|3\rangle}{\Omega_1/2 + \Delta_3 + i\Gamma/2}, \tag{3.51}$$

respectively. We note that

$$\langle +|\vec{d}|3\rangle = 1/\sqrt{2}(\langle 1| + \langle 2|)\vec{d}|3\rangle = +1/\sqrt{2}\,\langle 2|\vec{d}|3\rangle \tag{3.52}$$

$$\langle -|\vec{d}|3\rangle = -1/\sqrt{2}\,\langle 2|\vec{d}|3\rangle \tag{3.53}$$

(recall that there is no atomic dipole on the 1–3 resonance, $\langle 1|\vec{d}|3\rangle = 0$)). Therefore, if the probe field is tuned midway between the two resonances ($\Delta_3 = 0$), both induced atomic dipole moments contribute equally in magnitude but with opposite phase and thus cancel exactly giving zero absorption probability. The atom does not scatter light at frequency ω_3, in other words, it is transparent for this light, or viewed differently, the atomic sample remains dark.

Such a dark state was observed in sodium vapour[44] and in numerous other experiments. A review of related theoretical and experimental work is found in references[45,46] . Dark resonances in ionic spectra are reported, for example, using trapped Ba^+ ions,[47–49] with Ca^+,[50–53] with Sr^+,[54,55] and Yb^+,[56] and have been used for cooling of Ba^+[57] and Ca^+.[58]

A spectrum that displays a single resonance line for $\Omega_1 \to 0$ and two resonances for $\Omega_1 > 0$, as shown in Figure 3.8(a), is also known as an Autler–Townes doublet.[59]

So far, in order to obtain some insight into the physical mechanism of EIT, we considered particular combinations of parameters of the light field and the atom ($\Delta_1 = 0, \Omega_3 \ll \Omega_1$). The description of the atomic response can, of course, be generalized by allowing for arbitrary detuning Δ_1 and no longer requiring $\Omega_3 \ll \Omega_1$ when calculating the EIT absorption profile.[46,60]

In the considerations above, the field at frequency ω_3 was considered to be a weak field that probes the level structure induced by the field at frequency ω_1. Now, when the coupling strengths Ω_3 and Ω_1 are comparable in size, the field at frequency ω_3 itself alters the atomic response appreciably.

A characteristic feature of the atomic spectrum remains: a dark resonance (no scattered light) arises when $\Delta_3 = \Delta_1$.[60]

An example absorption profile that results when Δ_3 is scanned is displayed in Figure 3.8(b). For $\Delta_3 = \Delta_1$, a coherent superposition of state $|1\rangle$ and state $|3\rangle$ is created again as was discussed above, a dark state that does not absorb light.

We have seen that by employing two fields the absorption spectrum can be altered drastically. Such a modified absorption spectrum can be exploited for the cooling of ions: the absorption profile is positioned relative to the the sideband spectrum of the ion(s) such that a high absorption probability is obtained for the red sideband (leading to cooling) while the absorption probability for the carrier and the blue sideband (contributing to heating) remains small (see Figure 3.8(b)). With a Ca^+ ion, EIT cooling has been demonstrated.[58] For a collection of ions that form a Coulomb crystal it should be possible to make the absorption spectrum overlap with the sideband spectrum such that several vibrational modes are cooled simultaneously.

Already this relatively simple system (three atomic states and two fields) allows for modifying the atomic response considerably to achieve efficient laser cooling. Of course, in order to make this qualitative description of EIT cooling quantitative, terms that describe the atomic motion and spontaneous emission have to be added to the Hamiltonian. A detailed theoretical analysis of EIT cooling of a trapped atom for the case $\Delta_1 = \Delta_3$ (two-photon resonance) can be found in reference[61] . There, the cooling rate and the mean phonon number in steady state are given for an atom confined in a harmonic oscillator potential.

3.5. Cavity Cooling

The atomic absorption, emission, and coherent scattering characteristics can be modified strongly by placing an ion in a cavity[62–65] which could be exploited for cooling.[66–71] For example, enhancing atomic scattering into a cavity mode that is tuned to the blue sideband resonance of a trapped ion will lead to cooling of the ion when the incident photon had a lower frequency than the scattered photon.[72]

3.6. Cooling Scheme Combining Laser Light and RF

As outlined above, EIT cooling relies on shaping the atomic absorption and emission spectrum using laser fields that create light-induced states and shift them suitably. In reference[73] a different way of using the ac Stark shift is proposed that also eliminates the carrier transition that otherwise would slow down cooling and increase the steady-state temperature. One of the two fields employed there could be an RF field, while for repumping a laser field is used.

When laser radiation is applied to ions that are exposed to a magnetic field gradient, then the effective LDP, Eq. 3.43, has two components, one being η which is due to the momentum transferred to the ion, the other being κ, which stems from the state selective displacement (Figure 3.5(b)). This could be exploited for cooling: If, in addition to an RF field driving an ionic resonance, a laser field is used that induces a Raman transition, and the phase between these fields is adjusted properly, then (in a classical picture) the momentum kicks experienced by the ion due to light scattering and the displacement could constructively add up to (de-)excite the ion's motion. Here, only the relative phase between the fields is relevant. The proper quantum mechanical treatment shows that again the carrier and blue sideband transition could be eliminated by interference. A cooling scheme based on this that promises high cooling rates is proposed in reference 74.

A discussion of further interesting cooling techniques that have been or could be applied to trapped atomic and molecular ions (e.g.,[75–77]) is beyond the scope of this introduction. Let us mention a novel fast cooling scheme that relies on pulsed excitation of ionic resonances to exert properly timed momentum kicks to the ion in order to cool its motion.[78]

References

1. F. Herfurth et al., HITRAP-Heavy, highly charged ions at rest: Status and experimental opportunities, *J. Phys.: Conf. Ser.* **388**, 142009 (2012); F. Herfurth et al., Precision measurements with highly charged ions at rest: The HITRAP project at GSI, *International Journal of Mass Spectrometry* **251**, 266 (2006).
2. N. Madsen, (2013) "Antihydrogen formation and trapping" in M. Knoop, N. Madsen, and R. C. Thompson (eds), *Physics with Trapped Charged Particles*, Imperial College Press, London, pp. 219–238.
3. D. Hall and G. Gabrielse, Electron Cooling of Protons in a Nested Penning Trap *Phys. Rev. Lett.* **77**, 1962 (1996).

4. M. Holzscheiter, *Physica Scripta* **22T** 73 (1988).
5. M. König, G. Bollen, H.-J. Kluge, T. Otto, and J. Szerypo, Quadrupole excitation of stored ion motion at the true cyclotron frequency, *Int. J. Mass Spectrom. Ion Processes* **142**, 95 (1995).
6. D. Wineland and H. Dehmelt, *Bull. Am. Phys. Soc.* **20** (1975).
7. T. W. Hänsch and A. L. Schawlow, Cooling of gases by laser radiation, *Opt. Commun.* **13** 68 (1975).
8. W. Neuhauser, M. Hohenstatt, P. Toschek, and H. Dehmelt, Optical-sideband cooling of visible atom cloud confined in parabolic well, *Phys. Rev. Lett.* **41** (4), 233 (1978).
9. D. J. Wineland, R. E. Drullinger, and F. L. Walls, Radiation-pressure cooling of bound resonant absorbers, *Phys. Rev. Lett.* **40**, 1639 (1978).
10. D. J. Wineland and W. M. Itano, Laser cooling of atoms, *Phys. Rev. A* **20**, 1521 (1979).
11. W. Neuhauser, M. Hohenstatt, P. E. Toschek, and H. Dehmelt, Localized visible Ba+ mono-ion oscillator, *Phys. Rev. A* **22** 1137 (1980).
12. W. Itano and D. Wineland, Laser cooling of ions in harmonic and penning traps, *Phys. Rev. A* **25**, 35 (1982).
13. S. Stenholm, The semiclassical theory of laser cooling, *Rev. Mod. Phys.* **58**, 699 (1986).
14. W. M. Itano, J. C. Bergquist, J. J. Bollinger, and D. J. Wineland, Cooling methods in ion traps, *Physica Scripta.* (T59), 106 (1995).
15. C. Adams and E. Riis, Laser cooling and trapping of neutral atoms, *Progress in Quantum Electronics* **21** 1 (1997).
16. H. J. Metcalf, P. van der Straten, *Laser Cooling and Trapping.* (Springer, New York, 1999).
17. J. Eschner, G. Morigi, F. Schmidt-Kaler, and R. Blatt, Laser cooling of trapped ions, *J. Opt. Soc. Am. B.* **20** 1003 (2003).
18. A. Ostendorf, C. B. Zhang, M.A. Wilson, D. Offenberg, B. Roth, and S. Schiller, Sympathetic cooling of complex molecular ions to millikelvin temperatures, *Phys. Rev. Lett.* **97**, 243005 (2006).
19. S. Jefferts, C. Monroe, E. Bell, and D. Wineland, Coaxial-resonator-driven rf (paul) trap for strong confinement, *Phys. Rev. A* **51**, 3112 (1995).
20. R. Loudon, *The quantum theory of light.* (Oxford University Press, Oxford, 2000), third edition.
21. D. J. Wineland et al., Experimental issues in coherent quantum-state manipulation of trapped atomic ions, *J. Res. Natl Inst. Stand. Technol.* **103**, 259 (1998).
22. F. G. Major, V.N. Gheorghe, G. Werth *Charged Particle Traps.* (Springer, Berlin, 2005).
23. J. Cirac and P. Zoller, Quantum computations with cold trapped ions, *Phys. Rev. Lett.* **74** 4091 (1995).
24. R. Hendricks, J. L. Soerensen, C. Champenois, M. Knoop, M. Drewsen, Doppler cooling of calcium ions using a dipole-forbidden optical transition, *Phys. Rev. A* **77**, 021401(R) (2008).

25. E. Peik, J. Abel, T. Becker, J. von Zanthier, and H. Walther, Sideband cooling of ions in radio-frequency traps, *Phys. Rev. A* **60**, 439 (1999).
26. F. Diedrich, J. C. Bergquist, W. M. Itano, and D. J. Wineland, Laser cooling to the zero-point energy of motion, *Phys. Rev. Lett.* **62**, 403 (1989).
27. V. Letchumanan, G. Wilpers, M. Brownnutt, P. Gill, and A. G. Sinclair, Zero-point cooling and heating-rate measurements of a single $^{88}Sr^+$ ion, *Phys. Rev. A* **75** 063425 (2007).
28. F. Schmidt-Kaler et al., Ground state cooling, quantum state engineering and study of decoherence of ions in paul traps, *J. Mod. Opt.* **47** 2573 (2000).
29. C. Schwedes, T. Becker, J. von Zanthier, H. Walther, and E. Peik, Laser sideband cooling with positive detuning, *Phys. Rev. A* **69** 053412 (2004).
30. B. E. King et al., Cooling the collective motion of trapped ions to initialize a quantum register, *Phys. Rev. Lett.* **81** 1525 (1998).
31. T. Monz et al., 14-qubit entanglement: Creation and coherence, *Phys. Rev. Lett.* **106**, 130506 (2011).
32. C. Monroe et al., Resolved-sideband raman cooling of a bound atom to the 3d zero-point energy, *Phys. Rev. Lett.* **75** 4011 (1995).
33. G. Morigi, H. Baldauf, W. Lange, and H. Walther, Raman sideband cooling in the presence of multiple decay channels, *Opt. Commun.* **187**, 171 (2001).
34. F. Mintert and C. Wunderlich, Ion-trap quantum logic using long-wavelength radiation, *Phys. Rev. Lett.* **87** 257904 (2001).
35. A. Khromova et al., Designer Spin Pseudomolecule Implemented with Trapped Ions in a Magnetic Gradient. *Phys. Rev. Lett.* **108**, 220502 (2012).
36. C. Wunderlich, G. Morigi, and D. Reiss, Simultaneous cooling of axial vibrational modes in a linear ion trap, *Phys. Rev. A* **72**, 023421 (2005).
37. M. Johanning et al., Individual addressing of trapped ions and coupling of motional and spin states using rf radiation, *Phys. Rev. Lett.* **102** 073004 (2009).
38. S. X. Wang, J. Labaziewicz, Y. Ge, R. Shewmon, and I. L. Chuang, Individual addressing of ions using magnetic field gradients in a surface-electrode ion trap, *Appl. Phys. Lett.* **94** 094103 (2009).
39. P. Staanum and M. Drewsen, Trapped-ion quantum logic utilizing position-dependent ac stark shifts, *Phys. Rev. A* **66**, 040302 (2002).
40. C. Ospelkaus et al., Microwave quantum logic gates for trapped ions, *Nature* **476**, 181 (2011).
41. C. Cohen-Tannoudji and S. Haroche, Interprétation de diverses résonances magnétiques en termes de croisements et anticroisements de niveaux d'énergie du systéme global atome-photons de radiofréquence., *C. R. Acad. Sc. Paris* **262B**, 37 (1966).
42. C.Cohen-Tannoudji, *Atoms in strong resonant fields*, In eds. S. H. R. Balian and S. Liberman, *Frontiers in Laser Spectroscopy*, p. 1. North-Holland (1977).
43. C. Cohen-Tannoudji, J. Dupont-Roc, G. Grynberg *Atom-Photon Interactions.* (WILEY-VCH, Weinheim, 2004).
44. G. Alzetta, A. Gozzini, L. Moi, and G. Orriols, An experimental method for the observation of r.f. transitions and laser beat resonances in oriented na vapour, *Il Nuovo Cimento B (1971)* **36** 5 (1976).

45. E. Arimondo, *V Coherent Population Trapping in Laser Spectroscopy*, In ed. E. Wolf, *Prog. Opt.* **35**, 257 (1996).
46. M. Fleischhauer, A. Imamoglu, and J. P. Marangos, Electromagnetically induced transparency: Optics in coherent media, *Rev. Mod. Phys.* **77** 633 (2005).
47. G. Janik, W. Nagourney, and H. Dehmelt, Doppler-free optical spectroscopy on the Ba+ mono-ion oscillator, *J. Opt. Soc. Am. B.* **2**, 1251 (1985).
48. Y. Stalgies, I. Siemers, B. Appasamy, T. Altevogt, and P. E. Toschek, The spectrum of single-atom resonance fluorescence, *Europhys. Lett.* **35**, 259 (1996).
49. Y. Stalgies, I. Siemers, B. Appasamy, and P. E. Toschek, Light shift and fano resonances in a single cold ion, *J. Opt. Soc. Am. B.* **15**, 2505 (1998).
50. F. Kurth et al., Doppler free dark resonances for hyperfine measurements and isotope shifts in Ca+ isotopes in a Paul trap, *Z. Physik D.* **34**, 227 (1995).
51. M. J. McDonnell, D. N. Stacey, and A. M. Steane, Laser linewidth effects in quantum state discrimination by electromagnetically induced transparency, *Phys. Rev. A* **70**, 053802 (2004).
52. C. Lisowski, et al., Dark resonances as a probe for the motional state of a single ion, *Appl.Phys. B* **81** 5 (2005).
53. M. Albert, A. Dantan, and M. Drewsen, *Nature Photon.* **5**, 633 (2011).
54. G. P. Barwood, P. Gill, G. Huang, H. A. Klein, and W. R. C. Rowley, Sub-khz clock transition linewidths in a cold trapped 88sr+ ion in low magnetic fields using 1092-nm polarisation switching, *Opt. Commun.* **151**, 50 (1998).
55. T. Lindvall, M. Merimaa, I. Tittonen, and A. A. Madej, Dark-state suppression and optimization of laser cooling and fluorescence in a trapped alkaline-earth-metal single ion, *Phys. Rev. A* **86**, 033403 (2012).
56. H. A. Klein, A. S. Bell, G. P. Barwood, and P. Gill, Laser cooling of trapped Yb+, *Appl. Phys. B.* **50**, 13 (1990).
57. D. Reiss, K. Abich, W. Neuhauser, C. Wunderlich, and P. E. Toschek, Raman cooling and heating of two trapped Ba^+ ions, *Phys. Rev. A* **65**, 053401 (2002).
58. C. F. Roos et al., Experimental demonstration of ground state laser cooling with electromagnetically induced transparency, *Phys. Rev. Lett.* **85**, 5547 (2000).
59. S. H. Autler and C. H. Townes, Stark effect in rapidly varying fields, *Phys. Rev.* **100**, 703 (1955).
60. S. E. Harris, Electromagnetically induced transparency *Phys. Today* 26 (1997).
61. G. Morigi, Cooling atomic motion with quantum interference, *Phys. Rev. A* **67**, 033402 (2003).
62. G. M. Meyer, H.-J. Briegel, and H. Walther, Ion-trap laser, *Europhys. Lett.* **37**, 317 (1997).
63. G. R. Guthohrlein, M. Keller, K. Hayasaka, W. Lange, and H. Walther, A single ion as a nanoscopic probe of an optical field, *Nature* **414**, 49 (2001).
64. A. B. Mundt et al., Coupling a single atomic quantum bit to a high finesse optical cavity, *Phys. Rev. Lett.* **89** 103001 (2002).

65. P. F. Herskind, A. Dantan, J. P. Marler, M. Albert, and M. Drewsen, Realization of collective strong coupling with ion coulomb crystals in an optical cavity, *Nature Phys.* **5**, 494 (2009).
66. T. W. Mossberg, M. Lewenstein, and D. J. Gauthier, Trapping and cooling of atoms in a vacuum perturbed in a frequency-dependent manner, *Phys. Rev. Lett.* **67**, 172 (1991).
67. J. I. Cirac, M. Lewenstein, and P. Zoller, Laser cooling a trapped atom in a cavity: Bad-cavity limit, *Phys. Rev. A* **51**, 1650 (1995).
68. P. Horak, G. Hechenblaikner, K. M. Gheri, H. Stecher, and H. Ritsch, Cavity-induced atom cooling in the strong coupling regime, *Phys. Rev. Lett.* **79**, 4974 (1997).
69. V. Vuletic and S. Chu, Laser cooling of atoms, ions, or molecules by coherent scattering, *Phys. Rev. Lett.* **84**, 3787 (2000).
70. A. Beige, P. L. Knight, and G. Vitiello, Cooling many particles at once, *New J. Phys.* **7**, 96 (2005).
71. S. Zippilli and G. Morigi, Cooling trapped atoms in optical resonators, *Phys. Rev. Lett.* **95**, 143001 (2005).
72. D. R. Leibrandt, J. Labaziewicz, V. Vuletic, and I. L. Chuang, Cavity sideband cooling of a single trapped ion, *Phys. Rev. Lett.* **103**, 103001 (2009).
73. A. Retzker and M. B. Plenio, Fast cooling of trapped ions using the dynamical stark shift, *New J. Phys.* **9**, 279 (2007).
74. A. Albrecht, A. Retzker, C. Wunderlich, and M. B. Plenio, Enhancement of laser cooling by the use of magnetic gradients, *New J. Phys.* **13**, 033009 (2011).
75. G. Birkl, J. A. Yeazell, R. Räckerl, and H. Walther, Polarization gradient cooling of trapped ions, *Europhys. Lett.* **27**, 197 (1994).
76. J. Eschner, B. Appasamy, and P. E. Toschek, Stochastic cooling of a trapped ion by null detection of its fluorescence, *Phys. Rev. Lett.* **74**, 2435 (1995).
77. A. Beige, T. Freegarde, and F. Renzoni, New cooling mechanisms for atoms and molecules, *J. Mod. Opt.* **58**, 1297 (2011).
78. S. Machnes, M. B. Plenio, B. Reznik, A. M. Steane, and A. Retzker, Superfast laser cooling, *Phys. Rev. Lett.* **104**, 183001 (2010).

Chapter 4

Accumulation, Storage and Manipulation of Large Numbers of Positrons in Traps I – The Basics[a]

Clifford M. Surko
University of California, San Diego,
La Jolla CA 92093, USA
csurko@ucsd.edu

In this chapter, methods are described to create, store, manipulate and characterize positron plasmas. Emphasis is placed on the so-called buffer-gas positron trapping scheme for positron accumulation that uses positron-molecule collisions to accumulate particles efficiently. Manipulation and storage techniques are described that exploit use of the Penning–Malmberg trap, namely a uniform magnetic field with electrostatic confining potentials along the direction of the field. The techniques described here rely heavily on single-component-plasma research, and relevant connections are discussed. The use of rotating electric fields to compress plasmas radially (the so-called "rotating-wall" technique) is described; it has proven particularly useful in tailoring positron plasmas for a range of applications. The roles of plasma transport and available cooling mechanisms in determining the maximum achievable plasma density and the minimum achievable plasma temperature are discussed. Open questions for future research are briefly mentioned.

---

[a] This chapter, with corrections and minor updates as noted, is reprinted with permission from *Proceedings of the International School of Physics "Enrico Fermi"*, "Physics with Many Positrons", Course CLXXIV, in Brusa, R. S., Dupasquier, A., Mills, A. P., Jr. (eds), (IOS Press, Amsterdam, 2010), pp. 511–543.

4.1. Overview

The School on Physics with Many Positrons,[a] for which this and a companion[1,2] chapter were written, highlighted the fact that progress in the ability to accumulate and cool positrons and antiprotons is enabling new scientific and technological opportunities with low-energy antimatter. In a major sense, much of this work has its origins at the forefront of plasma physics research – the development of new ways to create and manipulate antimatter plasmas. These chapters describe the development of new plasma tools for this effort. Thus, they also fit well in this volume of lectures from the 2012 Les Houches Winter School on Physics with Trapped Charged Particles. The objective of these chapters is to provide a description of methods to efficiently accumulate, store and manipulate positrons in the form of single-component plasmas for use in a variety of applications. Aspects of these techniques are also relevant to the confinement and manipulation of antiprotons.[b]

Chapter 5 describes recently developed methods of creating positron beams with small transverse spatial extent. The prospects for accumulating and storing larger quantities of antimatter are also discussed in Chapter 5, namely a novel multicell positron trap capable of storing $\geq 10^{12}$ positrons for days or longer, as well as other selected topics. These chapters are intended to be tutorial in nature rather than a first description of research results. They borrow heavily from previously published material, sometimes repeating passages verbatim. This chapter relies heavily on the material in references[3–5]. The reader is requested to consult these and other original articles for further details.

Single-component plasmas (SCP) are the method of choice to accumulate, cool and manipulate large numbers of antiparticles. These collections of antimatter can be stored in a high quality vacuum for very long times using the suitably arranged electric and magnetic fields of a Penning–Malmberg trap[6] – this device functions as a nearly ideal

---

[b] There are, however, significant differences. Due to the antiproton's annihilation characteristics and heavier mass, the positron cooling techniques described here must be replaced, cooling the antiprotons sympathetically with cold electrons. Further, the accumulations of antiprotons to date have typically been gases of charged particles rather than plasmas.

electromagnetic bottle. Not only can these positron plasmas be made more or less arbitrarily free of annihilation, but, in addition, techniques are available to further cool, compress, and tailor them for specific applications. These antimatter plasmas now play an important role in science and technology and this is expected to continue.

Low-energy antimatter science relies upon many developments in positron technology. They include methods to cool plasmas rapidly using specially chosen buffer gases[7] or cyclotron emission in a large magnetic field;[8] the application of rotating electric fields for radial plasma compression;[4,5,7,9–11] the development of non-destructive diagnostics using plasma waves[11–13]; and the creation of beams of small transverse spatial extent by careful extraction from trapped and cooled antimatter plasmas.[14–16]

There are numerous applications of these positron plasmas and trap-based beams. Trapped positron plasmas and similarly confined clouds of antiprotons are the method of choice to make low-energy antihydrogen atoms.[11,17–20] One goal of that work is to test fundamental symmetries of nature by precision comparisons of hydrogen and antihydrogen. Attempts are being made to create and study electron–positron plasmas that are of interest in plasma physics and astrophysics.[21–24] Bursts of positrons from a trap-based beam were used to create the first positronium molecules (Ps_2). This represents an important step toward the creation of a Bose–Einstein condensate (BEC) of Ps atoms.[25] Positrons have been used extensively to study materials,[26–28] such as low dielectric constant insulators that are key components in high-speed electronics and chip manufacture.[28] An important focus of recent work is the further development of pulsed, trap-based positron beams that offer improved methods to make a variety of materials measurements. Commercial prototypes of these beam systems are now available.[29,30] Positrons are also important in medicine and biology; positron emission tomography is the method of choice to study metabolic processes in humans and animals, both to treat disease and to develop new therapies.[31] In the longer term, research in this area may well lead to the development of *portable* antimatter traps, and this, in turn, would facilitate many other uses of antimatter.[3,32]

Much of the following discussion relies on the physics of single–component plasmas in Penning–Malmberg (PM) traps, namely a plasma in a cylindrical set of electrodes immersed in a uniform magnetic field with electrostatic confinement along the direction of the field. Relevant parameters to describe these plasmas and the notation used here and CCII are listed in Table 4.1.[c] The book by R. C. Davidson[33] and the review article by Dubin and O'Neil[34] contain excellent, detailed discussions of the theoretical plasma physics concepts relevant to non-neutral plasmas, including those in PM traps.

4.2. Positron Trapping

Background and overview. In our world of matter, positrons are typically produced using accelerators or radioisotopes. To be trapped, they must be slowed to electron-volt energies from their initial, broad spectrum of energies, ranging from several kiloelectron-volts to ~ 0.5 MeV. Typically a "moderator" material is used to slow them down. This is done by either transmission through or reflection from a metal, such as single-crystal copper or tungsten (energy spread ~ 0.5 eV; efficiency ≤ 0.1 %),[26,35] or reflection from a frozen, solid rare gas such as neon (energy spread ~ 1 eV; efficiency ≥1 %).[36,37] These materials are chosen specifically for the characteristic that positrons do not readily bind to them or become trapped in voids or at defects. In particular, some metals have a negative positron work function and can be grown in large single crystals; they are thus well suited for positron moderation.

The accumulation and confinement of positrons in electromagnetic traps has a long history. In the early 1960s, Gibson, Jordan and Lauer injected radioactive neon gas in a vacuum chamber surrounded by magnetic mirror coils.[38] The emitted positrons were confined by the mirror field. The escape time, relative to the Ne gas puff, was used to measure the single-particle confinement time. Schwinberg, Van Dyck and Dehmelt confined small numbers of positrons in a Penning trap for very long times (weeks to months).[39] Their goal was to make precision

[c] Expressions in this chapter are in S. I. units, unless otherwise noted. In these units, ε_0 is the permittivity of free space.

comparisons of the properties of electrons and positrons. Mills and collaborators used a Penning trap to confine and bunch positrons from a radioisotope source[40] and from a microtron accelerator[41] for use in spectroscopic studies of Ps atoms. Brown, Leventhal, Mills and Gidley confined positrons in a Penning trap to measure the annihilation Doppler broadening spectrum of molecular hydrogen in order to model astrophysical annihilation spectra.[42] In all of these experiments, small numbers of positrons were confined at low densities (i.e., typically in the positron–gas regime rather than the plasma regime). Here we focus on the accumulation of large numbers of positrons in the plasma regime.

Table 4.1. Parameters used to describe single-component plasmas in PM traps.[2] See text for details. The symbols m and e are the positronic mass and charge, respectively, with the sign of e positive.

Quantity	Symbol	Formula	Units
temperature	T	–	eV
number density	n	–	m^{-3}
plasma length	Lp	–	m
thermal velocity	v_T	$(T/m)^{1/2}$	$m\ s^{-1}$
cyclotron frequency	ω_c	eB/m	rad s^{-1}
cyclotron cooling rate	Γ_c[d]	$B^2/4$	Hz
plasma frequency	ω_p	$(ne^2/\varepsilon_0 m)^{1/2}$	rad s^{-1}
cyclotron radius	r_c	$v_T/\ \omega_c$	m
Debye screening length	λ_D	v_T/ω_p	m
axial bounce frequency	f_b	$v_T/2L_p$	Hz
ExB rotation frequency	f_E	$\frac{ne}{4\pi\varepsilon_0 B}$	Hz
Brillouin density limit	n_B	$\varepsilon_0 B^2/2m$	m^{-3}

[d] Γ_c is used as the cooling rate, independent of the specific technique (e.g., for collisional cooling also). The formula displayed here is for cyclotron cooling with B in tesla.

While a number of devices and protocols have been used or proposed to trap antimatter, the device of choice is the PM trap because of its excellent confinement properties. Other variations of the Penning trap that have either been discussed or employed to trap antiparticles and antimatter plasmas include hyperboloidal,[39] orthogonalized cylindrical[43] and multi-ring electrode structures.[44] The PM trap is illustrated in Figure 4.1. It uses a uniform magnetic field to inhibit the diffusion of particles across the B field and an electrostatic potential well, imposed by the application of suitable voltages on a set of cylindrical electrodes, to confine the particles in the direction of the B field.[34,45,46] The extremely long confinement times that can be achieved in these traps[6,47] makes the accumulation of substantial amounts of antimatter feasible in the laboratory.

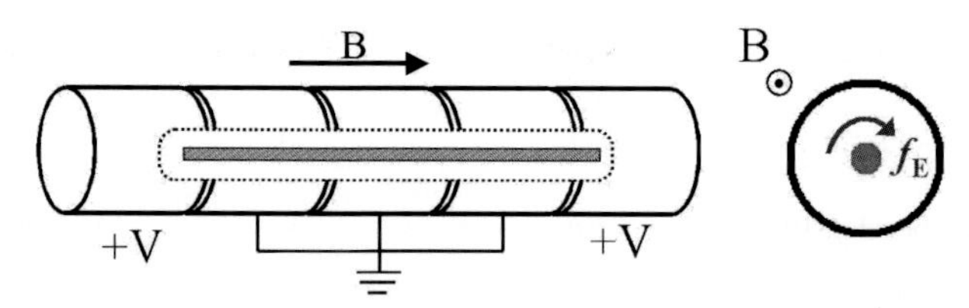

Fig. 4.1. Schematic diagram of a PM trap. The plasma is shown in the cutaway section. It is confined by a uniform axial magnetic field and by the electrostatic potential, V, at each end. As described below, the plasma rotates about its axis with an E x B frequency f_E, as illustrated in the end-on view (right).

An efficient accumulation scheme – the buffer-gas trap. Given this good trapping scheme, the challenge is to find an efficient method to fill the trap with positrons. A variety of trapping techniques have been developed to do this. If a pulsed positron source such as as a linear electron accelerator (LINAC) is used, the positrons can be captured by timed switching of the potential on one of the end confining electrodes. This end-gate switching technique has been employed extensively to condition positron beams from LINACs and other pulsed sources.[48] It has also been used to transfer positrons from one trap to another.[20,49,50] For high capture efficiency, the spatial extent of the incoming pulse must be smaller than twice the trap length, and the slew rate on the capture gate must be sufficiently rapid. In many circumstances, these conditions are relatively easy to meet.

When positrons are captured from a steady-state source, such as a radioisotope, energy can be extracted from the positrons to trap them, or energy can be transferred from the positron motion in the direction parallel to the magnetic field to the perpendicular direction by a variety of techniques. The latter effect results in "virtual trapping" in that the particles can subsequently be de-trapped by the reverse process. A variety of techniques have been developed to trap positrons using these approaches, including collisions with neutral gas atoms and molecules[51, 52], scattering from trapped ions,[39,53] scattering from trapped electrons in a nested potential well,[54] and trapping in a magnetic mirror.[55] Other methods used to trap positrons include using dissipation in an external resistor,[56] field ionization of weakly bound positronium (Ps) atoms[53,57] and the exchange of parallel and perpendicular momentum exploiting stochastic orbits.[58] Each of these techniques has its advantages, but it turns out that they are relatively inefficient.

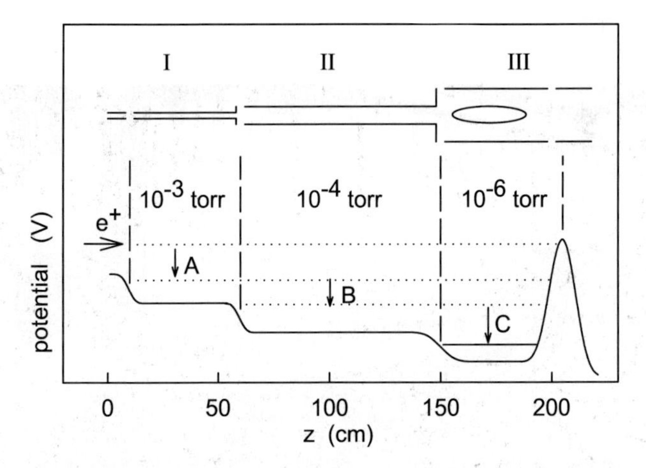

Fig. 4.2. Buffer-gas trapping scheme, showing the electrode geometry of a modified PM trap (above), the neutral gas pressure in each stage and the axial potential profile (below). There is an applied magnetic field, B ~ 0.15 T, in the z direction. Two-stage accumulators with B as small as 0.04 T have also been used successfully.[59]

The positron trapping method most widely used is the buffer-gas (BG) technique. It has the highest trapping efficiency and modest magnetic field requirements. Figure 4.2 illustrates the operating principle of such a buffer-gas positron accumulator which, in this example, has three stages.[52,60] Figure 4.3 shows the actual physical arrangement. Positrons are injected into a specially modified PM trap having a stepped potential profile, with each stage having a different pressure of buffer gas. Using a continuous gas feed and differential pumping, a high pressure ($\sim 10^{-3}$ mbar) is maintained in the small-diameter region at the left (stage I in Figure 4.2). Positrons are initially trapped in this region by inelastic collisions with buffer-gas molecules (marked "A" in Figure 4.2). The trapped positrons then make multiple passes back and forth in the trap.

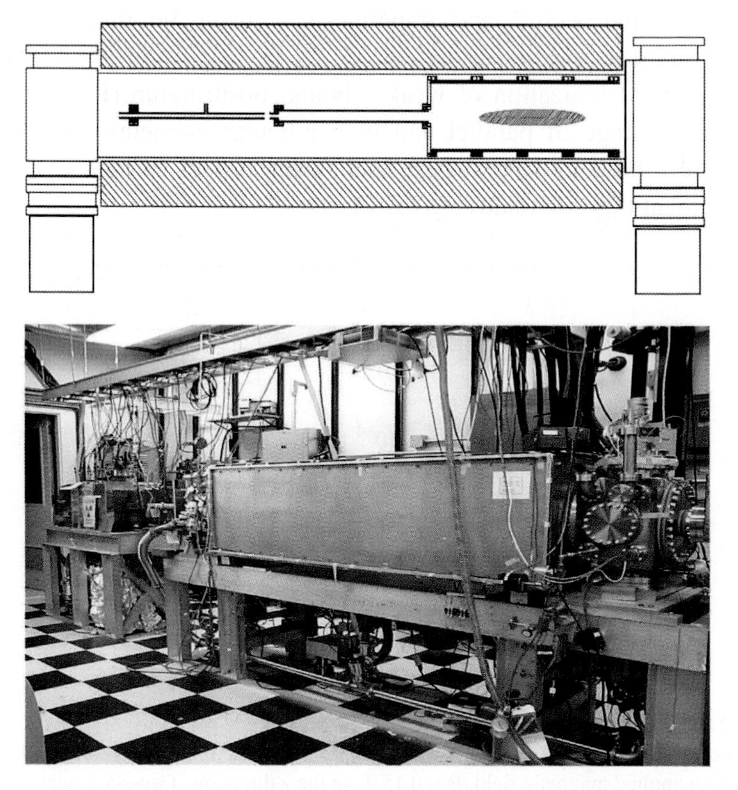

Fig. 4.3. A three-stage positron trapping apparatus with source and moderator at the University of California, San Diego (UCSD): (above) cutaway of a three-stage positron trap; (below) photograph of the positron source (a sealed ^{22}Na radioisotope source and solid neon moderator) in a lead enclosure at the left, and the three-stage trap in the large metal box on the right. For spatial scale, the floor tiles are ~ 0.3 x 0.3 m.

They lose energy by subsequent inelastic collisions ("B" and "C") in the successively lower pressure stages II and III, causing them to accumulate in stage III. Here, they cool to approximately the gas (i.e., the electrode) temperature, which is ~ 300 K.

This type of accumulator can be operated using a variety of gases including molecular nitrogen, hydrogen, carbon dioxide and carbon monoxide.[29] There are two considerations regarding the choice of buffer gas. One is to find a target species that has a relatively large cross-section for energy loss via inelastic scattering. The second is to avoid Ps atom formation, which results in loss of positrons through annihilation, either in the Ps atom or when the Ps strikes an electrode or the vacuum chamber.

It would be appealing to use the vibrational excitation of molecules for this energy loss process, however this results in a loss per collision ≤ 0.5 eV. In practice, this is too small to efficiently trap the spread of positrons from the moderator (e.g., energy spreads ~ 1 eV). An important effect is due to the fact that typical source/moderators are operated at a reduced magnetic field (typically $B \leq 0.03$ T). The quantity $E_\perp /B$ is an adiabatic invariant for these particles, where $E_\perp$ is the energy in motion in the plane perpendicular to B. Thus, when particles with a spread of $E_\perp$ values enter the higher magnetic field of the BG trap, the spread in parallel energies, $E_{||}$, increases significantly. This generally reduces the trapping efficiency since the inlet potential cannot be as carefully tuned so that incoming positrons just pass over it. The entire spread of $E_{||}$ must now pass over the inlet potential barrier of the trap. Positrons with larger values of $E_{||}$ must lose correspondingly more energy before they become trapped, and it is more difficult to tune the potential steps to optimize the energy loss per collision for all of the particles.

The highest trapping efficiency is obtained using molecular nitrogen. The reason it is superior is that, as shown in Figure 4.4, this species has a relatively large electronic excitation cross-section at positron impact energies ~ 10 eV, near the threshold for electronic excitation of the $a^1\Pi$ of N_2 at 8.8 eV,[61] while it also has a relatively small cross-section for Ps formation (i.e., a potent positron loss process) in this range of energies.

To our knowledge, molecular nitrogen is somewhat unique in this important characteristic. In most other molecules, the Ps formation threshold is below that for the lowest allowed inelastic electronic transition.

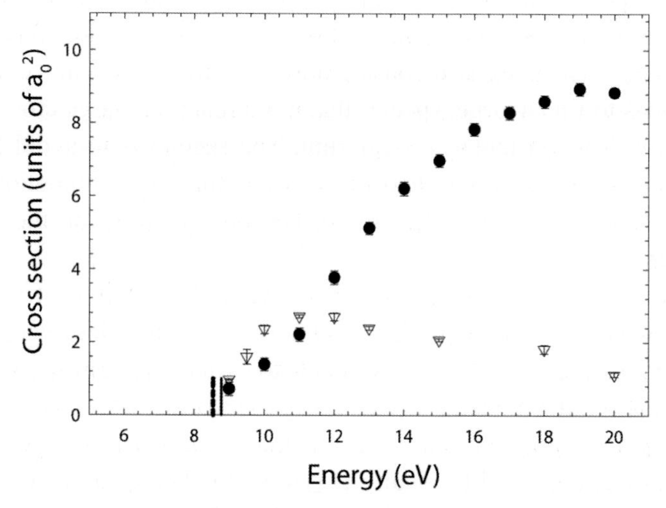

Fig. 4.4. Cross-sections in atomic units (a_0^2 = 2.8 x 10^{-21} m^2) for positron-impact excitation of the $a^1\Pi$ electronic state of N_2 (▽) and Ps formation (●). The dashed and solid vertical bars indicate the thresholds for electronic excitation and Ps formation, respectively. From reference[61].

The pressure in stage I is set so that the positrons make, on average, ~ one electronic-excitation collision in one transit through the trap and hence are confined in the potential well. This happens before they reflect off the potential barrier at the end of the trap opposite the source, exit the trap, and return to the moderator, where they would be lost to annihilation. Once trapped, the positrons move back and forth in the direction of the magnetic field. Additional stages with stepped potentials and correspondingly lower neutral gas pressures (i.e., two more stages in the trap illustrated in Figure 4.2) are arranged to trap the positrons in a region of low gas pressure in which the annihilation time is commensurately long. The positron lifetime in stage III of the trap illustrated in Figure 4.2 is typically $\geq$ 40 s. Longer lifetimes (e.g., hours or more) can be achieved by pumping out the buffer gas following positron accumulation.

While N_2 has a relatively large electronic excitation cross-section, its vibrational excitation cross-section is quite small. The addition of a low pressure (e.g., $\leq 10^{-7}$ mbar) of CF_4 or SF_6 in stage III is used to cool rapidly to room temperature.[9] The unusually large positron-impact vibrational cross section of carbon tetrafluoride,[62] which is discussed in more detail below, is responsible for rapid cooling to temperatures ≤ 0.16 eV, and SF_6 is believed to act similarly. For the typical pressure settings in the three-stage trap shown in Figure 4.3, operating with N_2 in stages I – III and CF_4 in stage III, the positrons are trapped in one transit back and forth through the trap. They lose additional energy by a second electronic excitation of N_2 and are thus confined in stages II and III in $\leq$ 100 μs. The positrons then make a similar collision in stage III and are confined to this stage in a few ms.[52] Finally, the positrons cool to room temperature by vibrational and rotational excitation of CF_4 in ≤ 0.1 s. A set of rate equations describing this cascade to lower positron energies is discussed in reference[52].

For accumulators with a solid neon moderator, the trapping efficiencies (i.e., defined as the fraction of positrons trapped and cooled relative to the number of incident slow positrons from the moderator) are typically in the range of 5–20%, and efficiencies of up to 30% have been observed under optimized conditions. Using a tungsten moderator, the efficiency can be as high as 50%. While not studied in detail, the trapping efficiency is likely limited by Ps atom formation and the small positron density in the first stage of the trap. This positron-density effect, which is discussed in more detail below, is due to $\tilde{E}xB$, asymmetry-induced radial transport, where $\tilde{E}$ is the (DC (direct current) in the laboratory frame) electrostatic field due to trap asymmetries. It is largest in the first trapping stage where the positron density is the smallest.

Using a 100 mCi ^{22}Na source and solid neon moderator, several hundred million positrons can be accumulated in a few minutes in the three-stage trap shown in Figure 4.3.[63] Once accumulated, the resulting positron plasmas can be transferred efficiently to another trap and stacked (e.g., for long-term storage).[20,49,50] Figure 4.5 shows the history of positron trapping using apparatus such as those described here using similar strength sources.

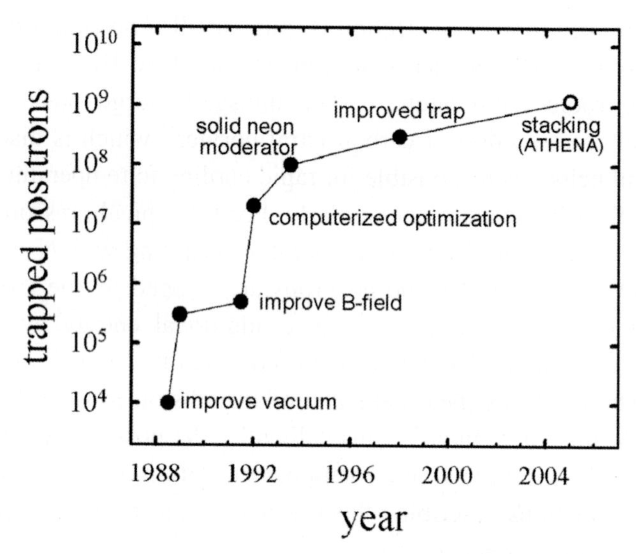

Fig. 4.5. Progress in creating positron gases and plasmas in PM traps using ^{22}Na positron sources with strengths ~ 50 – 100 mCi. For the data before 1993, tungsten moderators were used, while after that, solid neon moderators were used.

These buffer-gas traps are relatively efficient, arguably even efficient on an absolute scale. The difference between 5% and 30% efficiency is typically due to the fine tuning of the alignment of the incoming positron beam with respect to the electrode structure. In this regard, careful choice of the inner diameter of the first stage electrodes and operating pressure is likely of considerable importance. One area that, to the author's knowledge, has not been explored extensively is the extent to which elastic scattering on atoms or molecules (i.e., transfer of parallel energy to that perpendicular to the B-field), and the resulting process of virtual trapping could be used to advantage, particularly in the first stage of the buffer-gas trap. In this process, the particles will be trapped until another elastic scatter de-traps them, or an inelastic collision traps them absolutely.

Simpler, two-stage positron accumulators with correspondingly shorter positron lifetimes (e.g., $\leq$ 1 s) have now been developed.[59,64] Commercial two- and three-stage positron traps, such as that shown in Figure 4.6, are now sold commercially by R. G. Greaves at First Point Scientific, Inc., Agoura Hills, CA.

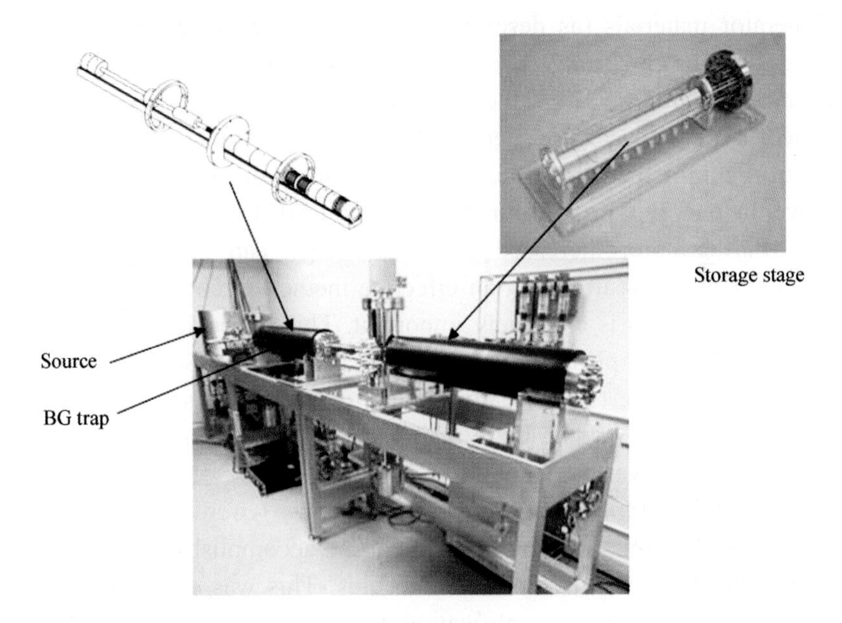

Fig. 4.6. A commercial two-stage buffer-gas trap and a separate storage stage (i.e., a three-stage system) with a ^{22}Na source and solid neon moderator. (below) Photograph of the system; with, left to right, the source/moderator (in the shiny cylinder), (in the black solenoids) the buffer-gas trap and the storage stage. Also shown are (top left) the buffer-gas trap electrodes, and (top right) the storage stage electrodes. Courtesy of R. G. Greaves, First Point Scientific, Inc., Agoura Hills, CA.

We end this discussion with a cautionary practical note about positron traps such as those described here. It is well known that positron annihilation rates on large hydrocarbon molecules can be extremely high. This arises from the fact that positrons tend to bind to these species (i.e., through a mechanism known as vibrational Feshbach resonances).[65] Oil molecules are particularly deleterious in this regard. Thus considerable care must be taken in achieving a good, oil-contaminant-free base vacuum in the accumulator (e.g., $\leq 5 \times 10^{-10}$ mbar) and/or trap. The vacuum system should be bakable (e.g., to 420 K or higher), if long confinement times are desired.

4.3. Positron Cooling

Moderator materials (as described above) are used to decelerate high-energy positrons from a source to electron-volt energies. Once accumulated in a PM trap, these collections of charged particles can be heated by small electric perturbations. This is quite deleterious to positron confinement for a number of reasons. For example, heating to above the energy threshold for Ps formation leads to a first-order particle loss. Furthermore, increased positron energy can also lead to de-confinement. Thus arranging an effective method to cool these positron gases and plasmas is extremely important. This is absolutely obligatory in cases where the plasma can be heated substantially. Such heating can occur, for example, when rotating electric fields are used to compress plasmas radially, or when there is a manipulation of positron gases and plasmas between various trapping regions.

Collisional cooling using atomic or molecular gases. At electron-volt energies and below, positron cooling can be accomplished by collisions with suitable gases of atoms or molecules. This was described briefly above, but a bit more amplification is in order. The cooling gas is selected to have a large inelastic scattering cross-section in order to achieve significant energy loss. However *positron loss* due to Ps atom formation must be avoided if possible. So-called "direct" annihilation of a positron with a bound electron in an otherwise elastic collision typically has a much smaller cross-section. Thus, where possible, one tries to work below the threshold for Ps formation (i.e., which can be several electron-volts or more). In fact, to avoid loss due to Ps formation with positrons on the tail of the Maxwellian distribution, the positron temperature should be kept a factor of three or more below the Ps formation threshold (e.g., $T \leq 2$ eV). For relatively low positron temperatures, direct annihilation then becomes an important factor in determining the lifetime of trapped positrons (e.g., ~ 40 s for N_2 at a pressure ~ 10^{-6} mbar).

Only recently have state-resolved inelastic positron-impact cross-sections been measured;[65] and so a general, quantitative understanding of the collisional positron cooling processes involving atoms and molecules is not available. Typically at energies in the electron-volt range, electronic transitions can be used to reduce the positron energy effectively. At energies in the range from 50 meV to those of the electronic transitions, vibrational transitions in molecules can be used, while below ~ 0.05 eV, one must rely on rotational transitions in molecules and momentum-transfer collisions with atoms to cool the positrons.

In the case where a single inelastic scattering channel is relevant (e.g., a vibrational mode j with energy ε_j), the cooling rate Γ_c will be,

$$\Gamma_c \equiv \frac{1}{T} dT/dt \approx -\frac{\nu_j \varepsilon_j}{T}, \tag{4.1}$$

where ν_j is the excitation rate for this transition. As we discuss in more detail below, this collisional cooling rate is applicable if there are no heating electric fields. If there are heating fields, then any momentum-transfer (e.g., an elastic) collision can convert the coherent field-driven component of the kinetic energy into heat, and this must be taken into account to determine the net cooling/heating rate. Such a detailed account of scattering processes is beyond the scope of this review. Likely Monte Carlo computer calculations would be useful in studying this balance of heating and cooling,[66] assuming the necessary collisional cross-sectional data are available.

Cross-sections for vibrational excitation of molecules have now been measured for several species,[67,68] and at least a semi-quantitative understanding of the magnitudes of these cross-sections is available.[69] Moreover, positron-cooling rates due to vibrational excitation have been measured for several molecules.[7,29,70] Cooling rates for selected molecules are given in Table 4.2. It turns out that SF_6 and CF_4 are particularly effective. In these species, there is a large amount of charge transfer to the F atoms. This results in a very large vibrational excitation

Table 4.2. Positron cooling rates in a PM trap using molecular gases at 2.6 x 10^{-8} mbar: time τ_a, for direct annihilation; measured cooling time, τ_c; and the energies of the vibrational quanta, ε_j. Data from references[7,9].

Gas	$\tau_a(10^3$ s)	τ_c(s)	ε_j(eV)
SF_6	2.2	0.36	0.076, 0.19
CF_4	3.5	1.2	0.16
CO_2	3.5	1.3	0.29, 0.083
CO	2.4	2.1	0.27
N_2	6.3	115	0.29

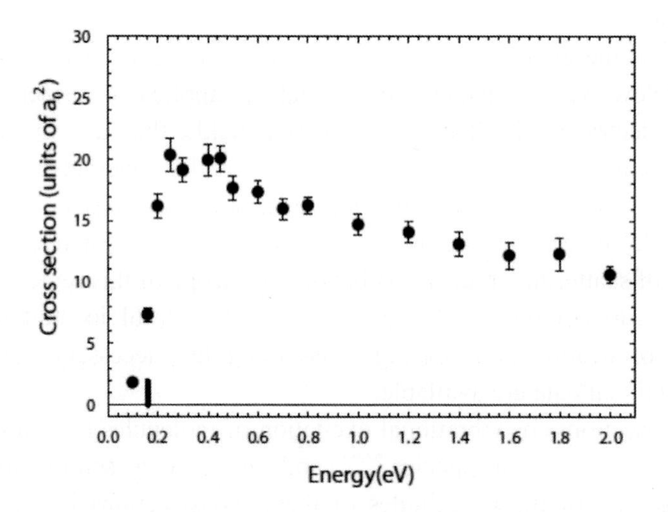

Fig. 4.7. Positron-impact cross-section for excitation of the ν_3 vibrational mode of CF_4 as a function of incident positron energy in atomic units (a_0^2 = 2.8 x 10^{-21} m^2). The relatively large and approximately constant cross-section above the threshold energy, ε_j = 0.157 eV, provides a very efficient and useful cooling mechanism. Reprinted from reference[62].

cross-section for the asymmetric stretch (i.e., ν_3) vibration in the molecule. Shown in Figure 4.7 is the cross-section for the ν_3 mode of CF_4.[62]

The current versions of buffer-gas positron traps typically use a mixture of N_2 and CF_4 or SF_6 in the final trapping stage for rapid cooling. There is very little information available on positron energy loss due to rotational excitation of molecules, save for an early study by Coleman *et al.* using a positron lifetime technique,[71] thus this would likely be a fruitful area for future work. As discussed below, CF_4 has also been used effectively for the cooling required for the radial compression of positron plasmas (i.e., to counteract the heating caused by the work done on the plasma by the applied torque).

Cyclotron cooling. A convenient method to cool electron-mass charged particles (positrons) in an ultra-high vacuum (UHV), is to arrange for them to emit cyclotron radiation in a strong magnetic field.[47] In this case, the positron temperature is typically a balance of heating (e.g., RF (radio frequency) electric fields are particularly effective in this regard) and the cyclotron cooling. In fact, cyclotron cooling at achievable magnetic fields is typically considerably less effective than cooling using gaseous collisions; so in this case, heat sources can produce quite significant effects. In the absence of a heating source, the particles will come to equilibrium at the temperature of the surrounding electrode structure. However, in the case in which parts of the vacuum system are at higher temperatures (e.g., when the electrodes are cooled cryogenically), this can result in heating the plasma above the temperature of the electrode structure.[e] The cyclotron-cooling rate for electron-mass charged particles is,[47,72]

$$\Gamma_c \approx B^2/4, \tag{4.2}$$

where B is in tesla and Γ_c is in s^{-1}. For example, the radiative cooling time, $1/\Gamma_c$ of positrons in a 5 tesla field is 0.16 s. Assuming an emissivity, $\varepsilon = 1$, for the electrodes at the cyclotron frequency, the surrounding electrode structure is at temperature T_w, and there is no

[e] J. Fajans, Private communication, 2009.

external heating, the time dependence of the positron temperature, $T(t)$, of a positron plasma at initial temperature T_1 will be:

$$T(t) = T_0 + (T_1 - T_w)\exp(-\Gamma_c t). \quad (4.3)$$

Shown in Figure 4.8 is a typical cooling curve for the thermal relaxation of an electron plasma confined in an apparatus at 300 K.

Two comments are in order regarding cyclotron cooling in an electrode structure. If one can arrange a resonant cavity at the cyclotron frequency, then the cooling rate is increased by the Q factor of the cavity.[73] The second comment is that the electrode structure must have a minimum size in order for cyclotron cooling to be effective. In particular, the structure must be at least large enough to accommodate the lowest-order resonant mode. For a long circular electrode structure, this means the inner diameter of the structure must be $D \geq \lambda_c$, where λ_c is the electromagnetic wavelength at the cyclotron frequency. For smaller values of D, the electrodes will act as a waveguide beyond cutoff, and radiation by the particles will be suppressed.

Sympathetic cooling using ions. The techniques described above are limited to producing a temperature equal to the temperature of the environment, (e.g., 4 K for cyclotron cooling in a trap cooled to liquid helium temperature). However, laser cooling of ions in traps permits cooling to temperatures much lower than their surroundings. This technique has been used[53] to reach positron plasma temperatures significantly below the ambient by cooling the positrons sympathetically using laser-cooled ions that were simultaneously confined in the same trap with the positron plasma. Using this technique, a high-density positron plasma ($n = 4\times10^{15}$ m^{-3}) was cooled to < 5 K in a room-temperature trap. This technique has the potential to produce positron plasmas with parallel energies less than 100 mK.[f]

[f] At very large B fields and low temperatures, the perpendicular energy of the particles will eventually be limited by the energy of the lowest Landau level.

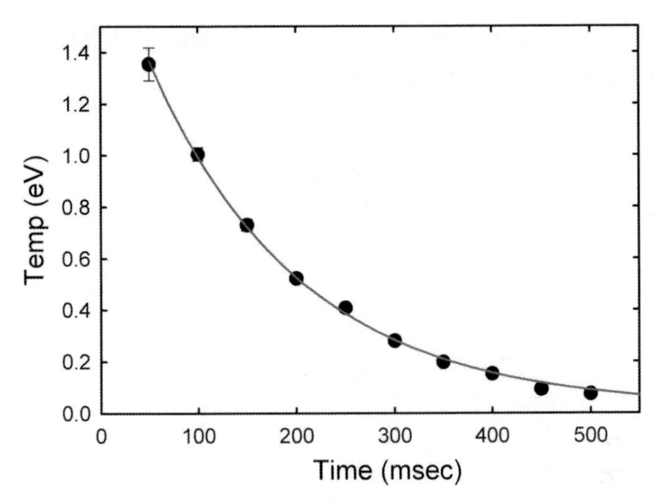

Fig. 4.8. Cyclotron cooling of an electron plasma in a magnetic field, B = 4.8 T, following heating with RF noise. Equation (2) yields $\Gamma_c = 6.5\ s^{-1}$, compared with a predicted value of $5.9\ s^{-1}$. Courtesy of T. R. Weber, UCSD, unpublished.

4.4. Confinement and Characterization of Positron Plasmas in Penning–Malmberg Traps

Basic concepts. A typical PM trap for positrons is shown in Figure 4.9. It consists of a set of cylindrical electrodes in a uniform magnetic field. The plasma is confined in the direction of the magnetic field by electrostatic potentials applied to electrodes at each end. A segmented electrode over a portion of the plasma is used to apply a rotating electric field to compress the plasma radially (this is discussed in more detail in Section 4.5, below). Also shown is a phosphor screen and a CCD (charge-coupled device based) camera for imaging the radial distribution of the plasma[7] and the RF circuitry to excite waves in the plasma (e.g., for temperature and density measurements).[74,75]

In a single-component plasma at temperature T in the PM trap, the particles make only small excursions in the plane perpendicular to B. They are characterized by the (average) cyclotron radius, $r_c = v_T/\omega_c$, where the cyclotron frequency ω_c is the angular frequency of gyration of the particle in the plane perpendicular to B, and $v_T = (T/m)^{1/2}$ is the average thermal velocity of the particle.

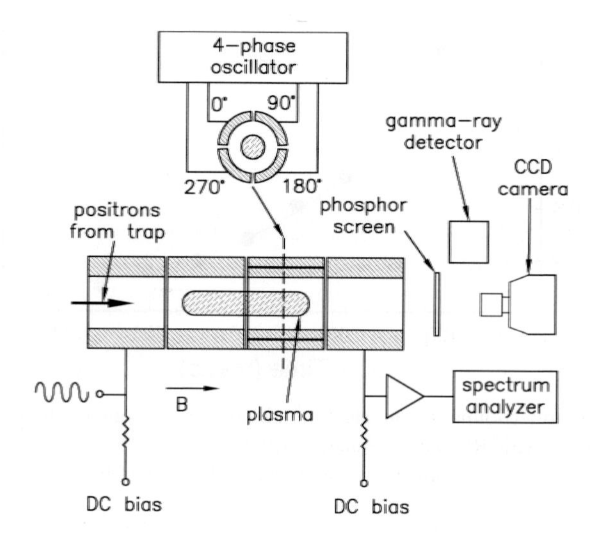

Fig. 4.9. Schematic diagram of a PM trap and associated apparatus for confining and manipulating positron plasmas. Shown is a segmented electrode for applying a rotating electric field for radial plasma compression, a phosphor screen and CCD camera for measuring radial density profiles and the electronics to excite plasma modes for diagnostic purposes.

They are subject to a confinement principle that arises from the fact that a charged particle in a B field has an angular momentum associated with it, beyond the ordinary mechanical momentum. As a consequence, at low temperatures where the thermal velocities of the particles are negligible, the canonical angular momentum P_θ[47] is approximately

$$P_\theta \approx -\frac{m\,\omega_c}{2}\sum_i r_i^2 , \tag{4.4}$$

where the r_i are the radial positions of the particles, and it is assumed that the particles are positively charged which fixes the sign of P.

In a PM trap with cylindrically symmetric electrodes, the angular momentum, P_θ, is constant. Thus the second radial moment of the particle distribution is also constant, and so the plasma cannot expand. In practice, these plasmas do expand slowly due to imperfections in the trap. In this case, the torque on the plasma is related to the outward transport rate, $\Gamma_0 = (1/n)(dn/dt)$, by

$$\tau = \frac{dP_\theta}{dt} = P_\theta \Gamma_0. \tag{4.5}$$

A single-component plasma in a PM trap is effectively a long cylindrical rod of charge. This collection of trapped particles will exhibit plasma behavior when the Debye length, $\lambda_D = v_{th}/\omega_p$, is such that $\lambda_D >> r_p$, L_p, where ω_p is the plasma frequency, r_p is the plasma radius and L the plasma length. In this case, potential perturbations in the plasma will be screened by motion of the particles in the direction parallel to B. Consequently, any remaining electric field in the plasma will be in the radial direction (i.e., neglecting end effects).

This radial electric field in and around the plasma results in a plasma potential that increases as one approaches the plasma center. From Gauss' law, for a long cylindrical plasma of N positrons with radius r_p in an electrode of radius r_w, the magnitude of this space-charge potential (i.e., the "plasma potential") at the plasma center is

$$\Phi = \frac{AN}{L}\left[1 + 2\ln(r_w / r_p)\right], \tag{4.6}$$

where $A = e/4\pi\varepsilon_0 = 1.4 \times 10^{-9}$ Volt–m. This value of Φ sets the minimum potential, V_c, required to confine the plasma, namely $V_c > \Phi$.

A key physical effect in PM traps and in other magnetized plasmas arises from the fact that the magnetic and electric space-charge fields are perpendicular to each other. This is illustrated in Figure 4.10 for a "slab" model that describes particle motion in two dimensions (i.e., ignoring the cylindrical symmetry of the PM trap). Charged particles in such fields undergo so-called "E x B drifts" at a velocity $v_E = E/B$, in the direction perpendicular to both E and B.[76,77] In terms of the cyclotron radius r_c and the characteristic distance $r_E = v_E/\omega_c$, the trajectories are "cycloids". Particles starting at the origin at time t = 0 orbit about a center *moving at velocity* v_E, located at (x, y) position ($r_E + v_E t$, r_E), with an associated radius,

$$\rho = \sqrt{r_c^2 + r_E^2}. \tag{4.7}$$

In particular, the secular motion is in the x direction (i.e., perpendicular to both E and B), and the oscillation amplitude 2ρ is dominated by the larger of r_E and r_c. This latter effect has very important consequences for particle transport, namely the transport step size (~ ρ) can be dominated by r_E. Thus the transport can become very large when E is large (i.e, $v_E > v_T$), giving rise to large and rapid excursions of the particles outward.

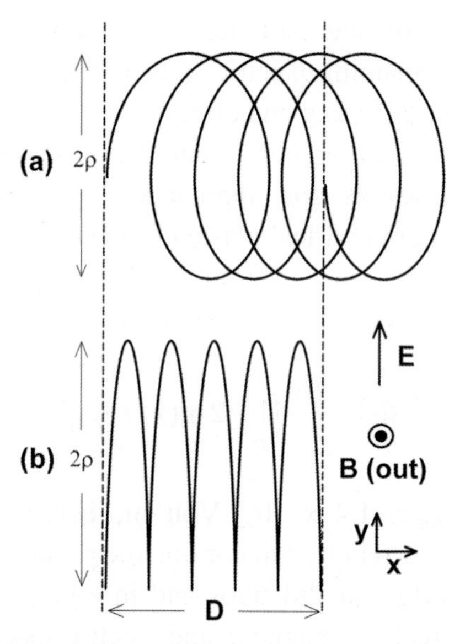

Fig. 4.10. E x B drift orbits for 5-cyclotron periods in a two-dimensional "slab" model (not to scale): (a) the usual drift-orbit case where the electric field is small, $r_E/r_c = 0.1$; and (b) a much larger electric field, $r_E/r_c = 10$. The corresponding radial excursions ρ and lateral distances (D) are approximately (a) r_c (πr_c), and (b) $10r_c$ ($100\pi r_c$). The particles make much larger excursions in the strong E field. (In the PM trap, y corresponds to the radial direction and x to the azimuthal direction.)

Due to the radial E field in the trapped plasma and the resulting E x B motion, the particles drift around the axis of symmetry at a frequency

$$f_E = \frac{ne}{4\pi \varepsilon_0 B}, \tag{4.8}$$

where n is the number density of the plasma. Note here the intimate connection between the rotation frequency, f_E, and the plasma density, n.

Transport due to neutral collisions. The angular momentum constraint of Eq. (4.4) implies that a single-component plasma confined by a magnetic field can expand only if there is a torque on it. In a perfect, azimuthally symmetric trap there would be no expansion. However in practice, this is not the case. Typically radial transport *is* observed. This can be due to trap imperfections, or, in the case where there is appreciable neutral background gas, this transport can be due to the drag on the plasma due to neutral collisions.[6,79]

The transport due to neutral gas collisions is reasonably well understood. In this case, the outward flux of particles (i.e., number of particles/area–time) J is,[6]

$$J = \frac{\partial}{\partial r}\left(\nu_p r_c^2 n\right) + \nu_p r_c^2 \left(\frac{eE}{T}\right) n, \tag{4.9}$$

where ν_p is the positron-neutral, momentum-transfer collision frequency (frequently dominated by elastic scattering), and E is the space-charge electric field. The two terms in Eq. (4.9) are respectively the flux due to collisional diffusion, and the flux induced by the electric field that involves the electrical mobility of the plasma. In the plasma regime, $(eEr_p)/T \gg 1$, and so the second term dominates the otherwise diffusive transport by a factor $\sim e\Delta\Phi/kT$, where $\Delta\Phi$ is the change in plasma potential across the plasma.[6] Assuming this is the case and inserting E for a rigid rotor, one finds for the outward transport rate, $\Gamma_0 \equiv (1/n)(dn/dt)$,

$$\Gamma_0 = \nu_p \left(\frac{r_C}{\lambda_D}\right)^2. \tag{4.10}$$

In the single-component positron (or electron) plasmas considered here, typically $r_c << \lambda_D$, and so the transport due to neutral-gas collisions is typically small.

Transport due to electric and magnetic asymmetries. In the case that gas collisions do not dominate the transport (e.g., a plasma in a UHV environment cooled by cyclotron radiation), a detailed, microscopic understanding of the transport has remained elusive in spite of 30 years research on the subject. It is believed to be due to azimuthal asymmetries. Recent studies point to the importance of so-called trapped particles and the influence of asymmetries upon them.[80] This somewhat subtle effect arises from the fact that particles trapped in electrostatic or magnetic wells (e.g., due to trap imperfections) do not experience the averaging effects that the bulk of the particles do, and so they can make larger radial excursions. When subsequently scattered out of this imperfection (i.e., trapping well), they can then cause greatly enhanced radial particle transport. From the perspective of antimatter-trap engineering, one typically relies upon empirical formulae based upon the now-extensive experimental studies. Data for the outward radial transport of particles (presumably due to asymmetries) from a plasma in a PM trap are shown in Figure 4.11.[81]

As shown in the Figure 4.11, there are two regimes of plasma transport. At sufficiently high plasma densities, Γ_0 is independent of n, whereas at lower densities $\Gamma_0 \sim (nL)^2$. The transition between these two types of behavior appears to occur when the axial bounce frequency, $f_b = v_{th}/2L_p$ is approximately three times the Coulomb collision frequency.[81] However, there is no theory at present for this effect, and it is unclear whether this result will hold in other experiments. The values of Γ_0 shown in Figure 4.11 are among the smallest values reported for the given parameters. In other experiments, Γ_0 can be as much as an order of magnitude greater, presumably due to larger trap asymmetries.[81] In practice, the best one can do to estimate the outward transport (and/or confinement time) is to use the reported values as order of magnitude estimates of the outward transport.

In considering the effects on transport due to electric asymmetries that are static in the laboratory frame, they are expected to be largest in plasmas with a small rotation frequency (i.e., in this case the transport is due to asymmetry-induced E x B flows). However at higher rotation frequencies, the rotation can also bring the asymmetry-induced fields (that are DC in the laboratory frame) into resonance with a plasma mode. This can act as a potent drag on the plasma and result in a high level of transport.

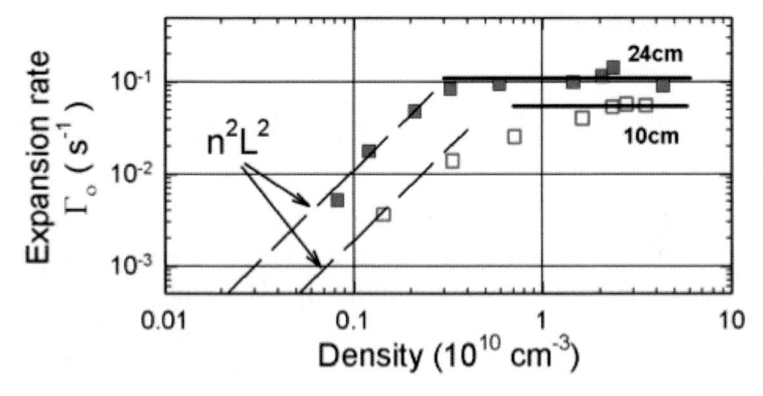

Fig. 4.11. The expansion rate, Γ_0, as a function plasma density for an electron plasma in a UHV PM trap in a 5-tesla field. The data show two regimes, including one in which Γ_0 is strongly density dependent. The transition occurs when the Coulomb collision frequency is ~ three times the axial bounce frequency. (In this figure, $L \equiv L_p$.) From reference[81]; see this reference for details.

The E x B transport at low rotation frequencies has important consequences for the operation of buffer-gas traps. In the first stages of a buffer-gas trap, or in traps confining small numbers of positrons, the rotation frequency will be small because the positron density is low. Thus, for example, a small static electrostatic asymmetry (e.g., arising from patch-voltages on the electrodes) can induce the rapid DC E x B transport of the particles to the wall. In the case of the buffer-gas trap, this means that one wants to get the particles out of these early trapping stages as quickly as possible and into the final stage where the plasma density (and hence the plasma rotation frequency) is higher. The small plasma rotation frequency in the first stage of buffer-gas traps can potentially play a significant role in limiting the trapping efficiency of these devices.

Plasma heating. Single-component plasmas in PM traps can be heated by various mechanisms, including ambient RF noise on the confining electrodes. One unavoidable heating source is the outward plasma expansion itself. Essentially, the radial, outward-directed electric field due to the plasma space charge preferentially gives the particles extra energy as they move outward radially. The heating rate, Γ_h, due to this effect can be written,[81–83]

$$\Gamma_h = \frac{1}{T}\frac{dT}{dt} = \left(\frac{e\phi_0}{2\eta T}\right)\Gamma_0, \tag{4.11}$$

where $1/\eta$ is the fraction of the space-charge potential that is dropped across the plasma, assuming $\phi = \phi_0$ at $r = r_w$. For a rigid-rotor plasma with a constant radial density profile, $\eta = \left[1 + 2\ln\left(a_w / r_p\right)\right]$, and ϕ_0 represents the potential drop across the plasma itself. Note that the plasma potential can be quite large (tens of volts are not atypical), so that in modestly cold plasmas, it can be the case that $\Gamma_h >> \Gamma_0$.

This heating must be mitigated by some type of cooling (e.g., cyclotron cooling or cooling due to collisions with gas molecules). In order for there to be a stable steady state, the heating rate must be smaller than the maximum cooling rate, i.e., $\Gamma_h/\Gamma_c < 1$, otherwise the temperature will increase in an uncontrolled manner.

If neutral collisions dominate both the transport and the cooling, we can combine Eqs (4.10) and (4.11) to find for the heating rate,

$$\Gamma_h = \frac{\nu_p}{4}\left(\frac{\omega p}{\omega c}\right)^2\left(\frac{r p}{\lambda D}\right)^2. \tag{4.12}$$

Neutral collisions with molecules can provide cooling via the excitation of vibrations (e.g., as is the case for CF_4). Considering the excitation of a single level, the cooling rate is given by Eq. (4.1). A measure of the effectiveness of this cooling can be obtained by forming the ratio, β, of the heating rate given by Eq. (4.12) to the cooling rate in Eq. (4.1). Thus,

$$\beta = \frac{\Gamma_h}{\Gamma_c} = \frac{\nu_p T}{4\nu_j \varepsilon_j}\left(\frac{\omega p}{\omega c}\right)^2\left(\frac{rp}{\lambda D}\right)^2. \tag{4.13}$$

As discussed above, the plasma temperature will be stable only for $\beta \leq 1$, and will "run away" for larger values of η. Since $\beta \propto (n/B)^2$. *This places an important constraint on the maximum achievable plasma density n.*

It is useful to express the density in terms of the Brillouin limit density n_B (i.e., the density at which $\omega_p^2 = \omega_c^2/2$; see Eq. (4.19) and related discussion below for details), in which case,

$$\left(\frac{n}{n_B}\right)^2 = \frac{8\nu_j \varepsilon_j}{\nu_p T}\left(\frac{rc}{rp}\right)^2. \tag{4.14}$$

Thus, to achieve high plasma densities, one would like a cooling gas with small ν_p and large ν_j. Carbon tetrafluoride fits this bill. As discussed above, it has an unusually large value of ν_j (cf., Figure 4.7[62]). It turns out that it also has a small value of ν_p,[84] making it a good choice for this purpose.

There are, however, two caveats regarding Eqs (4.12) – (4.14). Equation (4.14) is valid so long as the maximum density n is not very close to n_B. Close to the Brillouin limit, the cycloidal E x B orbits of the particles are very large and nearly unconfined, and a more careful calculation (not done here) is required. A practical criterion might be to set the amplitude of the cycloidal motion $\delta r = E/\omega_c B$ to be $\leq 0.1\ r_p$, which corresponds to $E/(\omega_c B r_p) \leq 0.1$ and $n/n_B \leq 0.1$. Further, we use particularly simple expressions for the collisional transport. Techniques such as Monte Carlo calculations would be very valuable in obtaining better estimates for the plasma expansion, heating and cooling.[66]

Diagnostic techniques. A variety of destructive and non-destructive techniques have been developed to measure the properties of non-neutral plasmas in traps, parameters such as plasma temperature, density, shape and the total number of particles. Destructive diagnostics involve releasing the particles from the trap and detecting them in various ways. Absolute measurements of the total number of particles can be made by dumping the particles onto a collector plate and measuring the total charge.[6] In the case of positrons, the annihilation gamma rays can be detected when the particles are dumped, and the total particle number can thus be extracted using a calibrated detector. Radial profiles can be measured using a phosphor screen biased at a high voltage (~ 5 – 10 kV). The resulting fluorescent light is measured using a CCD camera.[85] Plasma density can be inferred from the radial profiles and the total number of particles can be calculated using a Poisson–Boltzmann equilibrium code.[86] Plasma temperature can be measured by releasing particles slowly from the trap and measuring the tail of the particle energy distribution.[87]

Destructive diagnostics have been employed extensively in the development of new techniques to manipulate and trap antiparticles. However, for experiments where the particles are collected for long periods of time, such as antihydrogen production or the creation of giant pulses, destructive diagnostics are disadvantageous. Several non-destructive techniques have been developed, based on the properties of the plasma modes. For long cylindrical plasmas, the frequency of the diocotron mode yields the charge per unit length of the plasma, and hence provides information about the total number of particles.[88,89]

For spheroidal plasmas in harmonic potential wells, the frequencies of the axial Trivilpiece–Gould modes[90] yield the aspect ratio of the plasma and can be used to measure plasma temperature in cases where the aspect ratio is constant.[11–13,74,91] Such a mode spectrum is shown in Figure 4.12. The total number of particles can be determined by the Q factor of the response,[13] or by independently calibrating the amplitude response.[12,74] Passive monitoring of thermally excited modes can also be used to determine the plasma temperature.[92] Driven-wave techniques have also been used to monitor positron plasmas used for antihydrogen production.[11,13] They were also applied to characterize electron plasmas that are used to trap and cool antiprotons.[93]

An important technique for manipulating non-neutral plasmas is to compress the plasma radially using a rotating electric field to apply a torque on the plasma. This is the so-called "rotating-wall" (RW) technique. It has provided important new capabilities for single-component plasma research, such as counteracting outward plasma transport and permitting essentially infinite confinement times. It was first used to compress ion[94–97] and electron[82,98] plasmas. It has also been used to compress positron plasmas,[7,9] including those for antihydrogen production[11, 17, 99] and for the brightness-enhancement of positron beams.[14,15] This RW technique was also an important facet of the first successful creation of the positronium molecule, Ps_2.[25] It is expected to play a key role in work planned to produce giant pulses of positrons to create BEC of Ps atoms and the stimulated emission of annihilation radiation.[25]

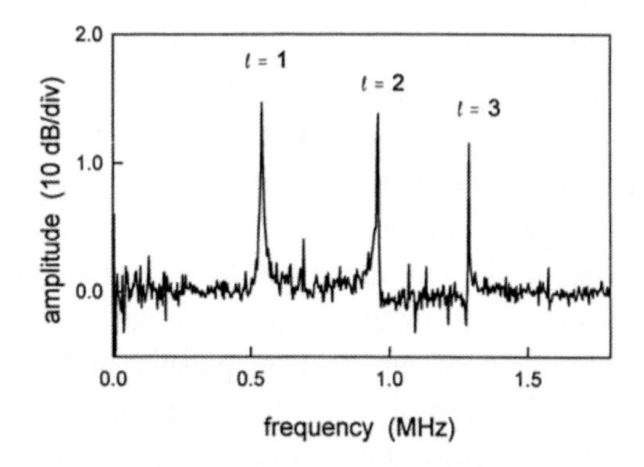

Fig. 4.12. Longitudinal compressional (Trivelpiece–Gould) modes of a positron plasma in a PM trap. From refence[74]; see this reference for details.

4.5. Radial Compression Using Rotating Electric Fields – the "Rotating-wall" (RW) Technique

The process of RW compression involves coupling a rotating electric field to the plasma to inject angular momentum. As described by Eq. (4.4), this then reduces the second moment of the radial particle distribution.[47] The arrangement for RW compression is shown schematically in Figure 4.9. Phased sine waves applied to a sectored electrode are used to generate a rotating electric field with a low-order azimuthal mode number (e.g., $m_\theta = 1$).[7,9,10,81] These fields produce a torque on the plasma, thereby compressing the plasma radially in a non-destructive manner.

Efficient cooling is required to counteract the heating caused by the torque-produced work done on the plasma. As described above, this cooling can be provided by cyclotron cooling (in the case of a strong confining magnetic field),[10,11,13,81] a buffer-gas (in the case of a weak magnetic field),[7,9,100] or by sympathetic cooling using laser–cooled ions.[95]

Early RW experiments relied on coupling to (Trivelpiece–Gould) plasma modes, which significantly limited the utility and flexibility of the technique. Two RW operating regimes were later discovered in which tuning to plasma modes is unnecessary. The first was in a plasma with buffer-gas cooling when the plasma radius is comparable to the Debye length, λ_D.[9] The second was in plasmas in a high-magnetic-field trap when the drive amplitude is sufficiently large (the "strong-drive" regime).[81] Most RW compression experiments now operate (or try to operate) in this second, strong-drive regime.

Shown in Figure 4.13 is an apparatus for studying PM plasmas cyclotron cooled in a high magnetic field. Shown in Figure 4.14 is an example of compression of an electron plasma in this device in the strong-drive regime. The protocol for these experiments is such that *the RW is applied at fixed values of both V_{RW} and f_{RW}*. Above a certain drive amplitude, the plasma evolves to a high-density steady state in which $f_E \approx f_{RW}$ (cf. Figure 4.14). As illustrated in Figure 4.15, the radial density profiles of these plasmas are "flat-top" in shape (i.e., a constant-density rigid-rotor in a state close to thermal equilibrium). Experiments at various values of f_{RW} are shown in Figure 4.16, illustrating the ability to access a broad range of high-density states in this strong-drive regime.

The ability to access the strong-drive regime depends upon overcoming the drag due to static asymmetries in the laboratory frame. These asymmetries drive waves (i.e., Trivelpiece–Gould modes) traveling backwards on the rotating plasma and thus act as a drag on it. This is illustrated in Figures 4.16 and 4.17 where a "step" appears in the data near the density $n = 0.4 \times 10^{10}$ cm^{-3}. The mode frequency is zero in the lab frame and referred to as a "zero-frequency mode" (ZFM).[81] The

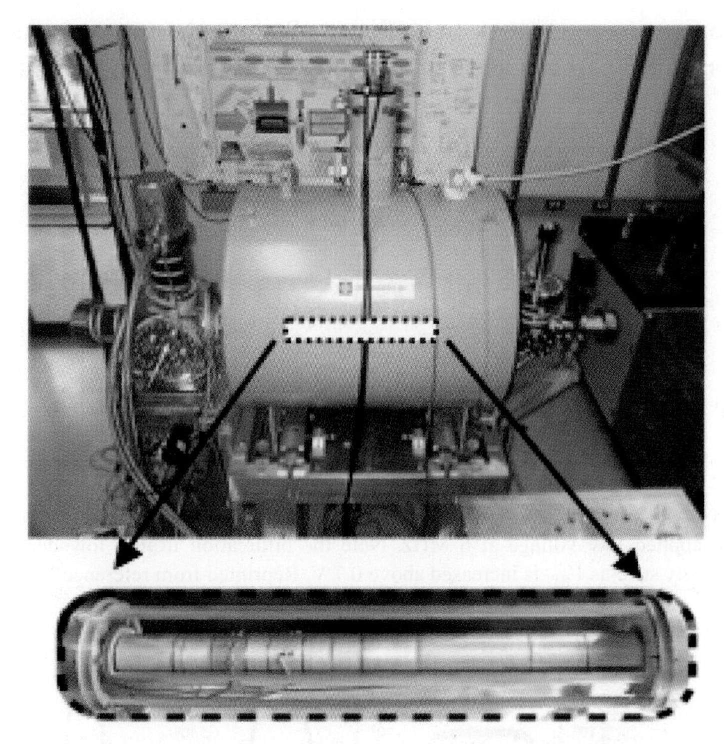

Fig. 4.13. A high-magnetic-field (5-tesla) UHV storage trap.[101] Also shown is a cutaway view of the electrode structure that contains two RW electrodes (left of center). The apparatus is also outfitted with a closed cycle pulsed-tube refrigerator for cooling the electrodes.

drag torques on the plasma have been modeled to include this ZFM effect,[5] namely the total torque on the plasma will be

$$\tau = \eta \frac{f_{RW} - f_E}{f_E} V_{RW}^2 - \frac{\beta f_E}{D^2 + f_E^{\,2}} - \frac{\gamma \delta f_0}{\left(f_E - f_0\right)^2 + \left(\delta f_0\right)^2}, \qquad (4.15)$$

where η, β, γ and D are constants. The terms in Eq. (4.15) represent the RW drive torque, τ_{RW} (first term) and the drag torques, τ_{drag}. The latter is the sum of the second and third terms, namely the background drag torque (second term, coefficient β) on the plasma due to trap imperfections, and the drag due to the ZFM (third term, coefficient γ).

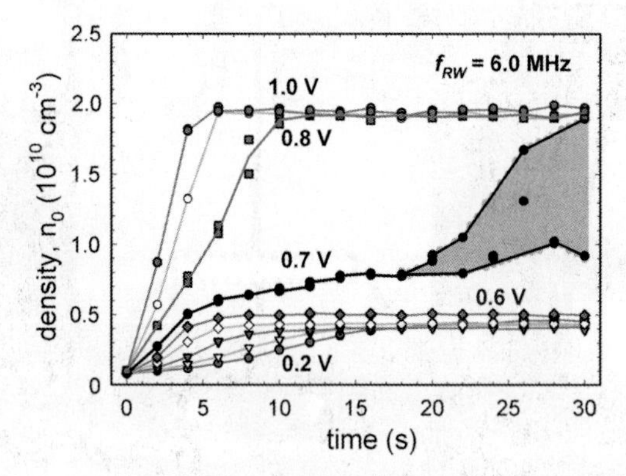

Fig. 4.14. Central electron density is shown as a function of time for various amplitudes of applied RW voltage at 6 MHz. Note the bifurcation from a low–density to a high density state as V_{RW} is increased above 0.7 V. Reprinted from reference[81].

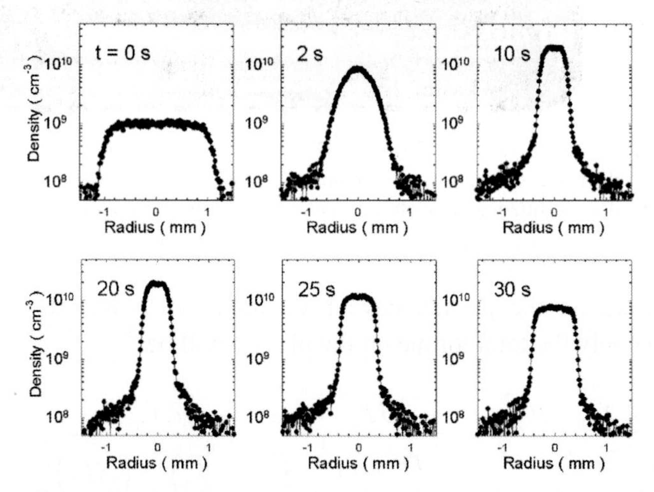

Fig. 4.15. Radial profiles for RW compression of an electron plasma in a 5-tesla magnetic field, using f_{RW} = 6 MHz, starting at t = 0 s. Steady-state compression is observed from t = 10 to 20 s, then the plasma expands with the RW off. Flat–top equilibrium profiles are observed, except at t = 2 s, where the plasma is hotter (i.e., $T \sim 3$ eV at that time). Reprinted from reference[81]; see this reference for details.

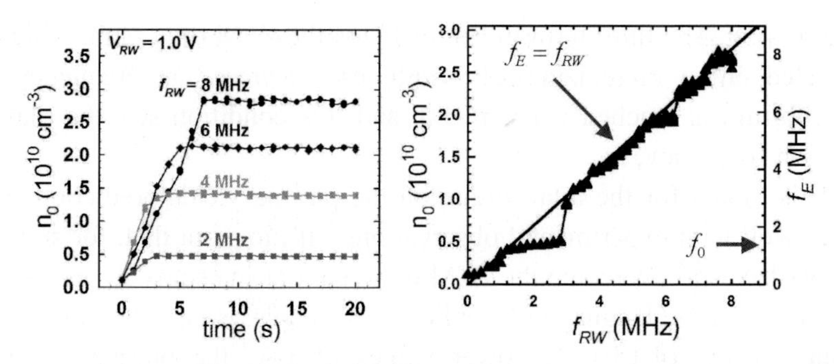

Fig. 4.16. (left) Central plasma density following application of the RW at various frequencies at V_{RW} = 1.0 V; and (right) steady-state density as a function of applied RW frequency, following the transition to the high-density state. The step near n = 0.4 x 10^{10} cm^{-3} is due to a so-called ZFM mode, which was key to understanding the high-density steady states. B = 5 T. Data from reference[81].

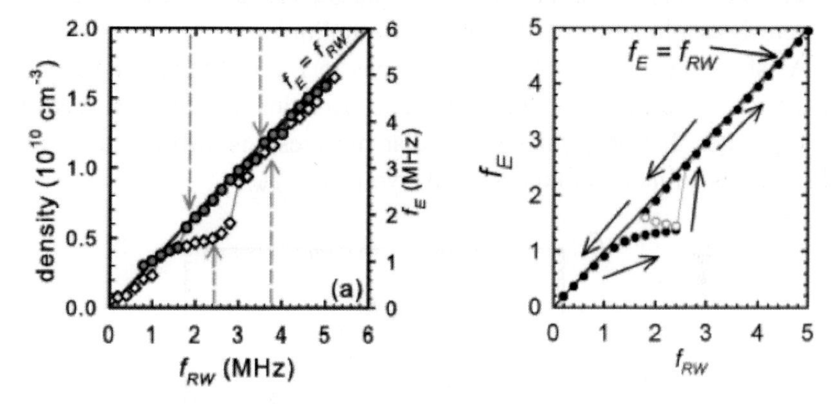

Fig. 4.17. (left) Density as a function of f_{RW}, when f_{RW} is fixed but the initial plasma density is smaller (upward arrow, ◇) or larger (downward arrow, ●) than that of the final, torque balanced steady states (i.e., the stable fixed points); (right) solutions of Eq. (4.13) for τ = 0, for the (●) stable and (O) unstable fixed points when approached varying f_{RW} in the directions shown by the arrows. The model exhibits the same qualitative behavior as the data. Analysis from reference[5]; see this reference for details.

The form of the second term in Eq. (4.15) was chosen empirically to model the observed outward transport data such as that shown in Figure 4.11. An example of the drag torque derived from that data is shown in

Figure 4.18. The third term in Eq. (4.15) is the ZFM drag term, which is modeled by a Lorentzian of width δf_0, *centered* at frequency f_0. Equilibrium is reached when $\tau = 0$, and this condition sets the plasma rotation frequency, f_E.

This model for the total torque on the plasma yields predictions that agree well with experimental observations.[5] It turns out that, for suitably strong drives to overcome the ZFM drag, the plasma spins up until $f_{RW} \approx f_E$ (which, in the language of nonlinear dynamics, is an "attracting fixed point" of Eq. (4.15)). At lower values of τ_{RW}, the plasma becomes "stuck" at a rotation frequency close to that of the ZFM (i.e., the "low-density fixed point" at f_0). The stable state to which the plasma relaxes depends upon which side of the ZFM the plasma starts: the fixed point is stable when $d\tau_{RW}/df_{RW} > d\tau_{drag}/df_{RW}$ and unstable when $d\tau_{RW}/df_{RW} < d\tau_{drag}/df_{RW}$. As a consequence, the plasma is predicted to exhibit hysteresis as a function of the RW drive amplitude.

As shown in Figure 4.17, the solutions to Eq. (4.15) provide a good qualitative description of this hysteretic behavior and the high-density steady states that are achieved. Similar hysteresis is also predicted and observed as a function of the RF drive voltage, V_{RW}.[5]

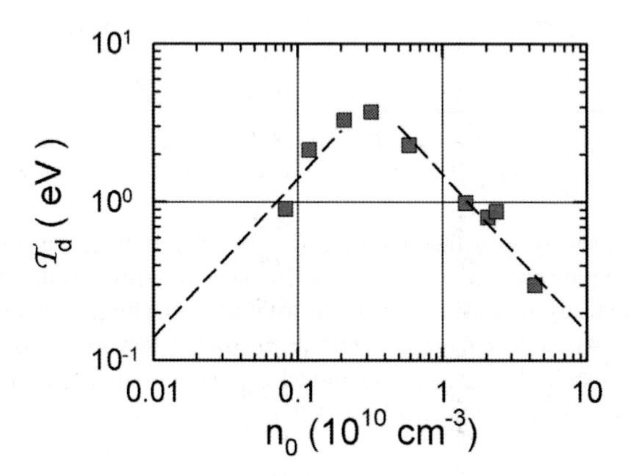

Fig. 4.18. Drag torque τ_d as a function of plasma density n_0 derived from the expansion data shown in Figure 4.11 for L_p = 24 cm. Dashed lines are guides to the eye. This dependence of τ_d on n_0 motivated the specific form of the second term in Eq. (4.15).

A key practical question is what limits the compression and the maximum achievable density. The UCSD experiments are routinely conducted with relative ease up to $f_{RW} \sim 8$ MHz and spottily up to ~ 18 MHz. This limit may be due to spurious resonances in the electronic circuitry or perhaps something more fundamental (e.g., the inability to couple effectively to the plasma at high frequencies); this will require further study to resolve.

RW compression in the single-particle regime. Low-density positron gases in Penning traps (i.e., collections of particles outside the plasma regime) have also been compressed using the RW technique with gas cooling.[100,102] For successful RW operation it was necessary that the particles be confined in a harmonic electrostatic potential well in the direction of the confining, uniform magnetic field. As shown in Figure 4.19, good compression was observed when $f_{RW} \leq \omega_z$, where ω_z is the axial bounce frequency in the harmonic well. In this case, it is believed that the particles couple to a rotating particle bounce resonance. As shown in Figure 4.19, at frequencies above ω_z, the particles are observed to heat rapidly and are de-confined.

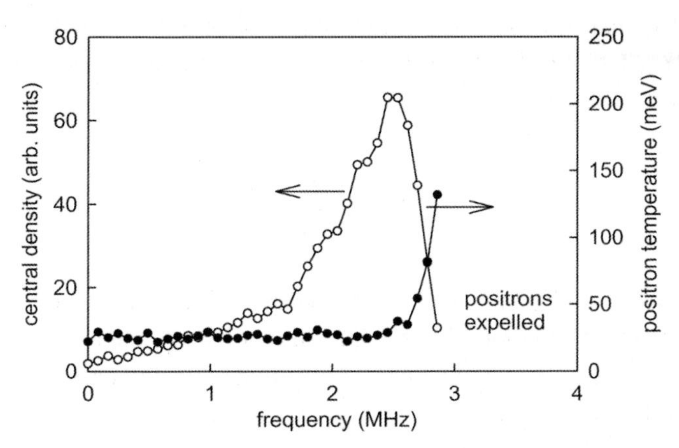

Fig. 4.19. Compression of a positron gas (i.e., in the single-particle regime) for an applied RW potential of 1.4 V. Good plasma compression is observed for f_{RW} at and below the axial bounce frequency of ω_z = 2.6 MHz. Above this frequency, plasma heating and particle expulsion are observed. The experiments use SF_6 for gas cooling. From reference[100]; see this reference for details.

The fact that the RW technique works in the single-particle regime is very useful in tailoring the charge clouds in buffer-gas traps, particularly ones that operate with fewer stages. In such traps, the cycle time must be kept short to avoid outward radial transport and annihilation, and hence the positron density is relatively low (i.e., the trapped positrons are in the single-particle, non-plasma regime).

Heating due to RW compression. Applying rotating electric fields to a plasma applies a torque τ_{RW} on it that heats the plasma by doing work on it. The heating rate can be written,[103]

$$P_H = \omega_{RW}\, \tau_{RW}, \tag{4.16}$$

where ω_{RW} is the angular frequency of the rotating electric field. In the strong-drive regime, the minimum power input to the plasma will be when the drive and drag torques are in balance, in which case $\omega_{RW} \approx \omega_E$ and $P_H = \omega_E\, \tau_{RW}$, where ω_E is the angular rotation frequency of the plasma.

The asymmetry-induced drag torque τ_a can be obtained by relating the time derivative of the plasma angular momentum (cf. Eq. (4.5)) to the outward expansion rate Γ_0.[81] Assuming a plasma of N particles with a flat-top density profile in surroundings at temperature T_W, the steady-state temperature T will be,[81]

$$T = T_W + \left(\frac{Ne^2}{3L_p}\right)\frac{\Gamma_0}{\Gamma_c}. \tag{4.17}$$

Illustrated in Figure 4.20 is the effect of plasma heating on RW compression. In this case the plasma is cooled by inelastic vibrational collisions with CF_4 molecules. Note that the temperature remains comparable to the ν_3 mode energy of 0.16 eV (i.e., the dominant positron-impact vibrational excitation) over an order of magnitude increase in the RW voltage. When it does break away from this value, as the RW voltage is increased further, the temperature rises rapidly and the maximum achievable compression decreases quickly.

Good compression is obtained as long as the collisional excitation of the ν_3 vibrational mode of CF_4 can control the plasma temperature. When the temperature increases much above the energy of this excitation (ε_3 = 0.16 eV), then the temperature runs away and the compression is much less efficient.

Note that the plasma temperature, given by Eq. (4.17), is that expected for the minimum heating rate, which was obtained when the RW drive and asymmetry drag torque τ_a are balanced in the strong-drive regime. If there is "slip" (i.e., if $\omega_{RW} > \omega_E$), the heating rate will be larger. In this case, the excess heating rate due to the slip will be,

$$\delta P = \tau_{RW}\left(\omega_{RW} - 2\pi f_E\right) = 2\pi\tau_{RW}\Delta f, \tag{4.18}$$

where Δf is the so-called slip frequency.

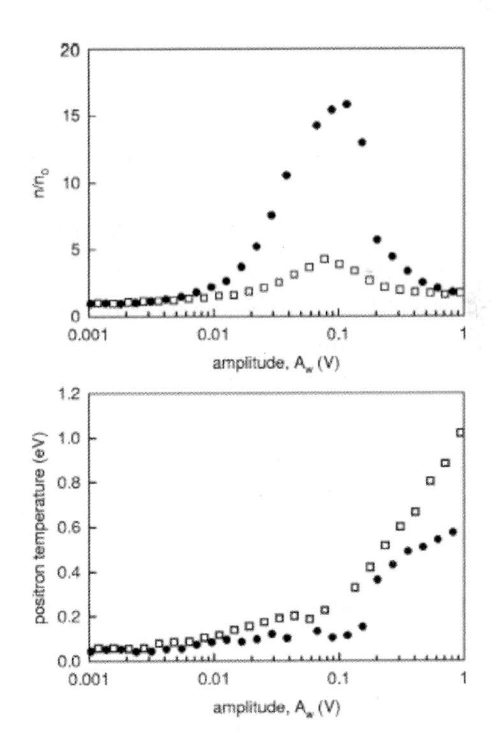

Fig. 4.20. Example of RW compression with gas cooling. Dependence of (above) the central density and (below) the positron temperature on RW drive amplitude, V_{RW} (labeled here as A_w), for cooling on CF_4 at pressures of (●) 3.5 x 10^{-8} mbar and (□) 8 x 10^{-9} mbar with f_{RW} = 2.5 MHz for 1 s. Reprinted from reference[7].

Maximum achievable density using RW compression. For many applications it is desirable to have as high a plasma density as possible. One constraint is the Brillouin limit. This limit arises from the fact that, for a particle in a PM trap rotating about the symmetry axis at frequency f_E, the v x B force acts both to provide the required inward centripetal force and to counteract the outward force due to the space-charge electric field. Due to the fact that the v x B force is proportional to the particle velocity v, and the centripetal force is proportional to v^2, this force balance is not possible above some maximum velocity v. And since the E x B rotation velocity, $v \propto n$, this imposes a maximum density limit, the so-called Brillouin limit.

The condition is[33]

$$\omega_p^2 = \omega_c^2/2, \tag{4.19}$$

where ω_p is the plasma frequency. The resulting Brillouin density limit is

$$n_B\left[m^3\right] = 4.8 \times 10^{18} B^2\left[T\right], \tag{4.20}$$

where n_B is in units of m^3 and B is in units of tesla. Above the Brillouin limit, particles at the plasma edge cannot be confined orbiting the axis of symmetry; they will move outward, unconfined.

However, if the plasma is in the presence of neutral gas molecules, even below this limit, any scattering will cause the particles to make relatively large cycloid-like orbits, moving outward on each collision an average distance, $E/\omega_c B$, where E is the space-charge electric field (cf. Figure 4.10). As the density increases, so will E, and hence the plasma will become more difficult to confine.

4.6. Concluding Remarks

The techniques described here have proven enormously useful in accumulating and manipulating positron, antimatter plasmas. They have played a central role in the quests to create low-energy antihydrogen and the positronium molecule, Ps_2. They have also proven crucial in studies of atomic physics processes such as positron scattering and annihilation in interactions with atoms and molecules.

That said, there are likely many opportunities for further improvement. While there are a myriad of possibilities, here we mention just a few of the obvious ones. There are likely a number of ways to make buffer-gas positron traps simpler, more compact and, perhaps, more efficient. This might be done by clever design of the neutral gas profile and the differential pumping arrangement. There are also questions as to what limits the maximum trapping efficiency and how this can be improved. Finally, the range of atomic and molecular gases explored for trapping and cooling, while extensive, has not been exhaustive; there may well be room for further improvement here too.

Regarding the RW technique, it is presently uncertain what limits the maximum density that can be achieved, and this is a crucial issue for many applications. One question is: can one approach the Brillouin limit, and, if not, why not? There is also a question as to whether one might use a resonant structure to enhance greatly the cyclotron cooling.[73] If so, this will likely permit a broader range of operating parameters and the ability to operate at lower magnetic fields.

Acknowledgments

I would like to acknowledge the contributions of A. Passner, M. Leventhal, T. J. Murphy, M. Tinkle, R. G. Greaves, J. R. Danielson, E. A. Jerzewski and T. M. O'Neil to the work described here. I would also like to thank M. Charlton for his careful reading of the manuscript and helpful suggestions. This chapter relies heavily upon the data and descriptions in references[3–5,83,104]. This work is supported by the U.S. DOE/NSF Plasma Initiative.

References

1. C. M. Surko, J. R. Danielson, T. R. Weber, "Accumulation, Storage and Manipulation of Large Numbers of Positrons in Traps II - Selected Topics" in Knoop, M., Madsen, N. and Thompson, R. C. (eds), *Physics with Trapped Charged*

Particles (Imperial College Press, London, 2013), pp. 129–172. (Reprinted with updates from reference[2].)

2. C. M. Surko, J. R. Danielson and T. R. Weber, "Accumulation, Storage and Manipulation of Large Numbers of Positrons in Traps II: selected topics" in Mills, A. P. and Dupasquier, A. (eds), *Physics with Many Positrons* (IOS press, Amsterdam, 2010), pp. 545–574.
3. C. M. Surko and R. G. Greaves, Emerging science and technology of antimatter plasmas and trap-based beams, *Phys. Plasmas* **11**, 2333–2348 (2004).
4. J. R. Danielson and C. M. Surko, "Plasma Compression using Rotating Electric Fields – the Strong Drive Regime", in Drewsen, M., Uggerhoj, U. and Knudsen, H. (eds), *Non-Neutral Plasma Physics* (American Institute of Physics Press, 2006), pp. 19–28.
5. J. R. Danielson, C. M. Surko and T. M. O'Neil, High-density Fixed Point for Radially Compressed Single-component Plasmas, *Phys. Rev. Lett.* **99**, 135005–135004 (2007).
6. J. H. Malmberg and C. F. Driscoll, Long-time containment of a pure electron plasma, *Phys. Rev. Lett.* **44**, 654–657 (1980).
7. R. G. Greaves and C. M. Surko, Radial compression and inward transport of positron plasmas using a rotating electric field, *Phys. Plasmas* **8**, 1879–1885 (2001).
8. J. Malmberg, T. M. O'Neil, A. W. Hyatt and C. F. Driscoll, presented at the Proceedings of the Sendai Symposium on Plasma Nonlinear Electron Phenomena, unpublished (1984) .
9. R. G. Greaves and C. M. Surko, Inward transport and compression of a positron plasma by a rotating electric field, *Phys. Rev. Lett.* **85**, 1883–1886 (2000).
10. J. R. Danielson and C. M. Surko, Torque-balanced high-density steady states of single component plasmas, *Phys. Rev. Lett.* **95**, 035001–035004 (2005).
11. R. Funakoshi, M. Amoretti, G. Bonomi, P. D. Bowe, C. Canali, C. Carraro, C. L. Cesar, M. Charlton, M. Doser, A. Fontana, M. C. Fujiwara, P. Genova, J. S. Hangst, R. S. Hayano, L. V. Jørgensen, A. Kellerbauer, V. Lagomarsino, R. Landua, E. L. Rizzini, M. Macrì, N. Madsen, G. Manuzio, D. Mitchard, P. Montagna, L. G. C. Posada, A. Rotondi, G. Testera, A. Variola, L. Venturelli, D. P. v. d. Werf, Y. Yamazaki and N. Zurlo, Positron plasma control techniques for the production of cold antihydrogen, *Phys. Rev. A* **76**, 012713–012718 (2007).
12. M. D. Tinkle, R. G. Greaves, C. M. Surko, R. L. Spencer and G. W. Mason, Low-order modes as diagnostics of spheroidal non-neutral plasmas, *Phys. Rev. Lett.* **72**, 352–355 (1994).
13. M. Amoretti, G. Bonomi, A. Bouchta, P. D. Bowe, C. Carraro and C. L. Cesar, Complete nondestructive diagnostic of nonneutral plasmas based on the detection of electrostatic modes, *Phys. Plasmas* **10**, 3056–3064 (2003).
14. J. R. Danielson, T. R. Weber and C. M. Surko, Extraction of Small-diameter Beams from Single-component Plasmas, *Appl. Phys. Lett.* **90**, 081503–081503 (2007).

15. T. R. Weber, J. R. Danielson and C. M. Surko, Creation of Finely Focused Particle Beams from Single–Component Plasmas, *Phys. Plasmas* **15**, 012106–012110 (2008).
16. T. R. Weber, J. R. Danielson and C. M. Surko, Energy Spectra of Tailored Particle Beams from Trapped Single–component Plasmas, *Phys. Plasmas* **16**, 057105–057108 (2009).
17. M. Amoretti, C. Amsler, G. Bonomi, A. Bouchta, P. Bowe, C. Carraro, C. L. Cesar, M. Charlton, M. Collier, M. Doser, V. Filippini, K. Fine, A. Fontana, M. Fujiwara, R. Funakoshi, P. Genova, J. Hangst, R. Hayano, M. Holzscheiter, L. Jorgensen, V. Lagomarsino, R. Landua, D. Lindelof, E. L. Rizzini, M. Macri, N. Madsen, G. Munuzio, M. Marchesotti, P. Montagna, H. Pruys, C. Regenfus, P. Riedler, J. Rochet, A. Rotondi, G. Rouleau, G. Testera, A. Variola, T. Watson and D. VanderWerf, Production and detection of cold antihydrogen atoms, *Nature* **419**, 456–459 (2002).
18. G. Gabrielse, N. Bowden, P. Oxley, A. Speck, C. Storry, J. Tan, M. Wessels, D. Grzonka, W. Oelert, G. Schepers, T. Sefzick, J. Walz, H. Pittner, T. Hansch and E. Hessels, Driven production of cold antihydrogen and the first measured distribution of antihydrogen states, *Phys. Rev. Lett.* **89**, 233401–233405 (2002).
19. G. Gabrielse, N. Bowden, P. Oxley, A. Speck, C. Storry, J. Tan, M. Wessels, D. Grzonka, W. Oelert, G. Schepers, T. Sefzick, J. Walz, H. Pittner, T. Hansch and E. Hessels, Background-free observation of cold antihydrogen with field-ionization analysis of its states, *Phys. Rev. Lett.* **89**, 213401–213404 (2002).
20. M. Amoretti, C. Amsler, G. Bonomi, A. Bouchta, P. D. Bowe, C. Carraro, M. Charlton, M. J. T. Collier, M. Doser, V. Filippini, K. S. Fine, A. Fontana, M. C. Fujiwara, R. Funakoshi, P. Genova, A. Glauser, D. Grögler, J. Hangst, R. S. Hayano, H. Higaki, M. H. Holzscheiter, W. Joffrain, L. V. Jørgensen, V. Lagomarsino, R. Landua, C. L. Cesar, D. Lindelöf, E. Lodi-Rizzini, M. Macri, N. Madsen, D. Manuzio, G. Manuzio, M. Marchesotti, P. Montagna, H. Pruys, C. Regenfus, P. Riedler, J. Rochet, A. Rotondi, G. Rouleau, G. Testera, D. P. v. d. Werf, A. Variola, T. L. Watson, T. Yamazaki and Y. Yamazaki, The ATHENA antihydrogen apparatus, *Nucl. Instrum. Methods A* **518**, 679–711 (2004).
21. V. Tsytovich and C. B. Wharton, Laboratory electron-positron plasma – a new research object, *Comments Plasma Phys. Controlled Fusion* **4**, 91–100 (1978).
22. R. G. Greaves and C. M. Surko, An electron-positron beam-plasma experiment, *Phys. Rev. Lett.* **75**, 3846–3849 (1995).
23. S. J. Gilbert, D. H. E. Dubin, R. G. Greaves and C. M. Surko, An electron-positron beam-plasma instability, *Phys. Plasmas* **8**, 4982–4994 (2001).
24. T. S. Pederson, A. H. Boozer, W. Dorland, J. P. Kremer and R. Schmitt, Prospects for the creation of positron-electron plasmas in a non-neutral stellarator, *J. Phys. B: At. Mol. Opt.* **36**, 1029–1039 (2003).
25. D. B. Cassidy, A. P. Mills, Jr., The Production of Molecular Positronium, *Nature* **449**, 195–197 (2007).
26. P. J. Schultz, K. G. Lynn, Interaction of positrons beams with surfaces, thin films, and interfaces, *Rev. Mod. Phys.* **60**, 701–779 (1988).

27. A. Dupasquier and A. P. Mills, *Positron Spectroscopy of Solids* (IOS Press, Amsterdam, 1995).
28. D. W. Gidley, D. Z. Chi, W. D. Wang and R. S. Vallery, Positron Annihilation as a Method to Characterize Porous Materials, *Ann. Rev. Mat. Sci.* **36**, 49–79 (2006).
29. R. G. Greaves and C. M. Surko, Positron trapping and the creation of high–quality trap–based positron beams, *Nucl. Instrum. Methods B* **192**, 90–96 (2002).
30. R. G. Greaves and J. M. Moxom, Recent results on trap-based positron beams, *Mat. Sci. Forum* **445–446**, 419–423 (2004).
31. R. L. Wahl, *Principles and Practice of Positron Emission Tomography* (Lippincott, Williams and Wilkins, Philadelphia, PA, 2002).
32. J. R. Danielson, T. R. Weber and C. M. Surko, Plasma Manipulation Techniques for Positron Storage, *Phys. Plasmas* **13**, 123502–123510 (2006).
33. R. C. Davidson, *Physics of Nonneutral Plasmas* (Addison-Wesley, Reading, MA, 1990).
34. D. H. E. Dubin and T. M. O'Neil, Trapped nonneutral plasmas, liquids, and crystals (the thermal equilibrium states), *Rev. of Mod. Phys.* **71**, 87–172 (1999).
35. A. P. Mills, Further improvements in the efficiency of low-energy positron moderators, *Appl. Phys. Lett.* **37**, 667–668 (1980).
36. A. P. Mills and E. M. Gullikson, Solid neon moderator for producing slow positrons, *Appl. Phys. Lett.* **49**, 1121–1123 (1986).
37. R. G. Greaves and C. M. Surko, Solid neon moderator for positron trapping experiments, *Can. J. Phys.* **51**, 445–448 (1996).
38. G. Gibson, W. C. Jordan and E. J. Lauer, Containment of positrons in a mirror machine, *Phys. Rev. Lett.* **5**, 141–144 (1960).
39. P. B. Schwinberg, J. R. S. Van Dyck and H. G. Dehmelt, Trapping and thermalization of positrons for geonium spectroscopy, *Phys. Lett.* **81A**, 119–120 (1981).
40. S. Chu and A. P. Mills, Excitation of the Positronium 1S –> 2S Two-Photon Transition, *Phys. Rev. Lett.* **48**, 1333–1336 (1982).
41. A. P. Mills, E. D. Shaw, R. J. Chichester and D. M. Zuckerman, Production of Slow Positron Bunches Using a Microtron Accellerator, *Rev. Sci. Instrum.* **60**, 825–830 (1989).
42. B. L. Brown, M. Leventhal, A. P. Mills and D. W. Gidley, Positron annihilation in a simulated low-density galactic environment, *Phys. Rev. Lett.* **53**, 2347–2350 (1984).
43. G. Gabrielse, L. Haarsma and S. L. Rolston, Open-endcap Penning traps for high precision experiments, *Int. J. Mass Spectrom.* **88**, 319–332 (1989).
44. H. Higaki and A. Mohri, Experiment on diocotron oscillations of spheroidal non–neutral electron plasmas in a multi-ring-electrode trap, *Phys. Lett. A* **235**, 504–507 (1997).
45. T. O'Neil, Plasmas with a single sign of charge (an overview), *Physica Scripta* **T59**, 341–351 (1995).
46. J. J. Bollinger, D. J. Wineland and D. H. E. Dubin, Non-neutral ion plasmas and crystals, laser cooling, and atomic, *Phys. Plasmas* **1**, 1403–1414 (1994).

47. T. M. O'Neil, A confinement theorem for nonneutral plasmas, *Phys. Fluids* **23**, 2216–2218 (1980).
48. D. Segers, J. Paridaens, M. Dorikens and L. Dorikens–Vanpraet, Beam handling with a Penning trap of a LINAC-based slow positron beam, *Nucl. Instrum. Methods* **A337**, 246–252 (1994).
49. R. G. Greaves, M. D. Tinkle and C. M. Surko, Creation and uses of positron plasmas, *Phys. Plasmas* **1**, 1439–1446 (1994).
50. L. V. Jørgensen, M. Amoretti, G. Bonomi, P. D. Bowe, C. Canali, C. Carraro, C. L. Cesar, M. Charlton, M. Doser, A. Fontana, M. C. Fujiwara, R. Funakoshi, P. Genova, J. S. Hangst, R. S. Hayano, A. Kellerbauer, V. Lagomarsino, R. Landua, E. L. Rizzini, M. Macrì, N. Madsen, D. Mitchard, P. Montagna, A. Rotondi, G. Testera, A. Variola, L. Venturelli, D. P. v. d. Werf and Y. Yamazaki, New Source of Dense, Cryogenic Positron Plasmas, *Phys. Rev. Lett.* **95**, 025002–025005 (2005).
51. C. M. Surko, M. Leventhal and A. Passner, Positron plasma in the laboratory, *Phys. Rev. Lett.* **62**, 901–904 (1989).
52. T. J. Murphy and C. M. Surko, Positron trapping in an electrostatic well by inelastic collisions with nitrogen molecules, *Phys. Rev. A* **46**, 5696–5705 (1992).
53. B. M. Jelenkovic, A. S. Newbury, J. Bollinger, W. M. Itano and T. B. Mitchell, Sympathetically cooled and compressed positron plasma, *Phys. Rev. A* **67**, 063406–063409 (2003).
54. N. Oshima, T. M. Kojima, M. Niigati, A. Mohri, K. Komaki and Y. Yamazaki, New scheme for positron accumulation in ultrahigh vacuum, *Phys. Rev. Lett.* **93**, 195001–195004 (2004).
55. H. Boehmer, M. Adams and N. Rynn, Positron trapping in a magnetic mirror configuration, *Phys. Plasmas* **2**, 4369–4371 (1995).
56. L. Haarsma, K. Abdullah and G. Gabrielse, Extremely cold positrons accumulated, *Phys. Rev. Lett.* **75**, 806–809 (1995).
57. J. Estrada, T. Roach, J. N. Tan, P. Yesley and G. Gabrielse, Field ionization of strongly magnetized Rydberg positronium: A new physical mechanism for positron accumulation, *Phys. Rev. Lett.* **84**, 859–862 (2000).
58. B. Ghaffari and R. S. Conti, Experimental evidence for chaotic transport in a positron trap, *Phys. Rev. Lett.* **75**, 3118–3121 (1995).
59. R. G. Greaves and J. Moxom, "Design and Performance of a Trap-Based Beam Source" in Schauer, M., Mitchell, T. and Nebel, R. (eds), *Non-Neutral Plasma Physics V* (American Institute of Physics, New York, 2003), pp. 140–148.
60. C. M. Surko, A. Passner, M. Leventhal and F. J. Wysocki, Bound states of positrons and large molecules, *Phys. Rev. Lett.* **61**, 1831–1834 (1988).
61. J. P. Marler and C. M. Surko, Positron-impact ionization, positronium formation and electronic excitation cross sections for diatomic molecules, *Phys. Rev. A* **72**, 062713–062710 (2005).
62. J. P. Marler and C. M. Surko, Systematic comparison of positron and electron impact excitation of the n_3 vibrational mode of CF_4, *Phys. Rev. A* **72**, 062702–062706 (2005).

63. C. M. Surko, S. J. Gilbert, and R. G. Greaves, "Progress in Creating Low-Energy Positron Plasmas and Beams" in Bollinger, J. J., Spencer, R. L., and Davidson, R. C. (eds), *Non-Neutral Plasma Physics* (American Institute of Physics, New York, 1999), pp. 3–12.
64. J. Clarke, D. P. v. d. Werf, B. Griffiths, D. C. S. Beddows, M. Charlton, H. H. Telle and P. R. Watkeys, Design and operation of a two-stage positron accumulator, *Rev. Sci. Instrum.* **77**, 063302–063305 (2006).
65. C. M. Surko, G. F. Gribakin and S. J. Buckman, Low-energy positron interactions with atoms and molecules, *J. Phys. B: At. Mol. Opt. Phys.* **38**, R57–R126 (2005).
66. A. Bankovic, J. P. Marler, M. Suvakov, G. Malovic and Z. L. Petrovic, Transport Coefficients for Positron Swarms in Nitrogen, *Nuc. Instrum. Methods B* **266**, 462–465 (2008).
67. S. J. Gilbert, R. G. Greaves and C. M. Surko, Positron scattering from atoms and molecules at low energies, *Phys. Rev. Lett.* **82**, 5032–5035 (1999).
68. J. Sullivan, S. J. Gilbert and C. M. Surko, Excitation of Molecular Vibrations by Positron Impact, *Phys. Rev. Lett.* **86**, 1494–1497 (2001).
69. J. P. Marler, G. Gribakin and C. M. Surko, Comparison of positron-impact vibrational excitation cross sections with the Born-dipole model, *Nucl. Instrum. Methods B* **247**, 87–91 (2006).
70. I. Al-Qaradawi, M. Charlton and I. Borozan, Thermalization times of positrons in molecular gases, *J. Phys. B* **33**, 2725–2732 (2000).
71. P. G. Coleman, T. C. Griffith and G. R. Heyland, Rotational-excitation and momentum-transfer in slow positron-molecule collisions, *J. Phys. B* **14**, 2509–2517 (1981).
72. B. R. Beck, J. Fajans and J. H. Malmberg, Measurement of collisional anisotropic temperature relaxation in a strongly magnetized pure electron plasma, *Phys. Rev. Lett.* **68**, 317–320 (1992).
73. T. M. O'Neil, Cooling of a pure electron plasma by cyclotron radiation, *Phys. Fluids* **23**, 725–731 (1980).
74. M. D. Tinkle, R. G. Greaves and C. M. Surko, Low-order longitudinal modes of single-component plasmas, *Phys. Plasmas* **2**, 2880–2894 (1995).
75. M. Amoretti, C. Amsler, G. Bonomi, A. Bouchta, P. D. Bowe, C. Carraro, C. L. Cesar, M. Charlton, M. Doser, V. Filippini, A. Fontana, M. C. Fujiwara, R. Funakoshi, P. Genova, J. S. Hangst, R. S. Hayano, L. V. Jørgensen, V. Lagomarsino, R. Landua, D. Lindelöf, E. L. Rizzin, M. Macrí, N. Madsen, G. Manuzio, P. Montagna, H. Pruys, C. Regenfus, A. Rotondi, G. Testera, A. Variola and D. P. v. d. Werf, Positron Plasma Diagnostics and Temperature Control for Antihydrogen Production, *Phys. Rev. Lett.* **91**, 05501–05505 (2003).
76. D. J. Griffiths, *Introduction to Electrodynamics* (Prentice Hall, Saddle River, NJ, 1981).
77. F. F. Chen, *Introduction to Plasma Physics and Controlled Fusion, Volume I: Plasma Physics*, Second edition (Springer, New York, 1984).
78. T. M. O'Neil and D. H. E. Dubin, Thermal Equilibria and Thermodynamics of Trapped Plasmas with a Single Sign of Charge, *Phys. Plasmas* **5**, 2163–2193 (1998).

79. C. F. Driscoll and J. H. Malmberg, Length-dependent containment of a pure electron-plasma column, *Phys. Rev. Lett.* **50**, 167–170 (1983).
80. A. A. Kabantsev and C. F. Driscoll, Trapped-particle Mediated Collisional Damping of Nonaxisymmetric Plasma Waves, *Phys. Rev. Lett.* **97**, 095001–095004 (2006).
81. J. R. Danielson and C.M. Surko, Radial Compression and Torque-balanced Steady States of Single-Component Plasmas in Penning-Malmberg traps, *Phys. Plasmas* **13**, 055706–055710 (2006).
82. E. M. Hollmann, F. Anderegg and C. F. Driscoll, Confinement and Manipulation of Non-neutral Plasmas Using Rotating Wall Electric Fields, *Phys. Plasmas* **7**, 2776–2789 (2000).
83. C. M. Surko and R. G. Greaves, A multi-cell trap to confine large numbers of positrons, *Rad. Chem. and Phys.* **68**, 419–425 (2003).
84. O. Sueoka, H. Takak, A. Hamada, H. Sato and M. Kimura, Total cross sections of electron and positron collisions with CHF_3 molecules: a comparative study with CH4 and CF4, *Chem. Phys. Lett.* **288**, 124–130 (1998).
85. X. P. Huang and C. F. Driscoll, Relaxation of 2D Turbulence to a Metaequilibrium Near the Minimum Enstrophy State, *Phys. Rev. Lett.* **72**, 2187–2190 (1994).
86. R. Spencer, G. W. Mason, and S. N. Rasband, "Numerical Non-Neutral Plasmas" in Schauer, M., Mitchell, T. and Nebel, R. (eds), *Non-Neutral Plasma Physics V* (American Institute of Physics Press, Melville, New York, 2003).
87. D. L. Eggleston, C. F. Driscoll, B. R. Beck, A. W. Hyatt and J. H. Malmberg, Parallel energy analyzer for pure electron plasma devices, *Phys. Fluids B* **4**, 3432–3439 (1992).
88. J. S. Degrassie and J. H. Malmberg, Wave-induced transport in pure electron-plasma, *Phys. Rev. Lett.* **39**, 1077–1080 (1977).
89. K. S. Fine and C. F. Driscoll, The finite length diocotron mode, *Phys. Plasmas* **5**, 601–607 (1998).
90. D. H. E. Dubin, Theory of electrostatic fluid modes in a cold spheroidal non-neutral plasma, *Phys. Rev. Lett.* **66**, 2076–2079 (1991).
91. C. S. Weimer, J. J. Bollinger, F. L. Moore and D. J. Wineland, Electrostatic modes as a diagnostic in Penning trap experiments, *Phys. Rev. A* **49**, 3842–3853 (1994).
92. F. Anderegg, N. Shiga, J. R. Danielson, D. H. E. Dubin, C. F. Driscoll and R. W. Gould, Thermally excited modes in a pure electron plasma, *Phys. Rev. Lett.* **90**, 115001–115001 (2003).
93. N. Oshima, T. M. Kojima, M. Niigaki, A. Mohri, K. Komaki, Y. Iwai and Y. Yamazaki, Development of a cold HCI source for ultra-slow collisions, *Nucl. Instr. Methods B* **205**, 178–182 (2003).
94. X. P. Huang, F. Anderegg, E. M. Hollmann, C. F. Driscoll and T. M. O'Neil, Steady state confinement of nonneutral plasma by rotating electric fields, *Phys. Rev. Lett.* **78**, 875–878 (1997).
95. X. P. Huang, J. J. Bollinger, T. B. Mitchell and W. M. Itano, Phase-locked rotation of crystallized non-neutral plasmas by rotating electric fields, *Phys. Rev. Lett.* **80**, 73–76 (1998).

96. T. B. Mitchell, J. J. Bollinger, W. M. Itano and D. H. E. Dubin, Stick-slip dynamics of a stressed ion crystal, *Phys. Rev. Lett.* **87**, 183001–183001 (2001).
97. J. J. Bollinger, T. B. Mitchell, X.-P. Huang, W. M. Itano, J. N. Tan, B. M. Jelenkovic and D. J. Wineland, Crystalline order in laser–cooled, non-neutral ion plasmas, *Phys. Plasmas* **7**, 7 (2000).
98. F. Anderegg, E. M. Hollmann and C. F. Driscoll, Rotating field confinement of pure electron plasmas using Trivelpiece-Gould modes, *Phys. Rev. Lett.* **81**, 4875–4878 (1998).
99. D. P. Vanderwerf, M. Amoretti, G. Bonomi, A. Bouchta, P. D. Bowe, C. Carraro, C. L. Cesar, "Transfer, Stacking, and Compression of Positronn Plasmas under UHV conditions" in Schauer, M., Mitchell, T. and Nebel, R. (eds), *Non-Neutral Plasma Physics V* (American Institute of Physics Press, Melville, New York, 2003), pp. 172–177.
100. R. G. Greaves and J. M. Moxom, Compression of Trapped Positrons in a Single Particle Regime by a Rotating Electric Field, *Phys. Plasmas* **15**, 072304–072306 (2008).
101. J. R. Danielson, P. Schmidt, J. P. Sullivan, C. M. Surko, "A Cryogenic, High-Field Trap for Large Positron Plasamas and Cold Beams" in Schauer, M., Mitchell, T. and Nebel, R. (eds), *Non-Neutral Plasma Physics V* (American Institute of Physics Press, Melville, New York, 2003) pp. 149–161.
102. C. A. Isaac, C. J. Baker, T. Mortensen, D. P. v. d. Werf and M. Charlton, Compression of Positron Clouds in the Independent Particle Regime, *Phys. Rev. Lett.* **107**, 033201–033204 (2011).
103. R. W. Gould, "Wave Angular Momentum in Nonneutral Plasmas" in Bollinger, J. J., Spencer, R. L. and Davidson, R. C. (eds), *Conference Proceedings No. 498, Non-Neutral Plasma Physics III* (American Institute of Physics, Melville, New York, 1999), pp. 170–175.
104. R. G. Greaves and C. M. Surko, Antimatter plasmas and antihydrogen, *Phys. Plasmas* **4**, 1528–1543 (1997).

Chapter 5

Accumulation, Storage and Manipulation of Large Numbers of Positrons in Traps II – Selected Topics[a]

Clifford M. Surko, James R. Danielson and Toby R. Weber

University of California, San Diego,
La Jolla CA 92093, USA
csurko@ucsd.edu; jdan@physics.ucsd.edu; tobyweber@gmail.com

This chapter describes research to create, manipulate and utilize positron, antimatter plasmas. One is the development of a method to extract cold beams with small transverse spatial extent from plasmas in a high-field Penning–Malmberg trap. Such beams can be created with energy spreads comparable to the temperature of the parent plasma and with transverse spatial diameters as small as four Debye screening lengths. Using tailored parent plasmas, this technique provides the ability to optimize the properties of the extracted positron beams. In another section, the design of a multicell positron trap is described that offers the possibility to accumulate and store a number of positrons that is several orders of magnitude more than is currently possible (e.g., particle numbers > 10^{12}). The device is scalable to even larger particle capacities. It would, for example, aid greatly in being able to multiplex the output of intense positron sources and in efforts to create and study electron–positron plasmas. This multicell trap (MCT) is likely to also be an important step in the development of portable traps for antimatter. The third topic is a discussion of possible ways to create and study electron–positron plasmas. They have a number of unique properties. These so-called "pair" plasmas are interesting both from the

---

[a] This chapter, with corrections and minor updates as noted, is reprinted with permission from *Proceedings of the International School of Physics "Enrico Fermi"*, "Physics with Many Positrons", Course CLXXIV, in Brusa, R. S., Dupasquier, A., Mills, A. P., Jr. (eds), (IOS, Amsterdam, SIF, Bologna, 2010) pp. 545–573.

point of view of fundamental plasma physics and for their relevance in astrophysics.

5.1. Overview

As discussed in the previous chapter,[b] the Penning–Malmberg (PM) trap is the method of choice to accumulate, store and manipulate antimatter plasmas. While that chapter and this one focus specifically upon applications to positron plasmas, the work described here is done with single-component electron plasmas for increased data rate. We are now fully aware how to prepare similar positron plasmas, and the experiments are done in an ultra-high vacuum (UHV) so that annihilation is not a problem. Thus, the extension from the electron to the positron case does not pose any particular difficulties. Chapter 4 described fundamental concepts in the accumulation and manipulation of antimatter plasmas in PM traps. Here we describe three topics that leverage those tools to create important new capabilities in several areas of positron research.

One goal is to develop an efficient way to create cold, bright, pulsed beams of antiparticles. It turns out that confining and cooling the particles in a PM trap allows considerable flexibility and significantly improved capabilities to form such beams. Applications include cold beams for spectroscopy that might be used, for example, to study positron interactions with atoms, molecules and atomic clusters.[2] Another application is the development of new tailored beam sources for materials studies.[3] The techniques described here could conceivably also be useful in producing cold antihydrogen atoms from cold antiproton and positron plasmas. This might, for example, be done by gently pushing a cold, finely focused beam of one species through the other.

A key goal of positron research is to develop methods to accumulate and store the maximum possible number of antiparticles for applications that require "massive" quantities of antimatter and intense bursts of antiparticles. Examples include the creation of electron–positron plasmas[4,5] and the development of portable traps for antimatter. We

[b] Table 4.1 in the previous chapter lists the standard parameters that describe these plasmas in the notation that will be used here.

describe here a scheme to do this, namely a "multicell trap" consisting of a suitable arrangement of PM traps contained in a common magnetic field and vacuum system.

Finally, we discuss possible scenarios to create and study electron–positron plasmas (so-called "pair plasmas"). Due to their characteristics, namely equal-mass particles with opposite signs of charge, they have a number of unique properties. In particular, nonlinear processes proceed very differently in these plasmas as compared to conventional plasmas in which the ion electron-mass ratio is three orders of magnitude larger. Relativistic pair plasmas are of particular interest in astrophysics. For example, copious amounts of this material are believed to be present in the magnetospheres of pulsars.

More generally, these projects are examples of the many potential contributions that research on the physics of single-component plasmas can make to the advancement of science and technology with antimatter.

5.2. Extraction of Beams with Small Transverse Spatial Extent

Specially tailored particle beams have found a myriad of applications in science and technology. This is especially true in studies involving antimatter.[5] Numerous examples involving antimatter are discussed in this volume and at the workshop upon which it is based. For many applications, it is desirable to have beams with a small energy spread and small transverse spatial extent. For a single-component plasma in a PM trap, the space-charge potential is largest on the axis of the plasma. Recently, we exploited this fact to create beams of small transverse spatial extent. This was accomplished by carefully lowering, in a pulsed manner, one of the confining end-gate potentials. This work is described in more detail in references[6–8]. The discussion presented here relies very heavily on that work. Here we describe the main results; glossing over the mathematical details. The experiments to test the predictions of the theory were done with electron plasmas for increased data rate.

A schematic diagram of the experimental arrangement and the electrostatic potential profile in the PM trap for pulsed extraction is shown in Figure 5.1. One quantity of interest is the radial profile of the extracted beam. Also of interest are the minimum possible beam diameter and the maximum number of particles that can be extracted in a pulse at this diameter.

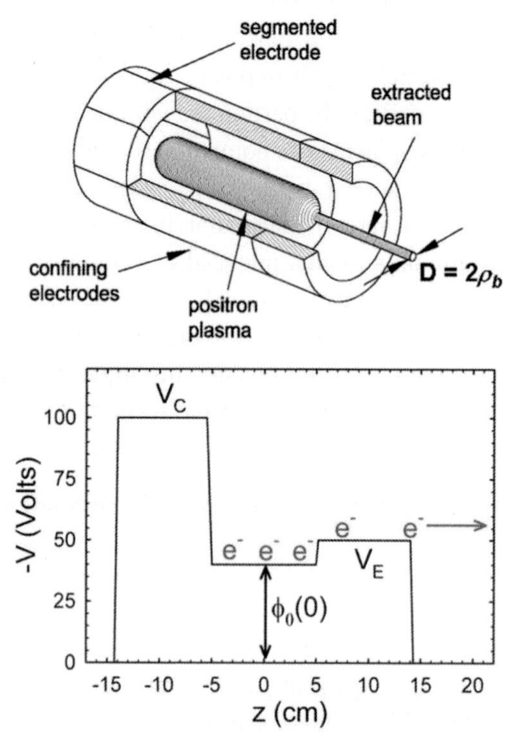

Fig. 5.1. (above) A cartoon of the experimental arrangement used to extract beams of small spatial extent from a single-component plasma in a PM trap; (below) potential profile in the vicinity of the trapped plasma (in this case electrons) in the direction along the confining magnetic field. A focus of the research described here is determining the minimum beam radius, ρ_b, and the other characteristics of the beam pulses.

The Penning–Malmberg trap used for these experiments is in a 4.8 T magnetic field. The corresponding cyclotron cooling time is $\tau_c = 0.16$ s. The trap electrodes had an inner diameter of 2.54 cm. The plasmas are typically in equilibrium, and thus they have a constant density n_0 (i.e., a flat-top radial distribution). They undergo an E x B rotation at the frequency f_E set by n_0. The plasma temperature T is set by the balance between heating sources due to background drag and/or rotating-wall

(RW) torques and the cyclotron cooling. Typical plasmas had a total number, N_0 = 4 x 10^8 electrons, a density n_0 = 1 x 10^9 cm^{-3}, plasma radius, r_p=0.1 cm, plasma length L_p =15 cm and T=0.05 eV (i.e., ~ twice the ambient temperature of 300 K). For these plasmas, the Coulomb collision time, τ_{ee} = 1 ms,[9,10c] is rapid compared to the cooling time, τ_c, thus ensuring that the plasmas are in a thermal equilibrium state.[11]

The plasma potential is largest at the (radial) center of the plasma and constant in the direction along the magnetic field due to plasma shielding (i.e., until one enters the small, "sheath" region at each end of the plasma). To extract a narrow beam, the confining potential, V_C, at one end of the plasma is carefully lowered to a predetermined value by applying a 10 μs square-wave pulse with amplitude ΔV. This extraction time is set by the fact that, as shown in Figure 5.2, the particles escape in a time ~ 5 μs.[7] Shown in Figure 5.3 is an example of the areal (two-dimensional) plasma density of the beam pulse as measured on a phosphor screen, together with charge-coupled device (CCD) images of the trapped plasma before and after beam extraction. As discussed below, the beam width depends upon the end-gate potential, V_E. Smaller-amplitude beams are narrower in diameter. Note that the beam extraction leaves a small "hole" (i.e., a region of decreased plasma density) at the center of the plasma. This hole moves coherently to the plasma edge and disappears in times ≤ 500 μs.[7]

---

[c] There are many considerations when calculating collision rates in cold, magnetized plasmas. These rates depend upon the plasma density n, the cyclotron radius r_c, the Debye length λ_D and the "distance of closest approach" $b \equiv e^2/(4\pi\varepsilon_0 T)$, where T is the plasma temperature. For the plasmas considered here, typically $b << r_c << \lambda_D$, in which case the rate of approach to thermal equilibrium is $\nu_{ee} = Cnb^2 v_T \ln(r_c/b)$, where v_T is the thermal velocity of the particles, and C is a constant of order of unity. The details and other interesting regimes are discussed in reference[10].

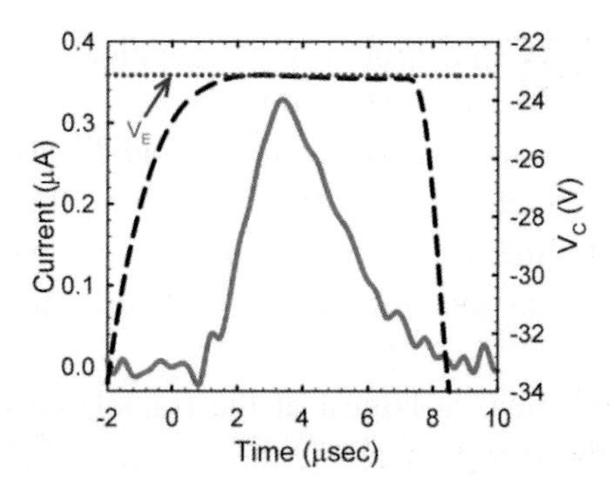

Fig. 5.2. (solid curve) Beam current as measured on a collector as a function of time in response to an approximately 10 μs reduction of the confining potential, V_C, (dashed curve) on one end of the plasma. Data from reference[7]; see this reference for details.

A simple theory was developed to describe the beam extraction process.[7,8] *A key assumption is that the particles do not scatter during beam extraction and that the fastest particles leave first*, with slower particles leaving sequentially depending upon their energies. While plausible, this is not strictly correct and should be checked experimentally, particularly in the case of cold, dense plasmas. Nevertheless, with this assumption, the energy and spatial distribution of the beam pulse can be calculated by simple integrals over the particle distribution in the plasma. Further, as shown below, these predictions are in good agreement with the results of experiments over a relatively wide range of plasma parameters.

The key piece of physics determining the properties of the extracted beam is that the exiting beam particles reduce the plasma potential in the extraction region, and this tends to inhibit more particles from leaving. The point is that the large value of the space-charge potential at the plasma center "pushes" the particles out of the trap. As particles are extracted, this potential decreases most at the radial center of the plasma, the radial potential profile is thus flattened near the plasma center and particles are then extracted over a region of larger radial extent. Since this perturbation in the potential depends upon radius, it can produce non-trivial radial beam profiles.

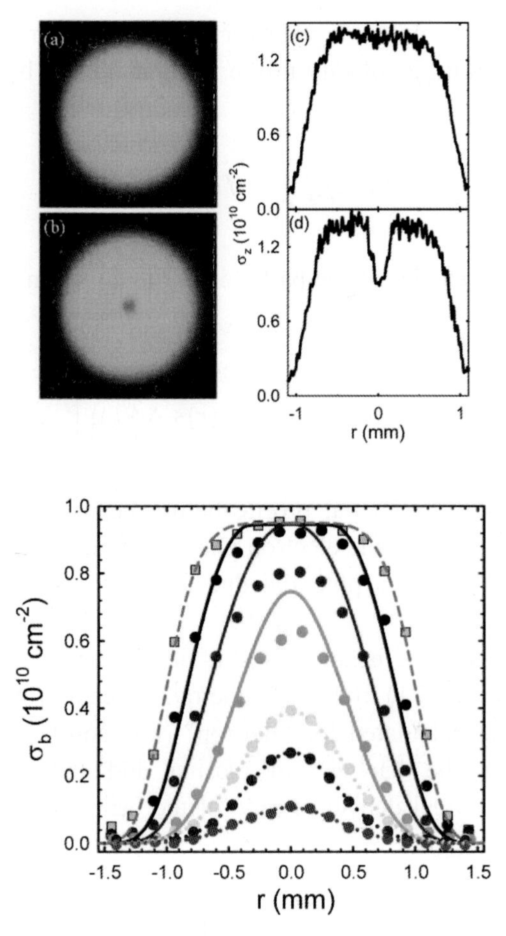

Fig. 5.3. (above) CCD camera images of the (two-dimensional) areal density profiles (a) before; and (b) 10 μs after beam extraction; shown in (c) and (d) are the corresponding radial profiles, $\sigma_z(r)$; (below) profiles, $\sigma_b(r)$, of extracted beams are shown for a selection of normalized beam amplitudes ξ, defined in Eq. (5.3) below: $\xi \approx 0.1$, 0.3, 0.5, 1.0, 1.9, 2.8. Shown for the three smallest beams are Gaussian fits (···), while the three largest beams are fit (–) to numerical solutions. The initial plasma profile, $\sigma_z(r)$, is also shown (■). Negative and positive values of r denote data along a chord through the plasma. Reprinted from reference[7].

As described in reference[7], assuming extraction from a long, cylindrical plasma, the areal density profile of the beam at a given radius can be written as a simple integral over the particle distribution function, with the lower limiting velocity in the integral v_{min} set by the extraction voltage, V_E. Thus

$$\sigma_b(r) = 2L_p \int_{v_{min}}^{\infty} f(r, v_{par}) dv_{par}, \tag{5.1}$$

where v_{par} is the velocity of the particles in the direction of the magnetic field, and

$$v_{min}(r) = \left[\frac{-2e}{m} \left(V_E - \phi_0(r) + \Delta\phi(r) \right) \right]^{1/2}, \tag{5.2}$$

where $-\phi_0(r)$ is the potential profile before beam extraction (i.e., the explicit minus sign is for the electron plasmas studied here), and $\Delta\phi(r)$ is the change in the plasma potential due to the extraction of the entire beam pulse.

For small-amplitude beam pulses, a simple analytic expression can be derived from Eq. (5.1) to describe the radial beam profile, taking $\phi_0(r)$ and $\Delta\phi(r)$ as that calculated for "flat-top" (i.e., constant-density) radial distributions for the background plasma and the extracted beam, respectively. The areal density profile of the beam is a Gaussian radial distribution,

$$\sigma_b(r) = \sigma_{bo} \exp\left[-\left(\frac{r}{\rho_b} \right)^2 \right]. \tag{5.3}$$

For the smallest-amplitude beams, where $\Delta\phi$ can be neglected, the characteristic radius, $\rho_b = 2\lambda_D$ [HW to 1/e], where λ_D is the Debye screening length. Qualitatively, λ_D is the smallest distance over which a change can be made in the plasma potential, and thus it sets the minimum transverse size of the beam. Since $\lambda_D \propto (n/T)^{1/2}$, extraction from a colder, higher-density plasma will produce narrow beams.

For larger beams, the expression for the width is found to depend upon the change in space charge due to the extracted beam. The critical parameter determining "small" and "large" beams is,[7]

$$\xi \equiv \frac{N_b}{N_0}\left(\frac{r_p}{2\lambda_D}\right)^2 = \frac{e\Delta\phi}{T}, \tag{5.4}$$

where N_b/N_0 is the fraction of the *total* plasma particles, N_0, extracted in the beam pulse, r_p is the plasma radius, and $\Delta\phi$ is the difference in space-charge potential across the beam. In terms of this parameter, and taking $\phi_0(r)$ and $\Delta\phi(r)$ as above,[d] the beam radius for Gaussian beams is found to be[7],

$$\rho_b = 2\lambda_D[1+\xi]^{1/2}. \tag{5.5}$$

When $\xi \geq 1$, the beam profiles depart significantly from the predicted Gaussian profiles. In this case, the profiles can be calculated numerically (i.e., again using the assumption that particles exit the plasma before scattering or radial diffusion occurs). As shown in Figure 5.3, the profiles of small-amplitude beams fit well with the Gaussian predictions, and the profiles of larger-amplitude beams are in reasonable agreement with the numerical calculations. As shown in Figure 5.4, the predicted *beam widths* from Eq. (5.5) are in good agreement with the data over the wide range of the beam amplitudes studied. In fact, the agreement for $\xi > 1$ cannot be justified by the theory and thus appears to be fortuitous.

Key results of this analysis are that the beam radius is limited to $\rho_b \geq 2\lambda_D$ and that this limiting value of $2\lambda_D$ can be readily achieved. Another important result is the identification of ξ as the parameter determining the beam widths. Equations (5.4) and (5.5) set a limit on the fraction of beam particles that can be extracted at the minimum diameter of $4\lambda_D$, namely that $N_b/N_0 << (2\lambda_D/r_p)^2$. Physically, this is the condition that the areal density within the pulse is small compared to the areal density of

---

[d] Assuming a flat-top, radial beam distribution *in lieu* of a Gaussian distribution makes a negligible error in the calculation of $\Delta\phi$ and ρ_b.

the parent plasma. Equations (5.4) and (5.5) also quantify the beam widths for larger beam pulses.

In practical applications, one would like to convert as much of a trapped plasma into a train of approximately identical pulses as possible. Shown in Figure 5.5 is an example where over 50% of the plasma was converted to a train of 20 nearly identical pulses. It is important to note that this was accomplished by maintaining the central plasma density constant throughout the extraction process using RW compression (cf., the previous chapter, Chapter 4) to keep the plasma density, and hence λ_D constant. After extraction of the beam pulses, the "holes" left by the extracted pulses propagate coherently to the plasma edge in a time < 1 ms. This returns the plasma to a rigid rotor, thermal equilibrium state. This would, in principle, allow pulse extractions at a kilohertz rate. Whether this is possible, given the fact that the plasma must be compressed with the RW to maintain constant n *and* λ_D, is an important topic for future research.

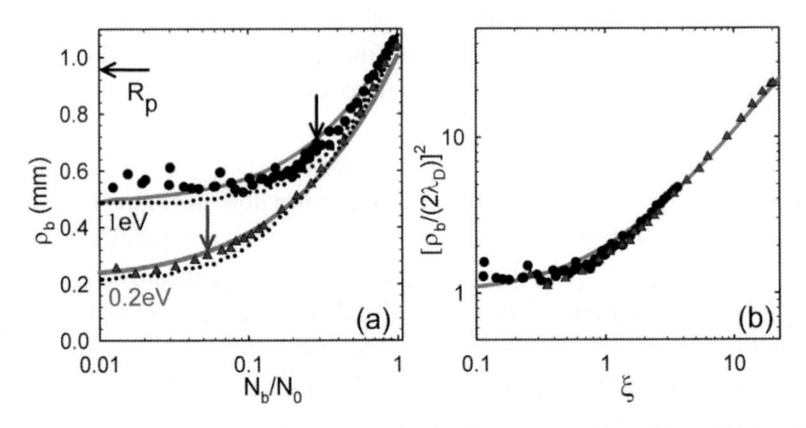

Fig. 5.4. (a) Beam-width, ρ_b, plotted vs. N_b/N_0 for $T = 1.0$ eV (●), and 0.2 eV (▲). The predictions (–) from Eqs (5.1) and (5.2) are also plotted, with no fitted parameters; (···) shows a numerical calculation from Ref.[7]; arrows correspond to beams with $\xi = 1$. (b) Data from (a) plotted as $(\rho_b/2\lambda_D)^2$ vs the dimensionless beam amplitude ξ, demonstrating the scaling predicted by Eq. (5.5). Reprinted from reference[7].

The energy distribution in the extracted beam is also of importance in many applications, such as when one wants to bunch the particles in time in positron-atomic physics (spectroscopic) applications or to study the energy dependence of certain processes such as vibrational or electronic excitation by positron impact. Recently, a detailed investigation was conducted of the energy distributions of the beam pulses formed in the manner described above.[8]

It is useful to define the dimensionless exit-gate voltage,[8]

$$\eta = (e/T)\left[V_E - \phi_0(0)\right]. \tag{5.6}$$

In terms of η, the beam amplitude can then be written,[8]

$$\xi = (1+\xi)\left[\frac{A\left(e^{-A^2}\right)}{\sqrt{\pi}} - \left(A^2 - 0.5\right) erfc(A)\right], \tag{5.7}$$

where

$$A = \left[\eta + \xi\left(\gamma + 2\ln\left(\frac{R_w}{\lambda_D}\right) + \Gamma\left(0, \frac{R_W^2}{\rho_b^2}\right)\right)\right]. \tag{5.8}$$

Here, R_w is the inner radius of the electrodes, and Γ is the upper, incomplete gamma function. Using the variables η and ξ, the distribution in energies $E_{||}$ of the beam particles in motion parallel to the magnetic field, can then be written

$$= \frac{1}{e}\frac{dN_b}{dE} = \frac{e^2}{L_p}\frac{d\xi}{d\eta}. \tag{5.9}$$

Thus, by measuring N_b as a function of the extraction voltage V_E (i.e., η in scaled variables), one can measure the parallel energy distribution of the beam. Typical data for N_b[8] are shown in Figure 5.6.

The beam parallel energy distributions, shown in Figure 5.7, are then obtained using Eq. (5.9).[12] Note that for a small number of extracted particles, $\xi = 0.02$, the distribution is fit well by the tail of a Maxwellian, whereas for $\xi = 0.4$, there is a distinct departure from this limit (i.e., due to the change in the plasma potential due to the extraction of the pulse). The dispersion in the total energy of the beam is defined by

$$\Delta E = \left(\langle E^2 \rangle - \langle E \rangle^2 \right)^{1/2} . \tag{5.10}$$

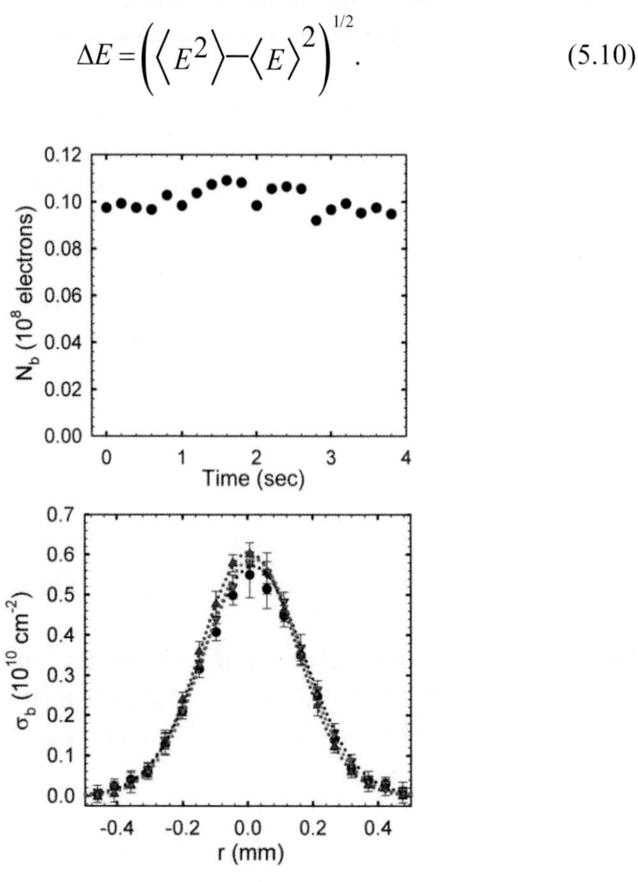

Fig. 5.5. (above) Amplitudes, N_b, for 20 pulses extracted consecutively with $\langle N_b \rangle = 1.0 \pm 0.05 \times 10^7$ and $\xi \approx 0.2$; (below) corresponding areal density profiles for the 1st (●), 10th (▲) and 20th (▼) extracted beams. The pulse amplitude and the radial beam profile remain constant, due to the fact that the density, and hence λ_D, is maintained constant by RW compression during the extractions; Gaussian fits to the profiles are also shown. Reprinted from reference[7].

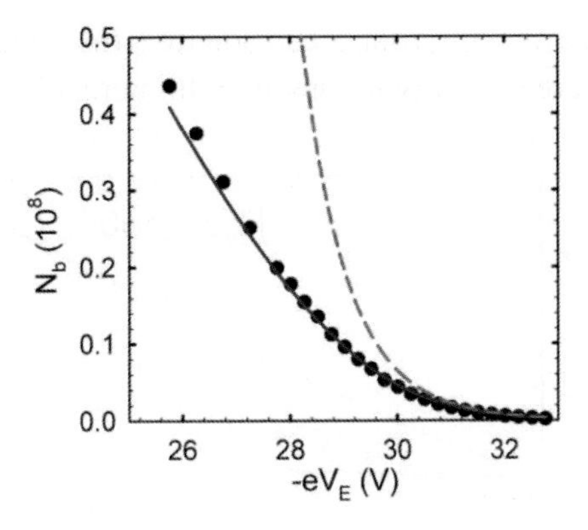

Fig. 5.6. (●) The number of beam particles, N_b, as a function of the extraction voltage, V_E; (solid line) the predictions of the theory of Eqs (5.7) – (5.9); and (——) the result using the small-beam approximation, $\xi \leq 1$. Reprinted from reference[8].

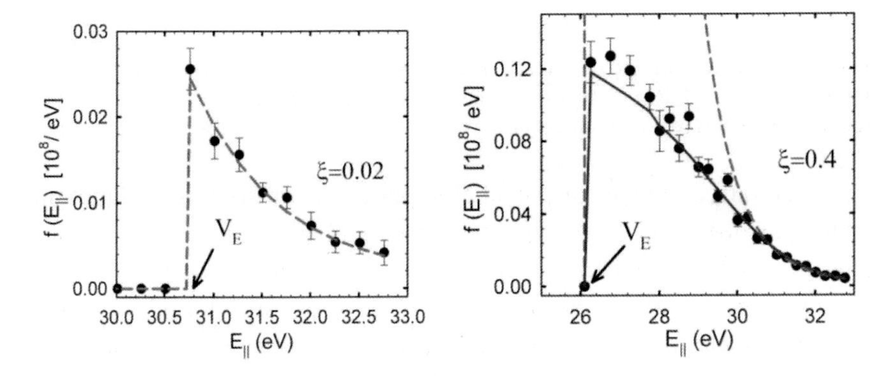

Fig. 5.7. Parallel energy distributions for two values of ξ: (——) prediction of the small–beam approximation; and (–) solutions of Eqs (5.7)–(5.9), compared with (●) the distribution obtained by taking the derivative of N_b as a function of V_E (i.e., $d\xi/d\eta$). Reprinted from reference[8].

Using Eqs (5.7)–(5.9), the parallel energy distribution function, $f(E_{||})$, and hence ΔE, can be calculated over a wide range of plasma parameters. Here it is assumed that the particle energy distribution in the motion in the plane perpendicular to B is a Maxwellian at temperature T.

As shown in Figure 5.8, the root mean square (rms) energy spread of the extracted beam, increases only modestly with increasing ξ.

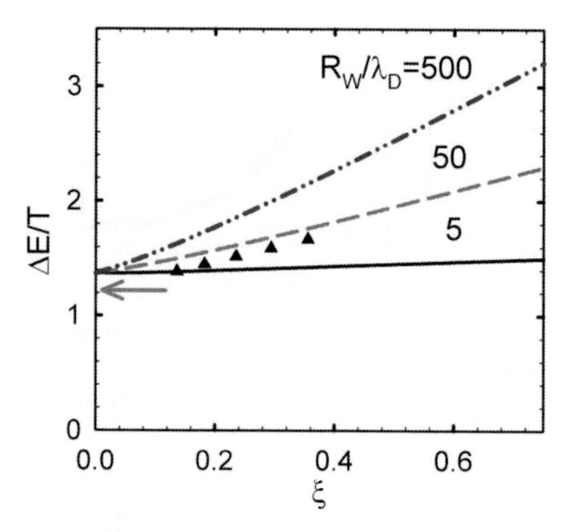

Fig. 5.8. Predictions for the rms spread in the total energy of the beam as a function of ξ for three values of R_W/λ_D (lines), compared with data (▲) for R_W/λ_D = 50. From reference[8]; see this reference for details.

There are a number of potential applications of the beam extraction technique described here. One example is in positron-atomic physics studies, where a magnetically guided beam (i.e., most compatible with extraction from parent plasmas in a PM trap) has distinct advantages.[2] However, there are also situations in which an *electrostatic,* as opposed to a magnetically guided, beam is desirable (i.e., a beam in a B-field free region). In particular, the electrostatic beams admit to the use of electrostatic focusing and remoderation for brightness enhancement.

In reference[7], an example is given of what could be done extracting a cold beam ($T \sim$ 10 meV) from a cryogenically cooled plasma in a 5-tesla field. Assuming it is guided to a region where $B \sim$ 5 gauss (5 x 10^{-4} T) and then extracted from the field, an electrostatic beam could be formed with a diameter of 1 mm and an energy spread of 10 meV (essentially all of which is in the perpendicular direction). Such a beam would be a considerable advance for positron-atomic physics scattering experiments and for certain types of materials analysis. For example, typical electrostatic beams used for atomic physics studies have energy spreads

≥ 0.5 eV and transverse extents ≥ 5 mm. There is a limit, however, on N_b for these cold, small-diameter beam pulses. To keep $\xi < 1$ and form a beam with an energy spread of 10 meV, for example, the number of particles per pulse is $N_b = 4 \times 10^3$.

In 2010, Weber and collaborators developed a method to extract particles tailored in a high-field PM trap (i.e., as described here) from the magnetic field to create a new class of electrostatic beams[13]. We refer the reader to reference[13] for details. Such beams are useful in a range of applications, including positron scattering studies involving atoms and molecules.[2] They are also important in that other methods to brightness-enhance beams such as "remoderation"[14] can be used once the particles are in a magnetic-field-free region.

5.3. Multicell Trap for Storage of Large Numbers of Positrons

Overview. Many applications require large numbers of positrons and/or long storage times. Furthermore, great benefits are anticipated if one could develop a *portable* antimatter trap. This would permit decoupling the end use of the antimatter from the source of the antiparticles, be it a particle accelerator, a nuclear reactor or even a compact, sealed radioisotope source. They are all awkward to deal with in many applications. While portable antimatter traps have been discussed previously,[5,15,16] none have yet been developed.

We describe here the next key step in research to develop a next–generation of positron storage devices. The near-term goal is to increase by orders of magnitude the number of positrons that can be accumulated and stored for long periods. Impediments to further progress include dealing with large plasma potentials. In addition, due to radial plasma transport in the presence of these large space-charge potentials, there are serious barriers to achieving very long confinement times (e.g., days) in these devices.

To achieve these objectives, we describe here the design of a novel *multicell* Penning–Malmberg trap.[17,18] In this discussion, we refer to a "cell" as a single-component plasma in an individual PM trap, in the case where more than one such plasma is arranged in the same magnetic field and vacuum chamber. The basic idea is that this MCT can confine and store antiparticles in separate plasma cells, shielded from one another by copper electrodes. These electrodes screen out the plasma space-charge potential, in turn reducing the required confinement voltages for a given number of particles by an order of magnitude or more. Such MCTs have been developed previously for other applications, namely arrays of quadrupole mass spectrometers used to increase sample analysis throughput.[19,20]

The initial goal is to develop a device in a common magnetic field and vacuum system that can store $\geq 10^{12}$ positrons for days or weeks without significant losses. As illustrated in Figure 5.9, this would increase the present state of the art by a factor $\sim 10^2 - 10^3$. Such a device would also represent a major step toward the development of a versatile, portable antimatter trap.

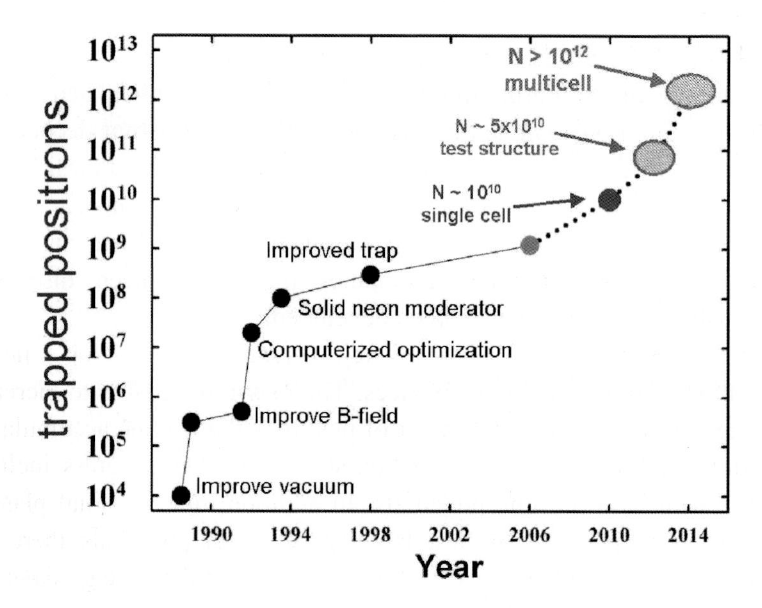

Fig. 5.9. Progress in positron trapping from similar strength sources of $\sim 50 - 100$ mCi ^{22}Na using a buffer-gas accumulator, including stacking positron plasmas in UHV;[5,21–23] (oval, UCSD) parameters achieved for an electron plasma, and (oval, multicell) the expected value for the MCT described here (updated in 2012).

The concept of the multicell PM trap is shown schematically in Figure 5.10.[17] There are several potential factors limiting long-term confinement of large numbers of positrons in PM traps. One is the Brillouin limit (cf., Section 4.4), which is the limiting density for plasma confinement in a uniform magnetic field. For electrons or positrons at tesla-strength magnetic fields, the Brillouin limit is beyond the capability of present-day experimental capabilities, and so (albeit unfortunately) it is not of immediate concern. A more important near-term limitation is the effect of plasma space charge, which is a key practical constraint for present-day positron traps. For large particle numbers, N, the space-charge potential of a cylindrical, single-component plasma of length L_p in a PM trap is proportional to N/L_p. For a fixed plasma length L_p, the number of particles, N, that can be stored in a trap is limited by the maximum potential, V_C that can be applied to electrodes in vacuum, in the presence of the plasma.

As described in Chapter 4, typical cylindrical plasmas in PM traps are space-charge limited at ~ $10^{11}\,\tilde{V}_C$ particles per meter of plasma length, where $\tilde{V}_C$ is the confining potential in kV (cf. Chapter 4).[17] For example, for a plasma of 10^{10} positrons with $L_p = 0.1$ m and $r_w/r_p = 8.8$, the plasma potential is 750 V, which, in turn, requires a value of $V_C >$ 750 V. While this value of V_C can be achieved relatively easily, the maximum possible operating potential for a compact PM trap, with closely spaced electrodes used to confine large numbers of electron-mass particles in a strong magnetic field, depends upon the specifics of the apparatus and must be demonstrated experimentally.

Another consideration arises from the fact that the heating due to outward diffusion is proportional to the plasma potential (cf. Chapter 4 and reference[17]). This heating can inhibit the ability to confine and compress positron plasmas. It can also lead to positronium (Ps) formation on background impurities in the vacuum system, and this represents a potentially serious positron loss process. For typical vacuum system contaminants, Ps formation has an energy threshold of several electron-volts and a relatively large cross-section ~ 10^{-20} m^2. The resulting neutral Ps atoms will annihilate quickly. To avoid this loss, the positron plasma must be kept relatively cool (e.g., $T \leq 2$ eV), and so unnecessary plasma heating must be avoided.

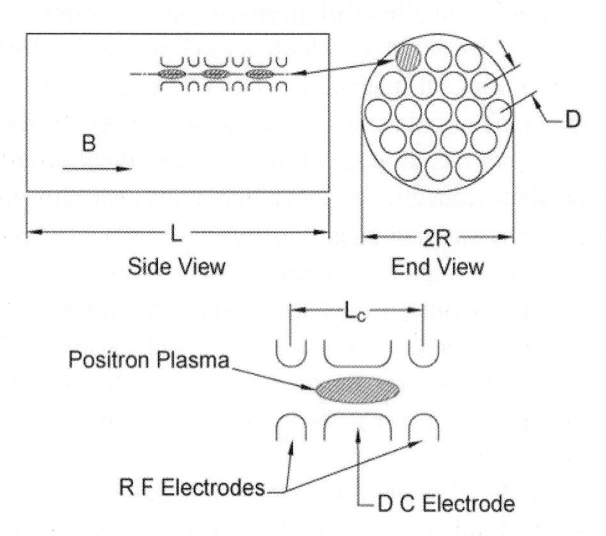

Fig. 5.10. Conceptual design of an MCT, showing the arrangement of cells parallel and perpendicular to the several-tesla magnetic field, *B*. This device consists of hexagonally close-packed cells perpendicular to the magnetic field and a number of in-line cells in the field direction. The RF (radio frequency) and DC (direct current) electrodes are shown in the lower diagram.

As shown in Figure 5.10, a key feature of the MCT is that large values of the plasma potential (i.e., due to the space charge of the plasma) can be mitigated by dividing the plasma into m, rod-shaped plasmas. Each plasma, of length L_p, is oriented along the magnetic field in a hexagonal-close-packed (HCP) arrangement transverse to the field. These rod-shaped plasmas are shielded from each other by closely fitting copper electrodes. For a given maximum confining electrical potential V_C, the number of stored positrons will be increased by a factor of m. Since the plasma heating rate due to outward expansion of the plasma is proportional to the potential drop across the plasma, the multicell design also reduces the requirements on plasma cooling. In the trap design considered here, cooling is accomplished by cyclotron radiation of the particles in a relatively large (e.g., several-tesla) magnetic field.

The multicell design also breaks up each long rod of plasma into p separate plasmas in the direction along the magnetic field (i.e., separated by electrodes at potential V_c). The plasma length is decreased by a factor L_p/p, and this reduces the effects of magnetic non-uniformities, since many of the cells are now both off the axis of symmetry and away from the mid-plane of the magnet. Breaking up the plasma longitudinally also reduces the rate of outward, asymmetry-driven radial transport (i.e., which is typically found to be proportional to L_p^2 (cf., Chapter 4 and reference[24]). The design parameters for a 21-cell MCT are summarized in Table 5.1. The electrode structure will be cooled to cryogenic temperatures to freeze out contaminant impurity molecules and to ensure a UHV environment. Using rotating electric fields in each cell for radial plasma compression, the positron loss is expected to be small on the design-goal time scale for confinement times of weeks.

Table 5.1. Design parameters of a 21-cell MCT.

Number of cells ($m \times p = 7 \times 3$)	21
Total positron number, N (10^{11})	≥ 5.0
Magnetic field (T)	5
Total electrode length, L (cm)	100
Electrode–package diameter, $2R$ (cm)	> 7.5
Plasma radius, R_P (cm)	0.2
Plasma length, L_p (cm)	20
Confinement voltage, V_c (kV)	1.0
Cell spacing (cm), D	2.0
Space-charge potential (V)	750
Rotating-wall frequency (MHz)	4

The plasma is expected to be considerably warmer than the electrode temperature (i.e., ~ 0.1 eV) due to plasma heating from the radio frequency fields used to achieve long-term plasma confinement. The work reported here used a confinement potential, $V_C = 1.0$ kV, which resulted in a maximum particle number of $N = 3 \times 10^{10}$ in a single PM cell. The design in Table 5.1 is conservative in this regard. If one could work with ~ 3 kV, which is likely, a trap for 10^{12} positrons would require only 14 cells. Alternatively, a 95-cell trap could confine ≥ 6 x 10^{12} positrons.

To fill the MCT, positrons will be accumulated in a buffer-gas trap using the arrangement shown in Figure 5.11. Typically N ~3 x 10^8 e^+ can be accumulated from a 100 mCi ^{22}Na radioactive source and noble gas moderator in a few minutes. The positron plasmas from the buffer-gas trap will be "stacked"[21,25,26] in UHV in the high-field trap[21] with a cycle time of several minute to achieve ≥ 10^{10} positrons in a single plasma cell. At these fill rates, trapping 10^{12} positrons would take several days to a week. However, stronger positron sources are currently in operation and/or under development in a number of laboratories around the world that could fill such a trap in a few hours or less.[5,27–29] A master plasma-manipulation cell (on the left side of the MCT) will receive plasmas from the buffer-gas trap, compress them and move them off axis radially by a technique described below, before depositing them in the multiple storage cells.

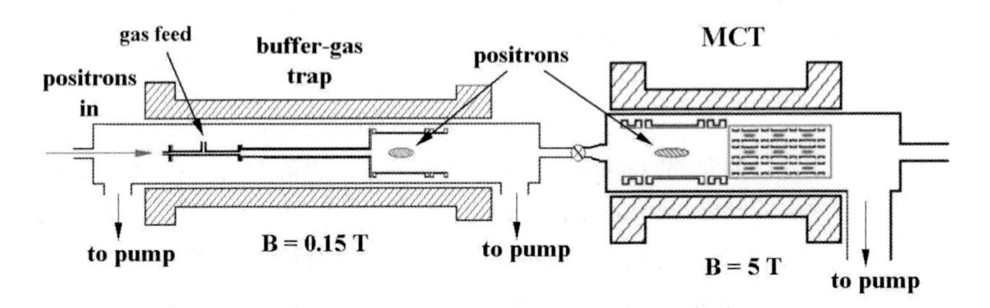

Fig. 5.11. Schematic diagram of the buffer-gas positron trap (left) used to accumulate positrons from a continuous source and shuttle them to the MCT. The buffer-gas trap is connected by a pulsed valve to the high-field UHV, multicell storage trap (right). Positrons from the source and moderator enter the buffer-gas trap from the left. The source could be a conventional radioisotope (^{22}Na) and a solid neon moderator, or a higher intensity source.

Validation of the MCT concept. A series of experiments were performed recently (using test electron plasmas) to validate key procedures necessary to operate an MCT successfully. The experiments were performed in the cylindrical high-field PM trap described in Chapter 4.[18] Plasmas are confined in various combinations of cylindrical electrodes ($r_w = 1.27$ cm) to achieve plasma lengths in the range $5 \leq L_p \leq 25$ cm. The electron plasmas are injected using a standard electron gun and confined radially by an applied 5-tesla magnetic field, with axial confinement provided by voltages applied to the end electrodes. In typical experiments, rotating electric fields (i.e., the RW technique described in Chapter 4) provided long-term confinement. The segmented RW electrodes are also used to excite and detect diocotron modes in the plasma that, as described below, were used to move plasmas across the magnetic field.

The plasmas are cooled by cyclotron radiation in the 5 T magnetic field at a rate, $\Gamma_c \sim 6\ s^{-1}$,[30] which is fast compared to the compression and expansion rates. Steady-state plasmas remain relatively cool (i.e. $T \leq 0.2$ eV; $T/e\phi_0 << 1$, where ϕ_0 is the plasma potential), even in the presence of strong RW fields. It was established that the steady-state density *could be maintained for more than 24 hours* with no loss of plasma.[18] The dependence of the plasma density on the total number of particles N is illustrated in Figure 5.12 for the three different confinement lengths L_c (i.e., the length of the potential well imposed by voltages on the electrodes) and a 1 kV confinement potential. The ability to create and manipulate two, in-line plasmas was also demonstrated.[18]

One of the key requirements for an MCT is development of a robust and compact method to move plasma across the magnetic field. While this could be accomplished by magnetic deflection or use of ***E x B*** plates, both of these techniques have disadvantages in terms of space requirements and the need to switch large magnetic fields and/or electrical potentials. A method to move plasmas across the field was developed using excitation of a so-called "diocotron mode" of the plasma.[31] Specifically, when a single-component plasma is displaced from the axis of the cylindrical electrode, the center of mass will exhibit an ***E x B*** drift due to the electric field of the image charge of the plasma induced in the electrodes that surround it. This uniform drift of the plasma center at frequency f_D about the axis is called a *diocotron mode.*

The amplitude of this mode is the displacement, D, of the plasma from the axis of symmetry of the confining electrodes.

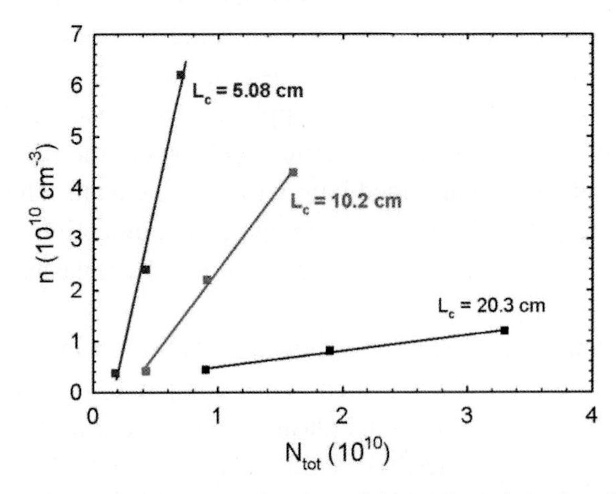

Fig. 5.12. The plasma density, n, as a function of the total number of confined particles, N, is shown for three different confinement lengths, L_c (i.e., the length of the electrodes forming the confinement well), using a confinement voltage, V_C = 1.0 kV. To vary N, three different electron-gun fill voltages were used, 0.3, 0.6 and 0.9 kV, at each value of L_c, represented by the three data points on each line. Reprinted from reference[18].

For a long plasma column with $L_p >> r_w$, the linear frequency of the $m_\theta=1$, $k_z = 0$ diocotron mode is approximately

$$f_D \approx (r_p/r_w)^2 f_E, \tag{5.11}$$

where $f_E = ne(4\pi\varepsilon_0 B)^{-1}$ is the ***E x B*** plasma rotation frequency (cf., Chapter 4 and reference[31]). In these experiments, $f_D \sim$ a few kHz $<< f_E$. The diocotron mode was excited by applying a sinusoidal voltage at a frequency near f_D to one sector of the four-sector electrode. As described below, the technique of "autoresonance" provides an effective and convenient method to control both the amplitude (i.e., radial displacement, D) and the azimuthal position of the plasma column as a function of time.

As the diocotron mode grows in amplitude and the plasma moves further off axis, the mode frequency changes. To lowest order, assuming $r_p \ll r_w$, the nonlinear diocotron frequency f_{NL}[32] is

$$f_{NL} = f_D \frac{1}{1-\left(D/r_w\right)^2}. \tag{5.12}$$

Figure 5.13 shows a comparison of the measured values of f_{NL}/f_D as a function of D compared to the predictions of Eq. (5.12).

In the experiments, plasmas could be imaged directly with CCD camera out to a displacement D ≈ 0.45 cm (r_w = 1.27 cm). Beyond that radius, D was measured using a pick-up signal on a segmented electrode.[18] A plasma displacement corresponding to 80% of the electrode radius was achieved. This has important, positive, implications concerning the ability to completely fill an electrode structure with multicell plasmas by addressing cells far from the magnetic axis.

The plasma could be phase-locked to a drive signal using the technique of "autoresonance". Autoresonance is the tendency of a driven nonlinear oscillator to stay in resonance with the drive signal, even when the system parameters are varied. This phenomenon was explored in some detail by Fajans *et al.*[33–35] for the case we are interested in, namely the diocotron mode. Using this technique, the diocotron mode is brought into autoresonance by sweeping the drive frequency from below the linear diocotron frequency to a selected, higher frequency. If the drive voltage is sufficiently strong, the excited diocotron mode amplitude (i.e., the displacement, D, of the plasma column from the symmetry axis) will grow as the mode increases in frequency with the drive frequency. In the autoresonant condition, the excited diocotron mode stays phase-locked to the applied signal. We plan to use this technique to inject trapped plasmas into the off-axis cells in the MCT, and we conducted experiments with electron plasmas to test this.

Figure 5.14 shows a model calculation to illustrate the autoresonant response of a plasma of relatively small spatial extent to a constant-amplitude sine wave, $V_D = V_o \sin(2\pi ft)$, as the drive frequency, f, is changed. The initial on-axis plasma is driven to a large displacement when the frequency of the drive is swept from below the linear mode frequency to a higher frequency. The final displacement, D, is determined by the final frequency of the applied signal. The angular position of the plasma in the plane perpendicular to the B-field is determined by the phase of this applied signal.

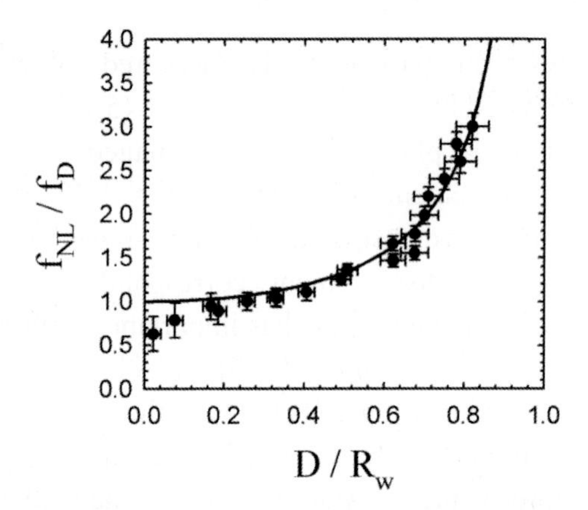

Fig. 5.13. The diocotron frequency (•) measured for plasmas displaced different distances D from the electrode center; and (—) the prediction of Eq. (5.12) with no fitted parameters. The linear diocotron frequency is $f_D = 2.9$ kHz. From reference[18].

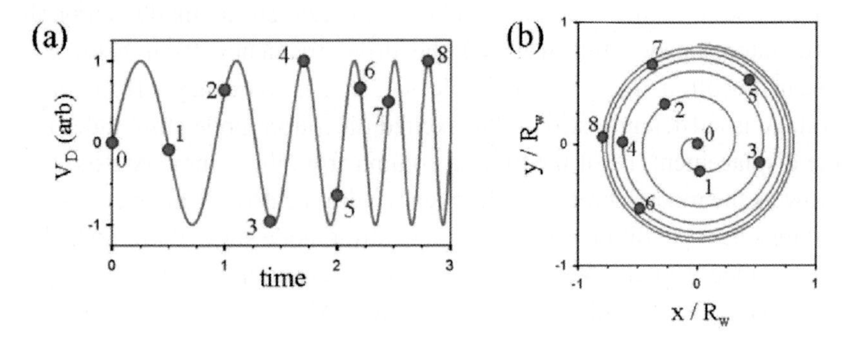

Fig. 5.14. Model calculation of the evolution of the plasma position during the excitation of a nonlinear diocotron mode in the condition of autoresonance: (a) the drive voltage $V_D(t)$ as a function of time in units of the period, τ_1, of the linear diocotron mode; and (b) the corresponding plasma orbit in the (x, y) plane perpendicular to the magnetic field at various times. Numbers correlate the plasma position with the phase of the drive signal. As the frequency increases, the plasma column moves to larger displacements.

Figure 5.15 (left panel) shows CCD images of autoresonantly excited plasmas that have been deposited on the screen at a fixed phase angle, $\phi = 0^{\circ}$, in the plane perpendicular to B. Note that these displacements are much larger than the plasma radius, demonstrating that $D >> R_p$ can be achieved. The ability of this technique to deposit plasmas at four predetermined *azimuthal* locations (90° apart) is shown in the right-hand panel of Figure 5.15. These experiments demonstrate that the autoresonance technique will enable depositing plasmas in cells at arbitrary locations in the plane perpendicular to the magnetic field. These results indicate that plasmas can be moved across the magnetic field in a few milliseconds or less and can be deposited in specific off-axis cells to a high degree of accuracy (e.g., +/– 0.2 mm in the radial and azimuthal directions).

The process of filling the MCT is illustrated schematically in Figure 5.16. Plasmas from a buffer-gas positron accumulator will be shuttled into a master plasma manipulation cell (left), then the dicotron mode will be excited to the appropriate values of displacement D and phase angle ϕ before the plasma is deposited into a specific off-axis cell. Shown in Figure 5.17 is a schematic illustration of the design of an electrode structure for a 21-cell trap (i.e., compatible with the design parameters summarized in Table I). It incorporates a master plasma manipulation cell for injection into off-axis cells. Each storage cell has a segmented electrode for RW radial plasma compression, an equal-length dc electrode, and confinement electrodes at each end.

Using the RW technique in the strong-drive regime (cf. Chapter 4), plasmas with a remarkably broad range of initial densities (e.g., varying by a factor of 20 or more) can be compressed to a given final-state density by the application of a single, fixed RW frequency. This results in considerable simplifications in the design of a practical MCT. In particular, in this regime of RW operation, active control and interrogation of individual plasma cells is unnecessary. This strong-drive regime also has important consequences in reducing plasma heating. As discussed in Chapter 4, this is due to the fact that the mismatch in frequencies between the plasma rotation and the RW drive is negligibly small, and this minimizes the RW heating. Thus, operation in the strong-drive regime approximates closely the minimum possible heating rate.

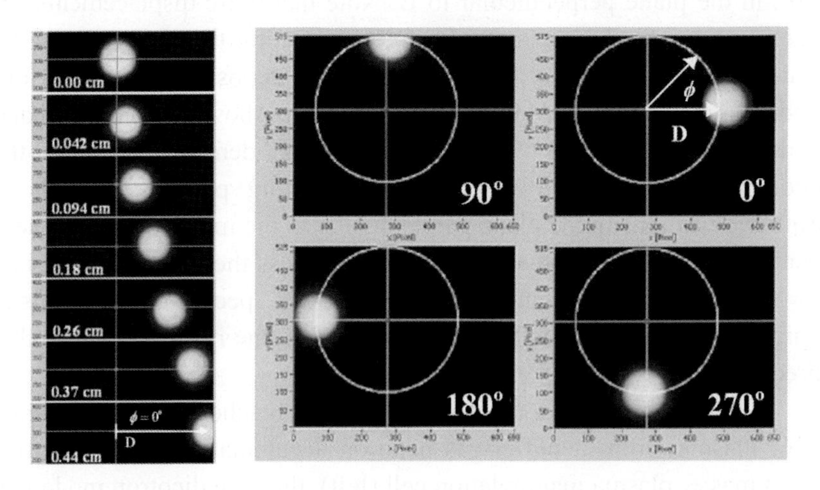

Fig. 5.15. CCD images of plasmas displaced from the axis by autoresonant excitation of the diocotron mode and dumped onto a phosphor screen: (left) six radial displacements at fixed azimuthal position; and (right) four azimuthal displacements at a fixed radial position. From reference[18].

The experiments indicate that it will be possible to access plasma parameters of $n \geq 5 \times 10^{15}$ m^{-3} at B = 5 T. In this regime, the outward transport rate, Γ_o, is independent of plasma density (i.e., instead of increasing as $\Gamma_o \propto n^2$ which is the case at lower plasma densities[11,36]). This too reduces the required RW drive torque and thus leads to considerably less plasma heating, and so the plasmas remain cool. Plasmas with parameters such as those listed in Table 5.1 can be created with T~ 0.1 eV. This is ideal for the multicell positron trap. As discussed above, one important consideration is keeping the plasma temperature sufficiently low so that one can avoid Ps formation by collisions of positrons with background gas molecules. The relatively low values of plasma temperature reported here, namely T ≤ 0.5 eV, easily fulfill this requirement.

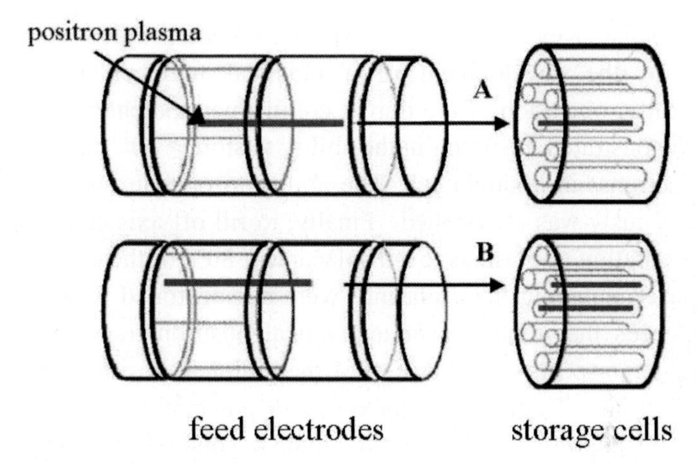

Fig. 5.16. Injection of plasmas into specific cells in the MCT: phase-dumped into an (A) on-axis, and (B) off-axis plasma cell.

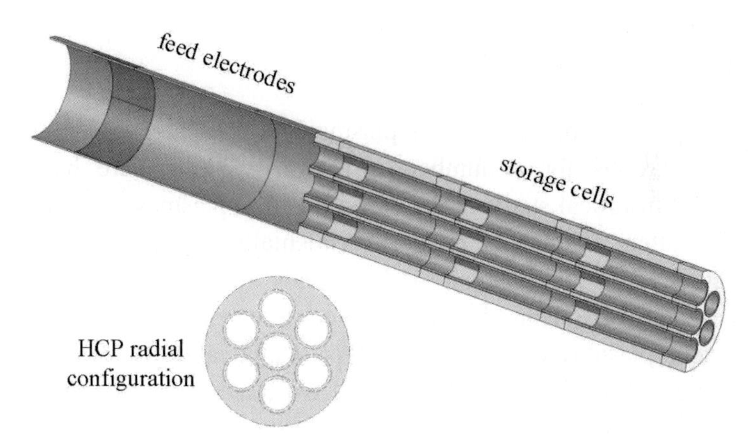

Fig. 5.17. Schematic diagram of the 21-cell multicell positron trap, showing three banks of seven cells in an hexagonally closed packed arrangement. Plasmas from the source will first enter the feed electrodes, then be moved off axis using autoresonant excitation of the diocotron mode to fill off-axis storage cells.

Summary. Key techniques have been demonstrated that will be critical to the development of a practical multicell positron trap. Specifically, operation of the trap at confinement potentials of 1 kV has been demonstrated, resulting in the ability to store $\geq 10^{10}$ particles (in this case electrons) in a single cell. The ability to operate two plasma cells simultaneously was established. Finally, to fill off-axis cells, diocotron-mode excitation of plasmas to a displacement 80% of the electrode radius and phased dumping these plasmas were demonstrated with a precision that exceeds that required for a practical positron trap. These results validate key aspects of the design of the multicell positron trap for $N \geq 10^{12}$ positrons. Further multiplexing can potentially increase the trap capacity by additional orders of magnitude beyond this benchmark goal.

In 2012, the University of California, San Diego (UCSD) Positron Group began to build a 21-cell MCT for 10^{12} positrons along the lines of that described in Table 5.1 and shown schematically in Figure 5.17.

5.4. Electron–Positron Plasmas

Electron–positron plasmas ("pair plasmas") are exceedingly interesting physical systems for a number of reasons. While there have been extensive theoretical studies of electron-positron plasmas,[37–43] there have been few attempts to study them experimentally. The major problem in conducting such experiments is that they are neutral plasmas (or approximately so). Thus the powerful confinement theorem for single-component plasmas (cf., Chapter 4) is inapplicable. Thus achievable confinement times for pair plasmas are expected to be many orders of magnitude shorter than those for pure positron plasmas, and this poses a huge challenge to experimentalists. However, with the advent of high-intensity positron sources and means to accumulate large numbers of positrons (e.g., the MCT described above), we are on the verge of creating these interesting plasmas in the laboratory.[e] Much of the

[e] For the (tenuous) pair plasmas considered here, Ps formation and positron annihilation with electrons can be neglected. The annihilation time for a positron in an electron plasma with density ~ 10^{20} m^{3} is approximately 1 s, independent of the plasma temperature. The rates of other processes, such as $e^+ + e^- \to Ps + h\nu$ and $e^+ + e^- + e^- \to Ps + e^-$, are also expected to be small.

following discussion follows closely to that presented in reference[44], albeit with revisions for recent progress. The reader is referred to this reference for further details.

In a seminal paper, Tsytovich and Wharton[45] pointed out that electron–positron plasmas possesses truly unique properties because of the equal-mass, opposite sign-of-charge nature of the plasma particles. For example, cyclotron radiation in these plasmas is linearly, rather than circularly, polarized. Furthermore, the nonlinear behavior of these plasmas is dramatically different than that in conventional electron–ion plasmas. For example, in an equal-mass plasma in which the species are in equilibrium at temperature T, the ion acoustic wave is very heavily damped; three-wave coupling processes (e.g., parametric decay) are absent; and nonlinear Landau damping is larger by the electron/ion mass ratio M/m, as compared to the conventional case. This completely changes the behavior of the plasma.

Relativistic electron–positron plasmas are thought to play a particularly important role in nature. For example, large quantities of this material are believed to be present in the magnetospheres of pulsars. As mentioned above, pair plasmas have been studied extensively theoretically,[39–43] but not experimentally, and thus this topic literally begs for experimental investigation. The first laboratory experiments to study something close to this situation were conducted by passing an electron beam through a positron plasma confined in a Penning trap (i.e., a study of the electron-beam, positron-plasma instability).[46,47] However, it is much more desirable to create an electron–positron plasma in which the two species are not drifting relative to each other.

Various techniques have been proposed for creating such simultaneously confined electron and positron plasmas. They include confinement in a magnetic mirror,[48] in toroidal a magnetic configuration known as a stellarator,[4] and in a combined Penning–Paul trap.[44] Due to the anticipated difficulties in simultaneous confinement of these plasmas, an intense positron source, such as that from a linear electron accelerator (LINAC), or an isotope-producing nuclear reactor such as the Munich fission reactor,[49] is virtually obligatory in order to achieve useful data rates in a laboratory experiment.

As mentioned above, it would be of great interest to study the relativistic regime. A magnetic mirror device is expected to provide good confinement for such a hot, electron-mass plasma. However, at the anticipated high temperatures, the Debye length is comparatively large for a given plasma density. Consequently, relativistic electron–positron plasma experiments will require *very* large numbers of positrons (e.g., $N \geq 10^{15}$ per plasma, with an expected lifetime $\leq$ 1s).[44] This is likely to challenge the capabilities of available positron sources for the foreseeable future.

An alternative approach to study relativistic electron–positron plasmas is the use of intense lasers. While these kinds of experiments are outside the scope of the present review, there has been great progress in this area in recent years, driven by the ever-increasing capabilities of high-intensity laser technology. We refer the reader to references[50–56] for further discussions of this promising new direction. As pointed out in reference[56], one can not only expect to generate very large numbers of positrons with these techniques (e.g., $N \sim 5 \times 10^{11}$), but one could also likely create strong confining magnetic fields using complementary laser-based techniques.

Combined trap for low-density electron–positron plasmas. One method to create and study a low-density, cool electron–positron plasma involves a hybrid trapping scheme.[44] The challenge of simultaneous confinement of both charge species can be overcome by the use of a combination of Penning- and Paul-trap technology. In a Paul trap, one can confine charged particles of both signs of charge by means of radio frequency (RF) fields. The basic concept is that in a rapidly oscillating high-frequency electric field ($\omega >> \omega_p$), the plasma is repelled from regions of large electric field. This concept gives rise to the notion of a ponderomotive force which can be written,[57]

$$\vec{F}_{pon} = -\frac{\omega_p^2}{2\omega^2}\vec{\nabla}\langle \varepsilon_0 E^2 \rangle. \tag{5.13}$$

Regions of large oscillating electric field act as potential hills and thus repel particles of both signs of charge. Paul traps have been used to confine quasi-neutral plasmas of positive and negative ions.[44] More

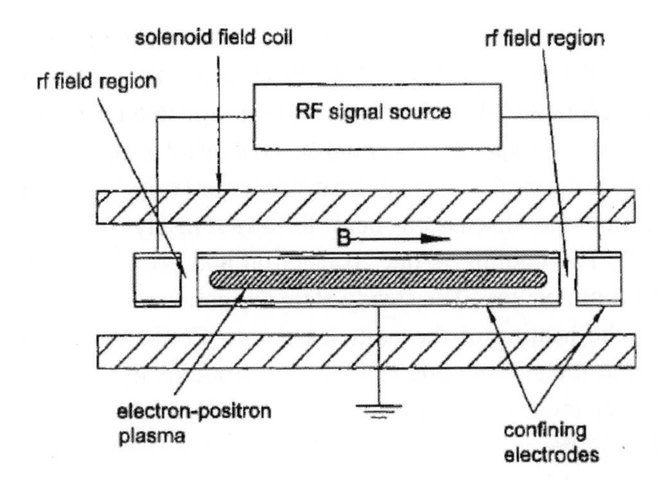

Fig. 5.18. A combined Penning–Paul trap for studying electron–positron plasmas. A uniform magnetic field B will provide confinement for both signs of charge in the direction perpendicular to B. RF fields at each end provide confinement in the direction along B. These large, high-frequency electric fields peak near the location at which the end and central electrodes meet. From reference[44].

recently, the simultaneous confinement of protons and electrons was demonstrated in a trap in which the electrons were confined by RF fields and the protons were confined in an overlapping Penning trap.[58]

In a natural extension of these experiments, we have proposed using a combined trap, illustrated in Figure 5.18, to confine an electron–positron plasma.[17] This would be done using a hybrid, Penning–Paul trap where radial confinement is provided by a magnetic field, as in a Penning trap; and confinement along the magnetic field is provided by RF electric fields E_{rf} (i.e., in place of the electrostatic potentials of the Penning trap). In this design, heating of the species by the RF is a problem. This can be overcome using the cooling provided by small amounts of a suitably chosen molecular gas (as described in Chapter 4).

The design parameters for the experiment are given in Table 5.2. The maximum achievable depth of the potential well using the Paul trapping technique is limited by practical considerations to a few electron-volts, thereby placing limits on the plasma temperature and the acceptable amount of charge imbalance. For these and other electron–positron plasma experiments, the advantage of using an intense positron source would be that the experiments could be conducted with more rapid cycle times, even if the confinement is relatively poor (which is likely).

Much of the discussion here follows closely that presented in reference[44]. The cooling is due to electron and positron collisions with neutral CO_2 molecules.[f] The steady-state temperature can be estimated by balancing the heating of the particles due to Coulomb collisions in the RF field and the cooling due to inelastic, vibrational, electron–molecule and positron–molecule collisions.[g] The heating rate due to Coulomb collisions will be

$$\frac{d\varepsilon}{dt} \approx 2m\nu_c \left\langle \left(\delta v\right)^2 \right\rangle, \tag{5.14}$$

where ν_c is the Coulomb collision frequency and $\delta v = eE_{rf}/m\omega$ is the particle velocity due to the RF field at frequency ω. It is useful to note

$$\frac{d\varepsilon}{dt} \propto U_{rf} = \frac{e^2 E_{rf}^2}{m\omega^2} \approx m\left\langle \left(\delta v\right)^2 \right\rangle, \tag{5.15}$$

that where U_{rf} is the RF trapping potential energy. The heating rate must be spatially averaged over the trajectory of the particles in the potential well. For a cylindrical plasma of length L, confined by cylindrical electrodes of radius $r_w << L$, with the confining electrode between $z = 0$ and $z = L$, the particles will be heated appreciably only in the regions of large RF electric field. This will occur only near the ends of the plasma in a region of extent $\delta z \sim 0.4\, r_w$ near the turning points of the motion.[44]

---

[f] It is assumed here that the molecular gas is CO_2, but other cooling gases could also be used (e.g., CF_4). It should be noted that SF_6 would not be a good choice because electrons attach to it.

[g] Consistent with the discussion in reference[44], the heating due to charged-particle/neutral, momentum-transfer collisions in the region of the rf fields was neglected. Recent estimates (2012) indicate that this does not significantly change the estimates made here.

To fix the design parameters, a trapping well depth, U_{rf} = 5 eV and a plasma temperature $kT \sim 0.5$ eV are assumed. Near the turning points, the particles experience an RF potential of strength ~ kT/e, and they spend more time near these locations than in other regions of the trap. Taking these factors into account and assuming $L = 30R_w$, the estimate made in reference[44] is that the time-averaged heating rate is,

$$\frac{d\varepsilon_{rf}}{dt} \approx 0.05\nu_c T\,.^{\mathrm{h}} \tag{5.16}$$

The rate of energy loss will be $d\varepsilon/dt \approx -\nu_{col}\varepsilon_j$, where ε_j is the average energy loss per collision and ν_{col} is the collision frequency for inelastic vibrational excitation of the molecule. We assume that the cooling gas is CO_2, which has a vibrational-quantum energy, $\varepsilon_j = 0.3$ eV, and a collision cross-section $\sigma \sim 10^{-16}$ cm^{-2} (cf., Table 4.1, Chapter 4)[44].[i] With these assumptions

$$\nu_{col} = n_n \sigma v_T \approx 2.0\times 10^5 P \text{ [Hz]}, \tag{5.17}$$

where n_n is the CO_2 number density and P is the pressure in millibars. Thus

$$\frac{d\varepsilon_{col}}{dt} = -7.0\times 10^4 P \text{ [eV/s]}. \tag{5.18}$$

In steady state, $d\varepsilon_{rf}/dt + d\varepsilon_{col}/dt = 0$. Thus $\nu_c \sim 2\times 10^6 P$ (Hz). For n = 10^7 cm^{-3} and T = 0.5 eV, $\nu_c \sim 2 \times 10^3$ Hz, which would require a CO_2 pressure of 1.3×10^{-6} mbar.

---

[h] The 0.05 factor in Eq. (5.16) accounts for the fact that the heating only occurs in the region of the RF fields. A similar weighting factor would be relevant to electron/positron neutral collisions (cf. footnote below).

[i] Simplifying assumptions were made in that analysis and are continued here. Experiments indicate that for positrons, the cross-section for vibrational excitation is twice the value used here,[2] while the vibrational excitation cross-section for positrons and electrons are taken to be the same. Further, heating due to electron/positron neutral collisions is neglected. Neither is expected to change substantially the estimates made here.

At this pressure, the annihilation time is ~ 80 s, the diffusion time due to collisions with neutral gas is ~ 500 s, the diffusion time due to electron–positron collisions is ~ 200 s, and the Bohm diffusion time[44] is ~ 100 μs. Thus the plasma can be expected to survive between 100 μs and several hundreds of seconds, depending on which transport mechanism dominates. This is an interesting issue in its own right and would likely be one of the first phenomenon to be studied. Since the plasma frequency is ~ 30 MHz, plasma wave phenomena could be studied, even if the confinement time was as short as 100 μs.

Table 5.2. Design parameters of an electron–positron experiment using a combined Penning–Paul trap. [Following reference[44], values here are in cgs (centimeter, gram, seconds), not S. I. units.]

Quantity	Design Value
density (cm–3)	10^7
plasma length (cm)	30
plasma radius (cm)	0.5
electrode inner diameter (cm)	1
particle number	5×10^8
RF frequency (MHz)	200
RF voltage (V rms)	100
rRFpotential well (V)	5
CO_2 pressure (mbar)	1.3×10^{-6}
plasma temperature (eV)	0.5

While this combined trap is suitable for low-density electron–positron plasma studies, it is not likely to be a viable geometry for confining high-density plasmas. This is due to plasma heating, which will increase rapidly with plasma density and the unavailability of a sufficiently rapid cooling mechanism to counteract it. While heating due to Coulomb collisions could be reduced at higher plasma temperatures, Ps formation (cf., Chapter 4) will quickly become a problem.

Magnetic-mirror confinement of hot electron–positron plasmas. Experimental studies of relativistic electron–positron plasmas will be much more challenging. The plasma limit requires $n\,\lambda_D^3 >> 1$, and $\lambda_D <$ L, where L is the characteristic dimension of the charge cloud. Thus, in order to have λ_D as small as 0.01 m at T_e > 200 keV (i.e., a mildly relativistic plasma), a density of $n = 10^{18}$ m^3 is required. At least, one must have $L = 10\ \lambda_D$ to study plasma wave phenomena, which in turn requires confining 10^{15} positrons. Beyond the challenge of accumulating such a large number of positrons, their confinement in a neutral plasma is expected to be a great challenge.

One possible geometry for such an experiment is a so-called magnetic mirror.[j] An experiment designed to test this confinement scheme for positrons is shown in Figure 5.19. In this experiment, positrons were accumulated from a 0.6 mCi ^{22}Na source and polycrystalline tungsten moderator.[59] It turns out that confinement in a magnetic mirror is better when the plasma is hot (i.e., thereby reducing the loss due to Coulomb collisions). In the mirror, it is also beneficial to arrange $T_\perp >> T_\parallel$, where $T_\perp$ and $T_\parallel$ are the perpendicular and parallel temperatures of the particles. Both conditions can be achieved relatively easily for electron-mass particles by heating at the cyclotron frequency using microwave radiation, and this is what was done in the experiment of reference[59]. The result was the confinement of ~ 10^4 positrons with confinement times of ~ 20 s and densities of ~ 5 x 10^8 m^{-3}. Given that the incident slow positron flux was low (~ 500/s), the results of this experiment are encouraging.

Confinement of the positrons can be further increased by placing electrodes on either end of the mirror, biased to as large a potential as possible. In this case, positrons exiting the usual "loss cone" in mirrors (i.e., particles with low values of $E_\perp/E_\parallel$ are not confined by the mirror fields) would be reflected back into the magnetic mirror. One unwanted side effect of the hot plasma is the intense X-ray and gamma-ray cyclotron emission from the hot particles.

---

[j] Particle confinement in a magnetic mirror relies on the fact that, in a slowly varying B-field, $E_\perp/B$ is a constant. Since the B-field does no work, this implies that, as B increases, $E_\parallel$ decreases. Thus, for a sufficiently strong field, particles are reflected (i.e., "mirrored"). See reference[57].

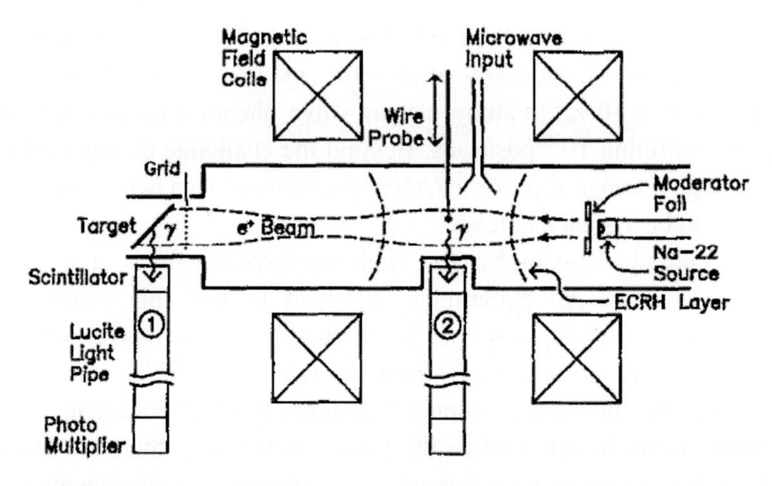

Fig. 5.19. Experimental arrangement to confine positrons in a magnetic mirror using a ^{22}Na source and moderator located in the mirror-field region.[59] In principle, such a configuration could be used to capture fast positrons from the source (i.e., eliminating the need for a moderator). Reprinted from reference[59].

Confinement in a stellarator. One of the simplest possibilities to confine a neutral plasma is to bend the field around into a toroidal (i.e., "donut") shape. However, plasma in a purely toroidal magnetic field is unstable to vertical drifts. In a tokamak, these drifts are mitigated by twisting the magnetic field lines using an induced toroidal current in the plasma. In the stellarator, this twist of the magnetic field is accomplished by external field coils. Both the stellarator and the tokamak were originally developed to confine hot fusion plasmas.[57,60]

The Columbia Non-neutral Torus (CNT), located at Columbia University was built specifically to conduct studies of non-neutral plasmas.[4,61] It has the capability to span the entire range of neutralization from pure electron to quasi-neutral plasmas, with a long-term goal of studying electron–positron plasmas. This device is a two-period, classical stellarator. It has the distinguishing feature, illustrated in Figure 5.20, that the required magnetic topology is created simply by four circular coils. The magnetic surfaces that this device generates are illustrated in Figure 5.21.

Fig. 5.20. The Columbia Non-neutral Torus plasma device showing a cutaway of the vacuum vessel, the four circular magnetic field coils that produce the stellerator field, and the calculated magnetic surfaces (faint deformed donut). For spatial scale, the vacuum vessel is approximately 1.8 m in diameter. Drawing courtesy of T. S. Pedersen.

This device is capable of confining stable, small-Debye-length plasmas with relatively long confinement times. Basic confinement and transport in the device are now broadly understood. Recently, a flux-surface-conforming electrostatic boundary mesh has been incorporated into the device. With this improvement, confinement times for pure electron plasmas are ~ 0.3 s. The design parameters for an electron–positron experiment in the CNT are a plasma volume of 0.1 m^3, T = 4 eV, n = 3 x 10^{12} m^{-3} and λ_D = 1 cm, with similar electron and positron inventories of ~ 10^{12} particles per species.

In 2012, a collaboration was formed to conduct such an electron–positron plasma experiment. It involves T. S. Pedersen and collaborators, who are now at the Max Planck Institute for Plasma Physics, Greifswald, Germany, Christoph Hugenschmidt and collaborators at the Munich FRM–II Research Reactor, and the UCSD positron group.[62] As described in reference[62], the experiment will be done using a new and more compact superconducting stellerator. The source of positrons will be the NEPOMUC facility (i.e., neutron induced positron source, Munich).

Fig. 5.21. The magnetic field topology in the CNT stellarator. Field lines on each of the (colored) surfaces remain on that surface as they transit both the long way and short way around the device. It is this twist of the field lines, which are induced by the external field coils shown in Figure 5.21, that stabilizes the confined plasma. Drawing courtesy of T. S. Pedersen.

5.5. Concluding Remarks

A method was described to extract beams of tailored width and brightness in a non-destructive, reproducible manner from plasmas in a PM trap. Simple analytical formulae predict the beam width and energy spread; key parameters of interest for a range of applications. The ability to extract multiple, nearly identical beams was demonstrated, utilizing over 50 % of a single trapped plasma with no loss of particles. Finally, a scenario was discussed in which the techniques described here can be used to produce high-quality electrostatic beams that are expected to be useful for a variety of positron applications. One major result of this work will likely be the ability to extract such beams from cryogenic plasmas. While challenging, this will offer the possibility of creating a new generation of bright, cold positron-beam sources with considerable potential for new physics and technology.

In a second area, key steps were described in the development of an MCT for the storage of large numbers of positrons. The availability of such large numbers of positrons opens up many new possibilities, providing bursts of positrons far larger than are available by any other means. The short-term goal of a trap for 10^{12} positrons is likely conservative, and we believe that it has a high probability of success. In the present design, this MCT could be made to fit in a volume of only a few cubic meters. It calls for a superconducting magnet and cryogens or a refrigerator. However, one can expect a rapid learning curve associated with the underlying science and technology. It is likely that further improvements in design can be made early in the development of such a multicell device, including increases in storage capacity and confinement time, decreases in the weight and size, and the reduction of other logistical requirements.

The third topic discussed here was the development of methods to study electron–positron plasmas. There are several possibilities in this area, each with advantages for specific kinds of studies. While it is fairly clear that such studies can be conducted, it is unclear what will be the optimum approach. One early focus will likely be study of the transition from single-component plasma confinement, to that in a partially compensated plasma, to that in a neutral plasma. The mechanisms of outward plasma transport in the three cases may well be quite different and interesting. Another topic, while challenging, is study of relativistic pair plasmas. It offers tremendous opportunities to make contact with the many theoretical studies that have been conducted to date.

There are presently world-class positron sources at the nuclear reactors at North Carolina State University[63] and in Munich, Germany[64] that could serve as sources for an electron–positron plasma experiment in a stellarator of the CNT design. Another strong source is being developed at the University of California Riverside.[65] These devices are, or will be, capable of slow positron fluxes ~ $5 \times 10^8 - 10^9$ e^+/s, and further increases in source strengths are under development. These intense positron sources would be ideal to fill an MCT. Further, they would be excellent for electron–positron experiments such as those described above. In this case, the MCT would be used to accumulate sufficient numbers of positrons (e.g., $N_{tot} \sim 10^{11} - 10^{12}$) from a strong source, which would then be injected in bursts into the CNT in times ~ 10 ms.

Not addressed here, but extremely interesting, is the desire to develop methods to provide ultra-intense, bright bursts of positrons for applications such as creating a Ps Bose–Einstein condensate (BEC), and eventually, an annihilation gamma-ray laser. The MCT can likely be dumped in tens of milliseconds, but much shorter bursts are desirable for the Ps BEC and other applications. It is likely that achieving these goals will be enabled by better general understanding and control of single-component antimatter plasmas.

At a more general level, the research thrusts discussed here provide examples of the potential of non-neutral plasma physics to impact upon antimatter physics.

Acknowledgments

We thank T. S. Pedersen for the material on the Columbia Non-neutral Torus, M. Bajpai for helpful discussions regarding an electron-positron plasma experiment, E. A. Jerzewski for his expert technical assistance in the experiments at UCSD, and M. Charlton for his careful reading of the manuscript and helpful suggestions. This work is supported by the U. S. DOE/NSF Plasma Initiative.

References

1. C.M. Surko, "Accumulation, Storage and Manipulation of Large Numbers of Positrons in Traps I. – The Basics" in Knoop, M., Madsen, N. and Thompson, R. C. (eds), *Physics with Trapped Charged Particles* (Imperial College Press, London, 2013) pp. 83–128.
2. C. M. Surko, G. F. Gribakin and S. J. Buckman, Low-energy positron interactions with atoms and molecules, *J. Phys. B: At. Mol. Opt. Phys.* **38**, R57–R126 (2005).
3. R. G. Greaves, J. M. Moxom, Recent results on trap-based positron beams, *Mat. Sci. Forum* **445–446**, 419–423 (2004).
4. T. S. Pedersen, A. H. Boozer, W. Dorland, J. P. Kremer and R. Schmitt, Prospects for the creation of positron–electron plasmas in a non–neutral stellarator, *J. Phys. B: At. Mol. Opt.* **36**, 1029–1039 (2003).

5. C. M. Surko and R. G. Greaves, Emerging science and technology of antimatter plasmas and trap-based beams, *Phys. Plasmas* **11**, 2333–2348 (2004).
6. J. R. Danielson, T. R. Weber and C. M. Surko, Extraction of Small–diameter Beams from Single-component Plasmas, *Appl. Phys. Lett.* **90**, 081503–081503 (2007).
7. T. R. Weber, J. R. Danielson and C. M. Surko, Creation of Finely Focused Particle Beams from Single-Component Plasmas, *Phys. Plasmas* **15**, 012106–012110 (2008).
8. T. R. Weber, J. R. Danielson and C. M. Surko, Energy Spectra of Tailored Particle Beams from Trapped Single-component Plasmas, *Phys. Plasmas* **16**, 057105–057108 (2009).
9. B. R. Beck, J. Fajans and J. H. Malmberg, Temperature and anisotropic-temperature relaxation measurements in cold, pure-electron plasmas, *Phys. Plasmas* **3**, 1250–1258 (1996).
10. D. H. E. Dubin, Collisional Transport in Nonneutral Plasmas, *Phys. Plasmas* **5**, 1688 (1998).
11. J. R. Danielson and C. M. Surko, Radial Compression and Torque-balanced Steady States of Single-Component Plasmas in Penning–Malmberg traps, *Phys. Plasmas* **13**, 055706–055710 (2006).
12. D. L. Eggleston, C. F. Driscoll, B. R. Beck, A. W. Hyatt and J. H. Malmberg, Parallel energy analyzer for pure electron plasma devices, *Phys. Fluids B* **4**, 3432–3439 (1992).
13. T. R. Weber, J. R. Danielson and C. M. Surko, Electrostatic Beams from Tailored Plasmas in a Penning–Malmberg Trap, *Phys. Plasmas* **17**, 123507–123510 (2010).
14. A. P. Mills, Brightness enhancement of slow positron beams, *Appl. Phys.* **23**, 189 (1980).
15. C. H. Tseng and G. Gabrielse, Portable trap carries particles 5000 kilometers, *Hyperfine Inter.* **76**, 381–386 (1993).
16. R. A. Lewis, G. A. Smith and S. D. Howe, Antiproton portable traps and medical applications, *Hyperfine Inter.* **109**, 155–164 (1997).
17. C. M. Surko and R. G. Greaves, A multi-cell trap to confine large numbers of positrons, *Rad. Chem. and Phys.* **68**, 419–425 (2003).
18. J. R. Danielson, T. R. Weber and C. M. Surko, Plasma Manipulation Techniques for Positron Storage, *Phys. Plasmas* **13**, 123502–123510 (2006).
19. O. J. Orient, A. Chutjian and V. Garkanian, Minature, high-resolution quadrupole mass-spectrometer array, *Rev. Sci. Instrum.* **68**, 1393–1397 (1997).
20. E. R. Badman and R. G. Cooks, Parallel minature cylindrical ion trap array, *Analytical Chem.* **72**, 3291–3297 (2000).
21. L. V. Jørgensen, M. Amoretti, G. Bonomi, P. D. Bowe, C. Canali, C. Carraro, C. L. Cesar, M. Charlton, M. Doser, A. Fontana, M. C. Fujiwara, R. Funakoshi, P. Genova, J. S. Hangst, R. S. Hayano, A. Kellerbauer, V. Lagomarsino, R. Landua, E. L. Rizzini, M. Macrì, N. Madsen, D. Mitchard, P. Montagna, A. Rotondi, G. Testera, A. Variola, L. Venturelli, D. P. v. d. Werf and Y. Yamazaki, New Source of Dense, Cryogenic Positron Plasmas, *Phys. Rev. Lett.* **95**, 025002–025005 (2005).

22. R. G. Greaves and C. M. Surko, Antimatter plasmas and antihydrogen, *Phys. Plasmas* **4**, 1528–1543 (1997).
23. C. M. Surko and T. J. Murphy, Use of the positron as a plasma particle, *Phys. Fluids B* **2**, 1372–1375 (1990).
24. E. M. Hollmann, F. Anderegg and C. F. Driscoll, Confinement and Manipulation of Non-neutral Plasmas Using Rotating Wall Electric Fields, *Phys. Plasmas* **7**, 2776–2789 (2000).
25. R. G. Greaves, M. D. Tinkle and C. M. Surko, Creation and uses of positron plasmas, *Phys. Plasmas* **1**, 1439–1446 (1994).
26. C. M. Surko, R. G. Greaves and M. Charlton, Stored positrons for antihydrogen production, *Hyperfine Inter.* **109**, 181–188 (1997).
27. C. Hugenschmidt, K. Schreckenbach, M. Stadlbauer and B. Straßer, First positron experiments at NEPOMUC, *Appl. Surf. Sci.* **252**, 3098–3105 (2006).
28. R. Krause-Rehberg, S. Sachert, G. Brauer, A. Rogov and K. Noack, EPOS – an intense positron beam project at the ELBE radiation source in Rossendorf, *Appl. Surf. Sci.* **252**, 3106–3110 (2006).
29. H. M. Chen, Y. C. Jean, C. D. Jonah, S. Chemerisov, A. F. Wagner, D. M. Schrader and A. W. Hunt, Intense slow positron production at the 15 MeV LINAC at Argonne National Laboratory, *Appl. Surf. Sci.* **252**, 3159–3165 (2006).
30. B. R. Beck, J. Fajans and J. H. Malmberg, Measurement of collisional anisotropic temperature relaxation in a strongly magnetized pure electron plasma, *Phys. Rev. Lett.* **68**, 317–320 (1992).
31. R. C. Davidson, *Physics of Nonneutral Plasmas* (Addison-Wesley, Reading, MA, 1990).
32. K. S. Fine, C. F. Driscoll and J. H. Malmberg, Measurements of a nonlinear diocotron mode in pure electron plasmas, *Phys. Rev. Lett.* **63**, 2232–2234 (1989).
33. J. Fajans, E. Gilson and L. Friedland, Autoresonant excitation of the diocotron mode in non-neutral plasmas, *Phys. Rev. Lett.* **82**, 4444–4447 (1999).
34. J. Fajans, E. Gilson and L. Friedland, Autoresonant excitation of a collective nonlinear mode, *Phys. Plasmas* **6**, 4497–4503 (1999).
35. J. Fajans, E. Gilson and L. Friedland, The effect of Damping on Autoresonant (nonstationary) Excitation, *Phys. Plasmas* **8**, 423 (2001).
36. J. R. Danielson and C. M. Surko, Torque-balanced high-density steady states of single component plasmas, *Phys. Rev. Lett.* **95**, 035001–035004 (2005).
37. M. A. Akhalkatsi and G. Z. Machabeli, Generation of Electromagnetic Waves in the Electron–Positron Plasma in the Vicinity of a Pulsar, *Astrophys.* **43**, 282–288 (2000).
38. U. A. Mofiz, Linear modes in the rotating neutron star polar-cap electron–positron plasma, *Phys. Rev. E* **55**, 5894–5900 (1997).
39. J. Zhao, J. I. Sakai, K. I. Nishikawa and T. Neubert, Relativistic particle acceleration in an electron–positron plasma with a relativistic electron beam, *Phys. Plasmas* **1**, 4114–4119 (1994).
40. J. Sakai, M. Eda and W. Shiratori, Wave generation and particle acceleration in an electron–positron plasma, *Phys. Scr.* **T75**, 67–71 (1998).

41. D. Gyobu, J. Sakai, M. Eda, T. Neubert and M. Nambu, Emission of electromagnetic waves from Langmuir waves generated by electron beam instabilities in pair plasmas, *J. Phys. Soc. Japan* **68**, 471–477 (1999).
42. A. D. Rogava, S. M. Mahajan and V. I. Berezhiani, Velocity shear generated Alfven waves in electron–positron plasmas, *Phys. Plasmas* **3**, 3545–3555 (1996).
43. T. Kitanishi, J. Sakai, K. Nishikawa and J. Zhao, Electromagnetic waves emitted from an electron–positron plasma cloud moving across a magnetic field, *Phys. Rev. E* **53**, 6376–6381 (1996).
44. R. G. Greaves and C. M. Surko, "Practical Limits on Positron Accumulation and the Creation of Electron-Positron Plasmas", in Anderegg, F., Schweikhard, L. and Driscoll, C. (eds), *Non-Neutral Plasma Physics IV* (American Institute of Physics, Melville, NY, 2002), pp. 10–23.
45. V. Tsytovich and C. B. Wharton, Laboratory electron–positron plasma-a new research object, *Comments Plasma Phys. Controlled Fusion* **4**, 91–100 (1978).
46. R. G. Greaves and C. M. Surko, An electron–positron beam-plasma experiment, *Phys. Rev. Lett.* **75**, 3846–3849 (1995).
47. S. J. Gilbert, D. H. E. Dubin, R. G. Greaves and C. M. Surko, An electron–positron beam-plasma instability, *Phys. Plasmas* **8**, 4982–4994 (2001).
48. H. Boehmer, Formation of electron–positron plasmas in the laboratory, *AIP Conference Proceedings* **303**, 422–434 (1994).
49. C. Hugenschmidt, Private communication, 2003.
50. D. Umstadter, Review of physics and applications of relativistic plasmas driven by ultra-intense lasers, *Phys. Plasmas* **8**, 1774–1785 (2001).
51. E. P. Liang, S. C. Wilks and M. Tabak, Pair production by ultraintense lasers, *Phys. Rev. Lett.* **81**, 4887–4890 (1998).
52. C. Gahn, G. D. Tsakiris, G. Pretzler, K. J. Witte, P. Thorolf, D. Habs, C. Delfin and C. G. Wahlström, Generation of MeV electrons and positrons with femtosecond pulses from a table-top lasesr system, *Phys. Plasmas* **9**, 987–999 (2002).
53. T. E. Cowan, M. Roth, J. Johnson, C. Brown, M. Christl, W. Fountain, S. Hatchett, E. A. Henry, A. W. Hunt, M. H. Key and A. MacKinnon, Intense electron and proton beams from Peta Watt laser-matter interactions, *Nucl. Instrum. Meth. Phys. Res. A* **455**, 130–139 (2000).
54. P. Zhang, N. Saleh, S. Chen, Z. Shen and D. Umstadter, An optical trap for relativistic plasma, *Phys. Plasmas* **10**, 2093–2099 (2003).
55. H. Chen, S. C. Wilks, J. D. Bonlie, E. P. Liang, J. Myatt, D. F. Price, D. D. Meyerhofer and P. Beiersdorfer, Relativistic Positron Creation Using Ultraintense Short Pulse Lasers, *Phys. Rev. Lett.* **102**, 105001–105004 (2009).
56. J. Myatt, J. A. Delettrez, A. V. Maximov, D. D. Meyerhofer, R. W. Short, C. Stoeckl and M. Storm, Optimizing electron–positron production on kilojoule-class high-intensity lasers for pair-plasma creation, *Phys. Rev. E* **79**, 066409–066410 (2009).
57. F. F. Chen, *Introduction to Plasma Physics and Controlled Fusion, Volume I: Plasma Physics*, Second edition (Springer, New York, 1984).
58. J. Walz, C. Zimmermann and L. Ricc, Combined trap with the potential for antihydrogen production, *Phys. Rev. Lett.* **75**, 3257–3260 (1995).

59. H. Boehmer, M. Adams and N. Rynn, Positron trapping in a magnetic mirror configuration, *Phys. Plasmas* **2**, 4369–4371 (1995).
60. N. A. Krall and A. W. Trivelpiece, *Principles of Plasma Physics* (San Francisco Press, San Francisco, 1986).
61. T. S. Pedersen, J. P. Kremer, R. G. Lefrancois, Q. Marksteiner, X. Sarasola and N. Ahmad, Experimental demonstration of a compact stellarator magnetic trap using four circular coils, *Phys. Plasmas* **13**, 102502–102506 (2006).
62. T. S. Pedersen, J. R. Danielson, C. Hugenschmidt, G. Marx, X. Sarasola, F. Schaue, L. Schweikhard, C. M. Surko and E. Winkler, Plans for the creation and studies of electron–positron plasmas in a stellarator, *New J. Phys.* **14**, 035010–035013 (2012).
63. A. G. Hathaway, M. Skalsey, W. E. Frieze, R. S. Vallery and D. W. Gidley, Implementation of a Prototype Slow Positronium Beam at the North Carolina State University PULSTAR Reactor, *Nucl. Instrum. Meth. in Phys. Res. A* **579**, 538–541 (2007).
64. C. Hugenschmidt, G. Kogel, R. Repper, K. Schreckenbach, P. Sperr, B. Strasser and W. Triftshauser, "NEPOMUC – the new positron beam facility at FRM II" in Hyodo, T., Kobayashi, Y., Nagasima, Y. and Saito, H., (eds), *Proceedings of the International Conference on Positron Annihilation, Volume 445–446* (Materials Science Forum, 2004) pp. 480–482.
65. D. B. Cassidy, R. G. Greaves, S. H. M. Deng, N. Lopez–Valdez, V. Meligne and A. P. Mills, "Development and Applications of an Accelerator Based Positron Source" in Mcdaniel, F. D. and Doyle, B. L. (eds.), *20th International Conference on Applications of Accelerators in Research and Industry* (AIP Conference Proceedings #1099, Toronto CA, 2009), pp. 866–869.

Chapter 6

Waves in Non-neutral Plasma

François Anderegg
University of California, San Diego, Physics Dept. 0319,
La Jolla CA 92093, USA
fanderegg@ucsd.edu

This chapter gives an elementary introduction to plasma waves in single-species non-neutral plasmas. The plasma is presumed to be trapped in an elongated cylindrical trap with an axial magnetic field; this geometry is referred to as a Penning–Malmberg trap. The first part concerns low frequency drift waves, called diocotron waves, followed by plasma waves of two different types, and the last part of the chapter is about cyclotron waves, which are the highest frequency electrostatic waves in Penning–Malmberg traps.

6.1. Diocotron Waves

The word diocotron comes from the Greek meaning to "pursue, chase". In 1952 researchers conducting experiments with hollow electron beams inside a conducting cylinder in a magnetic field observed that hollow beams of charge were unstable. The beam broke apart, and the pieces were rotating around the magnetic field axis and "chasing" each other. They named the instability the diocotron instability.

A comprehensive review of diocotron instability can be found in Chapter 6 of *Physics of Non-neutral Plasmas* by R. Davidson.[1] In this chapter, I will describe stable diocotron waves and not the diocotron instability observed on hollow electron beams. The properties of stable diocotron waves were first measured by deGrassie and Malmberg[2] in confined pure electron columns.

We will consider a simple case of a monotonically decreasing density profile of ions trapped in an elongated Penning–Malmberg trap. A simple introduction to Penning–Malmberg traps can be found in a *Physics Today*

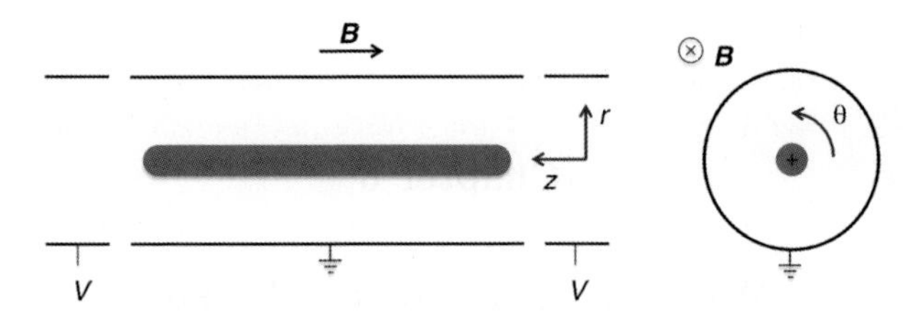

Fig. 6.1. Schematic of a Penning–Malmberg trap containing positive charges.

article.[3] Figure 6.1 shows the trap geometry and the coordinates used. The plasma is assumed to be "rigid", that is individual ions "bounce" rapidly along the magnetic field compared to their rotation around the axis of the trap. For low frequency drift waves, the rapid bouncing of the particles along the magnetic field axis averages any z-dependence, and we will assume here that diocotron waves have no axial variation, i.e. $k_z = 0$.

6.1.1. *Infinite length description*

To begin describing the diocotron wave, we will assume that the plasma column is "infinitely" long and that it can be described by a two-dimensional model. In general the density perturbation has the form $\delta n = \delta n(r) \exp\{i(m_\theta \theta + k_z z - \omega t)\}$; here we will discuss the case of $m_\theta = 1$ and $k_z = 0$.

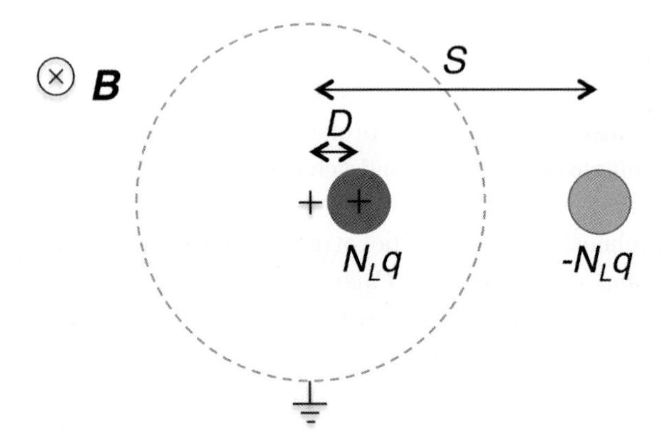

Fig. 6.2. Image charge model of the diocotron wave for an infinitely long line of charge.

Figure 6.2 shows an end view of a plasma column with line density $N_L[m^{-1}]$ of particles with (positive) charge q, displaced from the axis by a distance D. We can replace the wall by an equal and opposite "image charge" located at a distance S from the center of the trap such that the potential at $r = R_w$ is a constant: $\phi(R_w, \theta) = \text{const}$. For the sake of symmetry, the image charge has to be located at the same azimuthal angle θ as the real charges. The potential from two infinitely long lines of charge is:

$$\begin{aligned}\phi(r,\theta) &= -\frac{N_L q}{2\pi\varepsilon_0}\left[\ln\sqrt{r^2+D^2-2rD\cos\theta} - \ln\sqrt{r^2+S^2-2rS\cos\theta}\right] \\ &= -\frac{N_L q}{2\pi\varepsilon_0}\ln\left[\frac{r\sqrt{1+\frac{D^2}{r^2}-\frac{2D}{r}\cos\theta}}{S\sqrt{1+\frac{r^2}{S^2}-\frac{2r}{S}\cos\theta}}\right].\end{aligned} \tag{6.1}$$

By choosing $S/R_w = R_W/D$, that is, $S = R_w^2/D$, the potential at the location of the wall can be written as

$$\phi(R_W,\theta) = -\frac{N_L q}{2\pi\varepsilon_0}\ln\left(\frac{D}{R_w}\right), \tag{6.2}$$

which is independent of θ. The electric field from a line of charge can be calculated using Gauss' Law:

$$E(r) = \frac{\sum Q}{2\pi\varepsilon_0 r L} = \frac{N_L q}{2\pi\varepsilon_0 r}, \tag{6.3}$$

and therefore the image charge electric field at $r = 0$ is:

$$E_i(r=0) = -\frac{N_L q}{2\pi\varepsilon_0 S} = -\frac{N_L q D}{2\pi\varepsilon_0 R_w^2}. \tag{6.4}$$

The $E \times B$ drift velocity v_d of the (real) charge in the electric field of the image charge is:

$$\mathrm{v}_d = \frac{E_i}{B} = -\frac{N_L q D}{2\pi\varepsilon_0 B_z R_w^2}, \tag{6.5}$$

where we have assumed that the displacement D is small ($D/R_w \ll 1$). The infinite length small amplitude diocotron frequency is:

$$f_{\text{dio}}^{\infty} = \frac{\mathrm{v}_d}{2\pi D} = \frac{N_L q}{4\pi^2\varepsilon_0 B_z R_w^2}. \tag{6.6}$$

From an experimental point of view, the diocotron mode frequency gives a measure of the line density N_L. We have to be careful here since Eq. (6.6) is valid only for infinite length and small amplitude. If one measures the line density N_L with a dump technique[4] and also measures the diocotron frequency for a column with $L_p \sim 2R_p$, the infinite length equation gives a frequency too small by a factor of up to two or three.

In Sections 6.1.3 and 6.1.4 we will describe corrections to f_{dio}^{∞} to include the effect of a realistic plasma, that is, finite amplitude, finite length, temperature shift, etc.

6.1.2. *A negative energy mode*

The diocotron mode is a negative energy mode. This can be easily seen in the image charge model that we have used in this presentation. The image charge has the opposite sign of the "real" charge, therefore as the plasma is attracted towards its image charge, the electrostatic energy decreases as the mode amplitude increases. Here any kinetic energy is ignored. Let's calculate how much electrostatic energy is required to displace the plasma by a distance D in the image charge electric field:

$$\begin{aligned} W_{\text{ES}} &= \int_0^D Q \cdot E_i dx = \int_0^D N_L q L_p \left(\frac{-N_L q x}{2\pi\varepsilon_0 R_w^2} \right) dx \\ &= -\frac{(N_L q)^2}{4\pi\varepsilon_0} \frac{D^2}{R_w^2} L_p < 0. \end{aligned} \tag{6.7}$$

The electrostatic energy W_{ES} is negative; this means that the diocotron mode can be destabilized by dissipation. This destabilization is referred to as "resistive growth". Figure 6.3 illustrates how an azimuthal section of the wall (called a sector) can destabilize the dioctron mode. The power dissipated in the load connected to the sector is $P = \frac{1}{2} I^2 \text{Re}(Z)$, where I is the image current, and the electrostatic energy in the wave is W_{wave}.

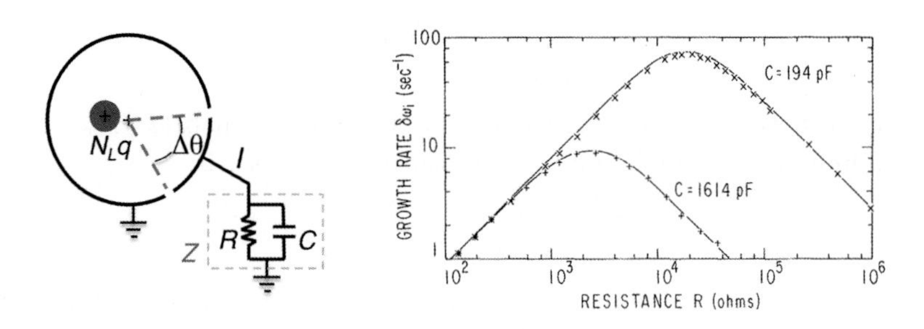

Fig. 6.3. Sectored electrode with model of impedance Z. Growth rate plotted versus resistance R.

Therefore the growth rate of the wave is:[5]

$$\gamma = \frac{P}{2W_{\rm wave}} = \frac{4\varepsilon_0}{\pi}\frac{\omega^2 L_s^2 \sin^2\frac{\Delta\theta}{2}}{L_p}\mathrm{Re}(Z)\,, \tag{6.8}$$

and for the RC circuit shown in Figure 6.3 the real part of the impedance is

$$\mathrm{Re}(Z) = \frac{R}{1+\omega^2 R^2 C^2}. \tag{6.9}$$

This growth rate was carefully verified experimentally[5] as shown in Figure 6.3. As the external resistance R is increased, the growth rate increases linearly up to a maximum corresponding to $\omega^2 R^2 C^2 = 1$; at the maximum growth rate, half of the current is flowing through the resistor and the other half through the capacitor. As R is further increased, the growth rate is reduced since more and more current flows through the dissipationless capacitor. Figure 6.3 also shows the growth rate for a different capacitor, further validating the model. Figure 6.4 illustrates how a feedback circuit can be used to damp a diocotron mode, by phase-shifting the sector voltage so as to obtain an effective "negative resistance". Changing the phase of the feedback controls the growth or damping rate of the diocotron mode.[6]

The diocotron mode provides a useful technique with which to control the position of the plasma in a trap. For example, Figure 6.5 shows a "phaser" picture of z-integrated density obtained by multiple dumps and measuring the charge passing through a collimator hole at varying azimuthal positions of the plasma.[7] Further, the diocotron mode is a good tool for the loading of off-axis multitrap cells[8] designed to store large numbers of particles such as positrons.

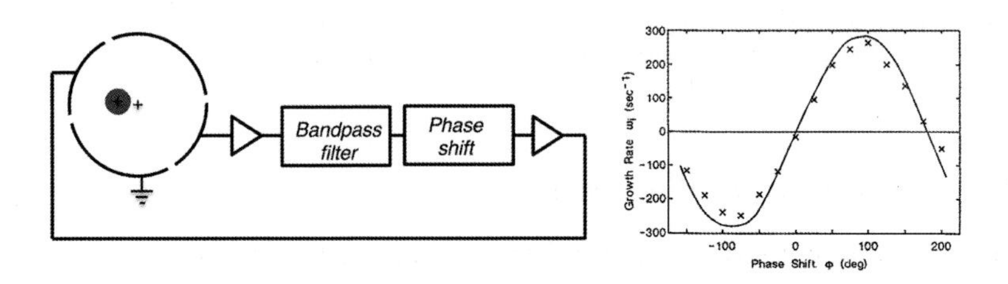

Fig. 6.4. Schematic of diocotron feedback circuit and feedback growth rate versus phase shift.

6.1.3. *Finite amplitude shift of diocotron mode*

For large displacement D, the plasma column distorts into an elliptical cross-section and the frequency of the diocotron mode increases, as[9]

$$f_{\text{dio}} = f_{\text{dio}}^{\infty} + f_{\text{dio}}^{\infty} \left(\frac{1 - 2(R_p/R_w)^2}{[1 - (R_p/R_w)^2]^2} \right) \left(\frac{D}{R_w} \right)^2 . \qquad (6.10)$$

The measured amplitude shift versus normalized displacement is shown in Figure 6.5 for three different plasma radii. Figure 6.5(b) shows phase-locked densities $n(r, \theta)$ for the diocotron at two amplitudes showing elliptical distortion; in both instances $R_p = 2.42$ cm and the wall radius was at $R_w = 3.81$ cm.

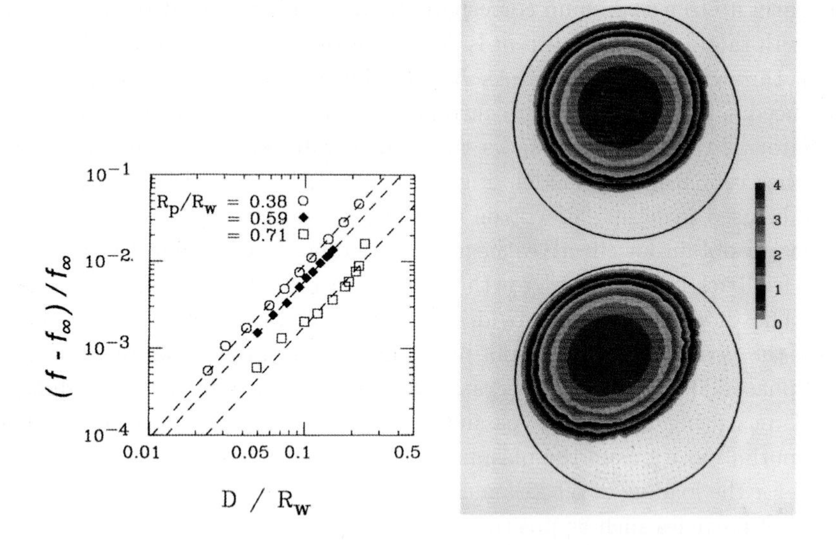

Fig. 6.5. (a) Measured frequency shift vs. amplitude for three different radius plasmas. (b) "Phaser" picture of z-integrated density; colors represent density on a linear scale of 10^6cm^{-3}, showing elliptical distortion of the plasma for large displacement.

6.1.4. *Finite length diocotron*

The confining potential at the end of the trap pushes the plasma in the z-direction, resulting in a radial force on an off-axis plasma. The force comes from the radial component of the confining potential and adds to the force due to the image charge. $F_{\text{tot}} = F_i + F_c$, where F_i is due to the image and

F_c is due to the confinement. A careful calculation of the frequency change has been conducted[10] and gives:

$$\frac{f_{\text{dio}}}{f_{\text{dio}}^{\infty}} = \frac{F_{\text{tot}}}{F_{i,\infty}} = 1 + \left[\frac{j_{01}}{2}\left(\frac{1}{4} + \ln\left(\frac{R_w}{R_p}\right) + \frac{k_B T 4\pi\varepsilon_0}{N_L e^2}\right) - 0.671\right]\left(\frac{R_w}{L_p}\right). \tag{6.11}$$

The plasma electrostatic pressure (term: $\frac{1}{4} + \ln(R_w/R_p)$) pushes on the end confining potential, increasing the diocotron frequency. The plasma kinetic pressure (term: $k_B T 4\pi\varepsilon_0/(N_L e^2)$) also similarly increases the frequency. Finally the finite length of the image charge reduces the force (term 0.671); the numerical factor comes from the specific shape of the vacuum potential of cylindrical electrodes. The finite length diocotron frequency equation has been extensively tested experimentally.

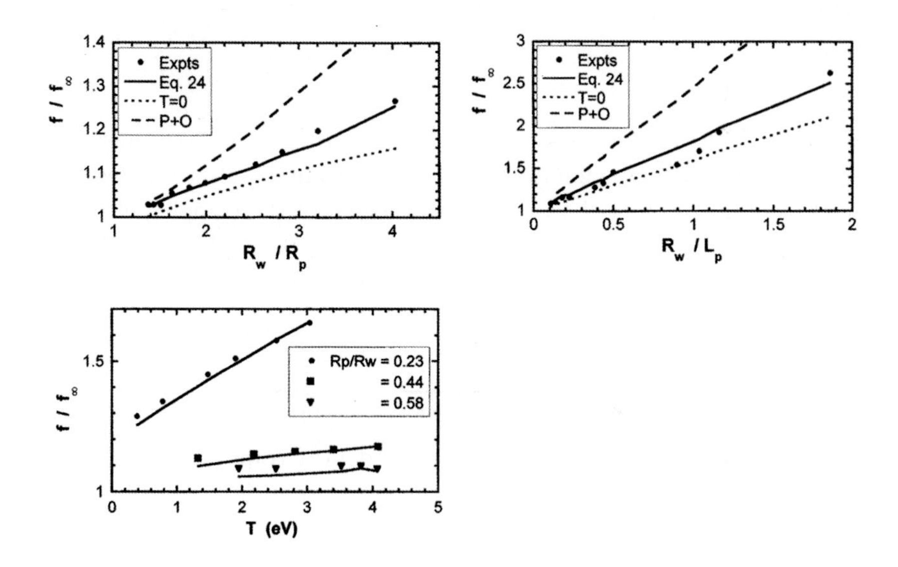

Fig. 6.6. Symbols show experimental measurement of diocotron frequency versus plasma radius, plasma length, and temperature showing agreement with Eq. (6.11). The long dashed line is a theory for "non-rigid" plasmas.[11]

Figure 6.6 shows the measured frequency shift plotted against R_w/R_p and R_w/L_p. In both cases the solid line is the theory prediction of Eq. (6.11) with no adjustable parameters. The kinetic pressure term $k_B T 4\pi\varepsilon_0/(N_L e^2)$ can also be written as $4\lambda_D^2/R_p^2$; it is small for plasmas which have many Debye lengths across, but it is large for clouds made out of a few particles with a large Debye length. The third graph of Figure 6.6 shows the measured kinetic pressure shift for three different plasma radii. Note that the kinetic pressure shift is much larger for small plasmas (large λ_D/R_p) than for large plasmas. It is important that these finite length corrections are accounted for when using the diocotron frequency to measure the line density.

6.1.5. *Magnetron regime*

For short and low density plasmas, the radial component of the confinement potential is substantially larger than the potential due to the image charge. In this regime the "diocotron mode" is called the "magnetron mode". In this regime the frequency of the mode is almost independent of the amount of charge in the trap, contrasting sharply with the diocotron mode described before. The frequency of the mode is:[10]

$$f = -\frac{E_r}{2\pi D B_z} = \frac{1}{2\pi D B_z}\left[\frac{\partial \phi_c}{\partial r} + \frac{\partial \phi_i}{\partial r}\right]_{r=D}, \tag{6.12}$$

where ϕ_c is the confinement potential and ϕ_i is the potential due to the image charge. For a cylindrical trap with a trapping electrode length L and a confining potential V_c applied to the end of the trapping electrode, one gets[10]

$$f = \frac{1}{2\pi B_z}\left[\underbrace{1.15\frac{V_c}{R_w^2}\frac{L}{R_w}}_{\text{magnetron}} - \underbrace{1.0027\frac{Q}{R_w^3}}_{\text{diocotron}}\right]. \tag{6.13}$$

For example, if 10^5 electrons are contained in a trap with $V_c = 10$ V, $R_w = 1$ cm and $L/R_w = 0.2$, the magnetron term is 230 kHz and the diocotron term is 1.4 kHz.

6.1.6. *Higher-order diocotron modes*

The image charge model we have used so far has the advantage of giving a physical intuition of the dioctoron mode but is limited to $m_\theta = 1$. The standard linear theory of diocotron modes can be found in References 1, 11, giving a diocotron mode frequency:

$$f_{\text{dio}}^{m_\theta} = f_{E\times B}\left[m_\theta - 1 + \left(\frac{R_p}{R_w}\right)^{2m_\theta}\right], \tag{6.14}$$

where $f_{E\times B}$ is the plasma rotation frequency and m_θ is the azimuthal mode number. For a square profile $f_{E\times B} = qn/(4\pi\varepsilon_0 B)$. For $m_\theta = 1$ this result is identical to the image charge model for infinite length and small amplitude, as in Eq. (6.6).

One can see from Eq. (6.14) that for the $m_\theta = 1$ mode, the diocotron frequency is the plasma rotation frequency reduced by $(R_p/R_w)^2$. Also for a small radius plasma, the $m_\theta = 2$ diocotron frequency is almost at the plasma rotation frequency.

6.2. Plasma Waves

We are considering a long plasma in a conducting cylinder with a magnetic field B aligned with the trap axis. The magnetic field is strong enough to make the cyclotron frequency much larger than the plasma frequency, but the exact magnitude of the magnetic field is not important.

To derive the frequency of plasma waves in a trap, we start by writing the continuity equation,

$$\frac{\partial n}{\partial t} + \frac{\partial}{\partial z} n \cdot \mathrm{v}_z = 0\,, \tag{6.15}$$

and Newton equation,

$$m\frac{\partial \mathrm{v}_z}{\partial t} = qE_z = -q\frac{\partial}{\partial z}\phi. \tag{6.16}$$

Note that we have kept only the z-dynamics in these two equations, but we will keep all components for the Poisson equation:

$$\nabla^2\phi = -4\pi q\delta n. \tag{6.17}$$

We will now assume that the density perturbation is $\delta n(r)\exp\{i(m_\theta\theta + k_z z - \omega t)\}$, and the above equations become

$$-i\omega\delta n + n\, ik_z\delta \mathrm{v}_z = 0; \tag{6.18}$$

$$-mi\omega\delta \mathrm{v}_z = -qik_z\delta\phi; \tag{6.19}$$

$$-k^2\delta\phi = -4\pi q\delta n. \tag{6.20}$$

Combining these three equations, one gets the Trivelpiece–Gould[12] mode dispersion relation for a cold plasma,

$$\omega^2 = \frac{k_z^2}{k^2}\frac{4\pi q^2 n}{m} = \frac{k_z^2}{k^2}\omega_p^2 \tag{6.21}$$

where ω_p is the plasma frequency.

When the thermal pressure is included, the dispersion relation becomes:

$$\omega^2 = \frac{k_z^2}{k^2}\omega_p^2 + 3\bar{v}^2 k_z^2 . \tag{6.22}$$

Here the thermal pressure increases the mode frequency. The wave vector k is equal to

$$k^2 = k_z^2 + k_\perp^2 , \tag{6.23}$$

and in cylindrical geometry for $R_p \ll R_w$:

$$k_\perp = \frac{1}{R_p}\left(\frac{2}{\ln(R_w/R_p)}\right)^{1/2} . \tag{6.24}$$

It is interesting to note that if we had kept all k (not only k_z), we would get

$$\omega^2 = \omega_p^2(1 + \frac{3}{2}k^2\lambda_D^2) , \tag{6.25}$$

which is the standard plasma wave dispersion relation for an infinite unmagnetized plasma, known in plasma physics as a Langmuir wave. The dispersion relation of these two types of plasma waves is shown in Figure 6.7.

In an infinite, unbound plasma, when we have a little bit of extra charge δq at one location and a small deficit of charge at another location, an

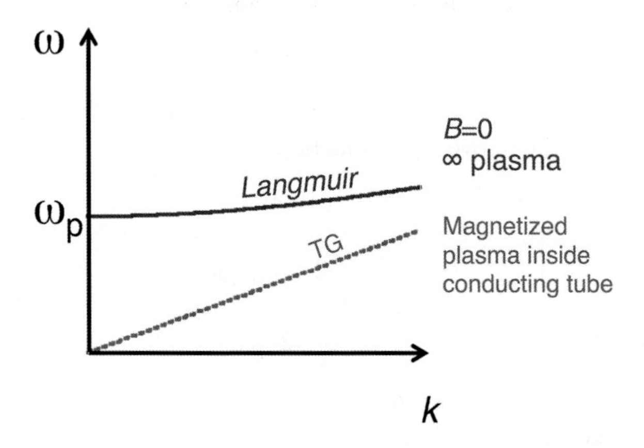

Fig. 6.7. Dispersion relation for Langmuir wave and Trivelpiece–Gould (TG) wave.

electric field E is established. In contrast, when a plasma is inside a conducting wall, the electric field E_z is reduced by the conducting wall. At the wall $E_z = 0$ and only E_r is not zero. This reduction of E_z reduces the restoring force and therefore lowers the oscillation frequency; the frequency of plasma waves, called Trivelpiece–Gould waves, is lower than the frequency of plasma waves in an unbounded system, called Langmuir waves ($f_{\mathrm{TG}} < f_{\mathrm{Langmuir}}$).

6.2.1. *Finite length Trivelpiece–Gould modes*

For trapped plasmas, the wavelength has to fit in the plasma such that the wave is reflected at each end of the trap, generating a standing wave with a parallel wave vector:

$$k_z = \frac{m_z \pi}{L_p}, \tag{6.26}$$

where m_z is the parallel mode number: $m_z = 1$ corresponds to one half wavelength in the plasma, $m_z = 2$ corresponds to a full wavelength in the plasma. In the limit of a long column ($k_z \lambda_D \ll 1$ and $R_p k_z \ll 1$), the frequencies of the axisymmetric ($m_\theta = 0$) Trivelpiece–Gould (TG) mode are:

$$\omega \cong \omega_p \left(\frac{R_p}{R_w}\right)(R_w k_z)\left[\frac{1}{2}\ln\left(\frac{R_w}{R_p}\right)\right]^{1/2}\left[1 + \frac{3}{2}\left(\frac{\overline{\mathrm{v}}}{\mathrm{v}_{\mathrm{ph}}}\right)^2\right]. \tag{6.27}$$

For non-axisymmetric modes $m_\theta \neq 0$.

$$\omega - m_\theta \omega_E = \omega_p \left(\frac{R_p}{R_w}\right) R_w k_z \frac{1}{j_{m_\theta, m_r}}\left[1 + \frac{3}{2}\left(\frac{\overline{\mathrm{v}}}{\mathrm{v}_{\mathrm{ph}}}\right)^2\right]. \tag{6.28}$$

Equation (6.28) assumed the mode frequency ω is of the same order as $m_\theta \omega_E$, which is valid for the ions. In contrast, for electron plasmas, ω is large compared to the rotation $m_\theta \omega_E$. For electron plasmas the Bessel function zero $\gamma m_\theta, m_r$ of Eq. (6.28) is replaced by $\gamma m_{\theta-1}, m_r$.[13] These dispersion relations are "acoustic", with $\omega \propto k_z$, and all frequencies have the same phase (and group) velocity. The $m_\theta \neq 0$ modes are Doppler shifted by the plasma rotation frequency; these can be useful for some rotating-wall applications.[14]

Experimental observations indicate that $k_z = m_z \pi / L_p$ is too large by about 10% and that the effective wavenumber corresponds to a longer wavelength extending past the end of the plasma. A theoretical model similar

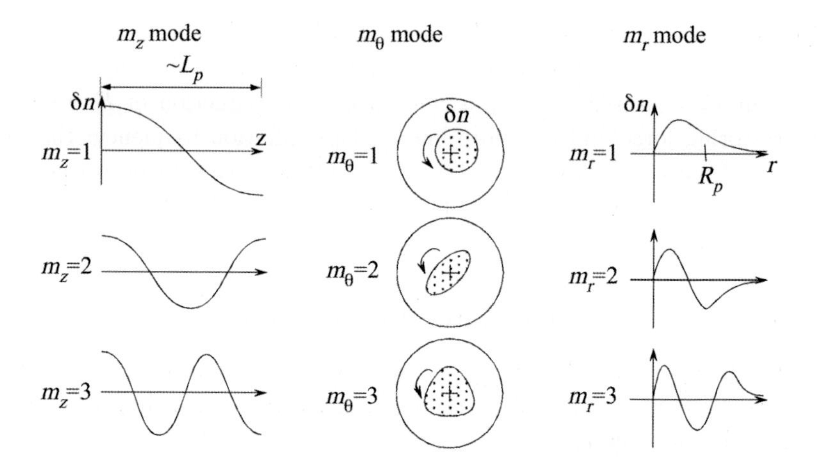

Fig. 6.8. Summary of longitudinal, azimuthal and radial wavenumber physical meanings.

to an organ pipe wavenumber calculation gives: $k_{\text{eff}} = k_z + \alpha_1 R_p + \alpha_2 R_W$, where α_1 and α_2 are given in reference 15.

Figure 6.8 summarizes all the ms used to describe the TG plasma modes. Higher m_z results in higher mode frequency; higher m_θ results in lower Doppler shifted mode frequency $\omega - m_\theta \omega_E$ and higher radial mode number m_r results in lower mode frequency.

6.2.2. *Thermally excited TG modes*

TG modes are easy to excite. Figure 6.9 shows the results of a "transmission" experiment in a pure electron plasma; $m_z = 1$ to 5 are shown with a -80 dBm drive. As the drive amplitude is reduced, the received amplitudes are reduced, and $m_z = 2$ disappears into the noise, since the antenna geometry was not effective at detecting $m_z = 2$. More interestingly, when the drive amplitude is turned off, TG modes are spontaneously excited at low level by thermal fluctuations.[17]

As we will see, the thermal excitation of plasma modes provides an effective diagnostic tool; since non-neutral plasmas can relax to a state of thermal equilibrium in the rotating frame, the "tools" of thermodynamics can be used. The plasma mode is excited by thermal electron motion in the plasma and by noise in the load (i.e. measuring circuit). At the same time, the plasma mode is damped due to Landau damping and by dissipation in the load.

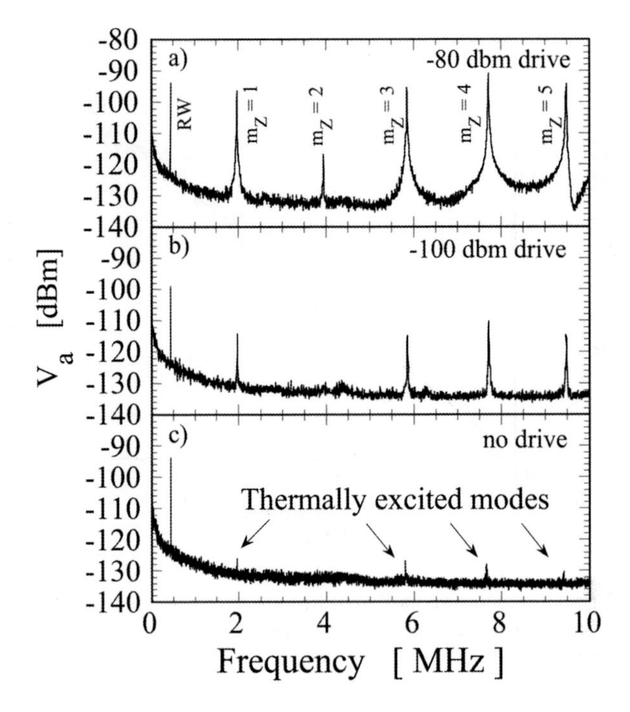

Fig. 6.9. Spectrum of $m_\theta = 1, 2, ..., 5$ Trivelpiece–Gould modes for three drive amplitudes including no drive, i.e. thermally excited.

Nyquist's theorem quantifies the amount of thermal noise in a circuit element:

$$\frac{V^2}{df} = 4k_BT\,\mathrm{Re}(Z). \tag{6.29}$$

Figure 6.10 shows that both temperature diagnostics agree to within 15%, which is typical of the accuracy of the dump temperature diagnostic. Figure 6.11 shows the electronic detection circuit attached to the electrodes of trap, and the lump circuit element model. Applying the Nyquist theorem to our trap, we get

$$\frac{V_a^2}{df} = \underbrace{4k_BT_p\,\mathrm{Re}(Z_p)\left|\frac{Z_L}{Z_p + Z_L}\right|^2}_{\text{plasma}} + \underbrace{4k_BT_L\,\mathrm{Re}(Z_L)\left|\frac{Z_p}{Z_p + Z_L}\right|^2}_{\text{load}}, \tag{6.30}$$

where V_a stands for the antenna voltage. The load impedance from a resis-

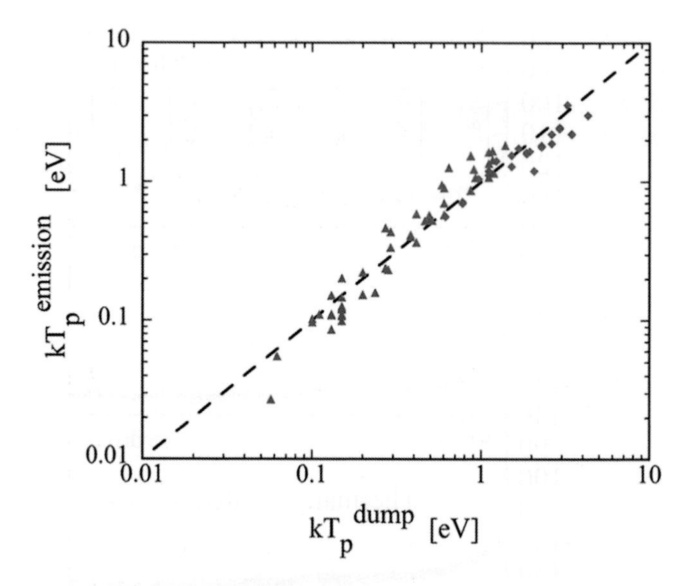

Fig. 6.10. Plasma temperature measured by emission technique versus standard dump temperature measurement. The two types of symbol represent measurements performed on two apparatuses.

tor and capacitor in parallel is $Z_L^{-1} = R_L^{-1} + i\omega C_L$. The plasma impedance around one mode is given by[16] $Z_p = G\omega_m^2/[i(\omega - \omega_m) + \gamma_m]$, where G is a coupling coefficient due to the geometry, and γ_m is the mode damping.

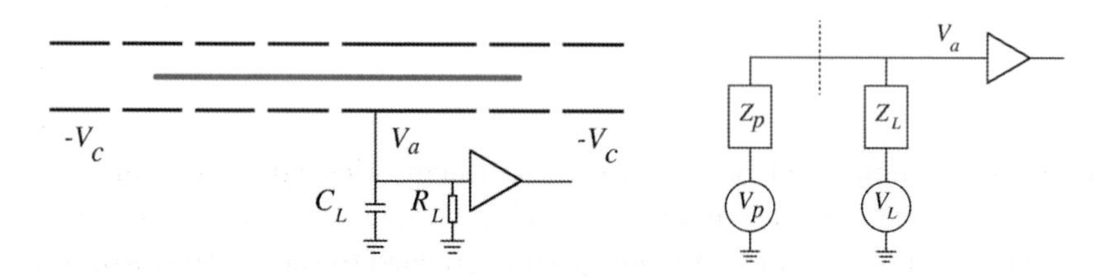

Fig. 6.11. Schematic diagram of Penning–Malmberg trap electrodes and electrical circuit analogue to plasma mode and receiver.

The measured voltage fluctuations on the antenna are shown in Figure 6.12; the fluctuations are decomposed in two parts, a Lorentzian from the plasma, and a dip-and-step from the load. The red trace is from the Lorentzian plasma contribution and the green trace is from the thermal noise of the load. The temperatures of the plasmas and of the load are obtained from this decomposition. The temperatures of the plasmas measured by this "emission" technique are plotted on the vertical axis of Figure 6.10 against the temperatures measured by a standard slow dump of the particles contained in the trap. Here we used a room-temperature amplifier to measure the thermally excited mode at plasma temperatures as low as 2.5 times room temperature.

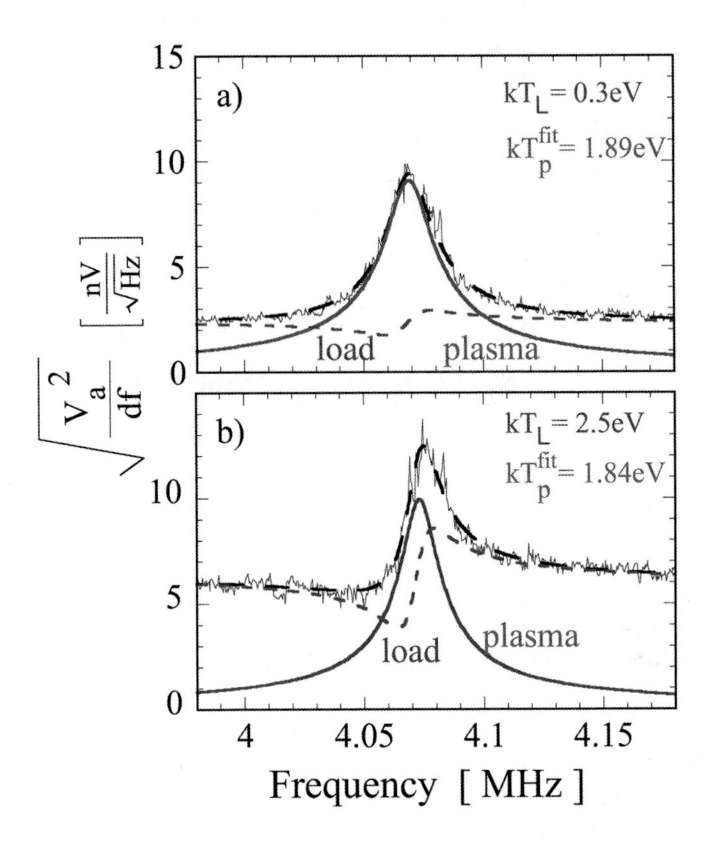

Fig. 6.12. Spectra of thermally excited TG mode for (a) $kT_{\rm plasma} = 1.89$eV and $kT_{\rm load} = 0.3$eV; (b) $kT_{\rm plasma} = 1.84$eV and $kT_{\rm load} = 2.5$eV. The long dashed line is Eq. (6.30) fitted to the data, the solid line is the plasma component and the short dashed line is the load noise filtered by the plasma.

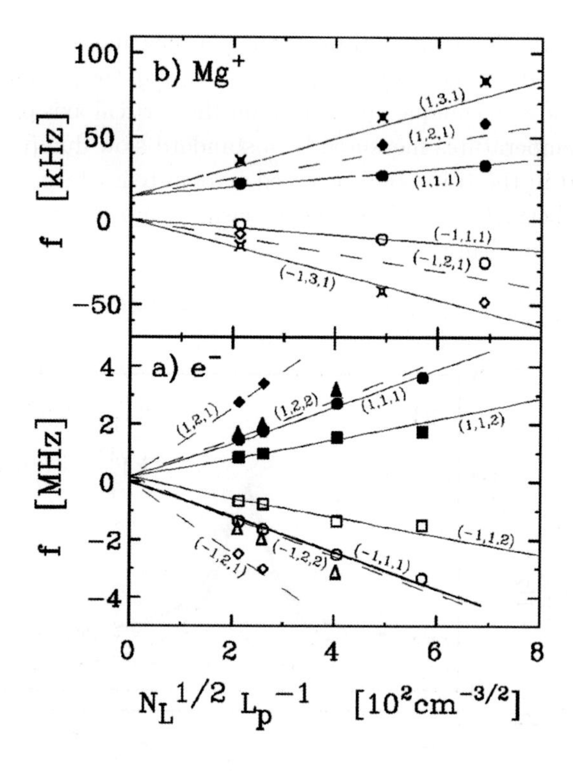

Fig. 6.13. Observed mode frequencies of $m_\theta = \pm 1$ as a function of plasma length L_p.

6.2.3. *Higher-order Trivelpiece–Gould modes*

So far we have discussed axisymmetric plasma waves. In this section, I will briefly show some experimental results of plasma modes with azimuthal dependance. Figure 6.2.3 shows the observed mode frequencies for azimuthal mode $m_\theta = \pm 1$ in a magnesium ion plasma and in a pure electron plasma. The modes are Doppler shifted according to Eq. (6.28) by the plasma rotation frequency. The shift is clearly visible on the Mg^+ data, but is hard to see on the electron data, since the electron mode frequency is larger by a factor $\sqrt{m_i/m_e}$, whereas the rotation frequencies are comparable. Figure 6.2.3 demonstrates that the Doppler shifted mode frequencies are proportional to $N_L^{1/2} L_p^{-1}$.

6.2.4. *Electron acoustic waves*

Electron acoustic waves (EAW)[18,19] are plasma waves with a slow phase velocity, typically $\omega \approx 1.4k\overline{v}$; in contrast, TG modes have $\omega \gtrsim 3k\overline{v}$. The name comes from the neutral plasma community: the ions are stationary, as for electron plasma waves, but the mobile electrons give an acoustic response. Analogous EAW modes are observed in pure electron and pure ion plasmas (where the name is somewhat misleading). The EAW is nonlinear so as to flatten the particle distribution to avoid strong Landau damping, but it can exist at small amplitude. Figure 6.14 shows the dispersion relation of plasma waves in an infinite size plasma where the dispersion relation has the shape of a thumb.

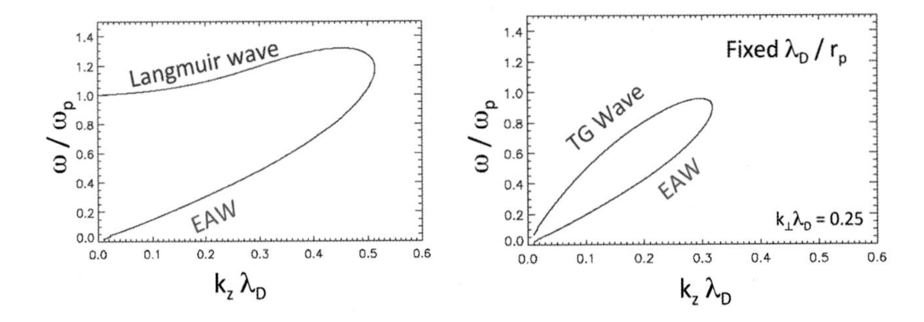

Fig. 6.14. Plasma wave dispersion relation in homogeneous infinite plasma and plasma of finite radial size.

In a trapped non-neutral plasma with a finite radial size R_p resulting in $k_\perp$ given by Eq. (6.24), the dispersion curve looks like a tear drop when λ_D/R_p is fixed. Experimentally it is easier to fix k_z and R_p, measuring the dispersion curve as a function of temperature T as shown in Figure 6.15. The squares correspond to TG waves and the dots to EAW. At small amplitude ($A_{\rm exc} = 50$ mV), no waves are observed for $T > 1.3$ eV corresponding to $R_p/\lambda_D < 2$. The TG wave is easily excited with bursts as short as 3–10 cycles; in contrast the EAW requires typically hundreds of cycles to be excited. However, at larger amplitude the waves are excited over a range of frequencies. The bar at $T = 0.8$ eV shows the range of frequencies over which a 100 cycle burst with $A_{\rm exc} = 300$ mV results in a wave at frequency $f = f_{\rm exc}$ ringing for hundreds of cycles. This means that at $T = 0.8$ eV, a wave can be excited at "any frequency" within the vertical extent of the grey bar. Similarly, waves at $T = 1.4$ eV are excited past the "end of the

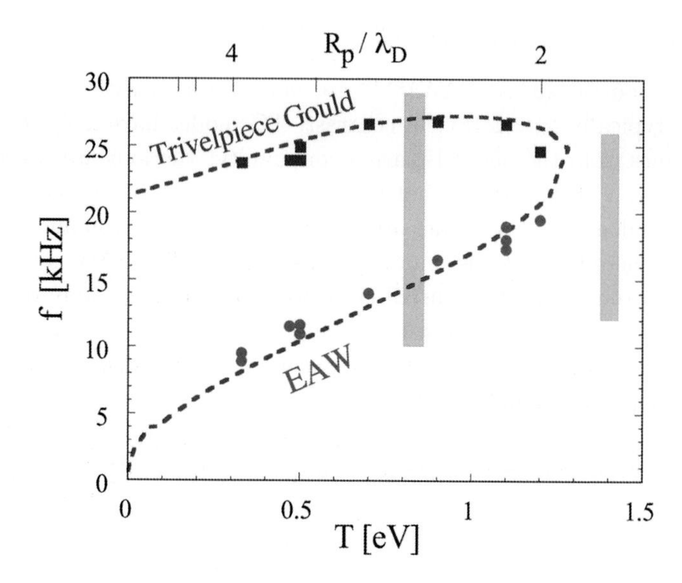

Fig. 6.15. Plasma wave dispersion relation for finite radial size plasma with fixed k_z plotted versus temperature.

thumb", where no near-linear solution exists. For these off-dispersion relation waves, the drive modifies the particle distribution until the distribution becomes resonant with the drive.

We note that similar modes may occur in laser–plasma interactions, in which case they are called KEEN (kinetic electrostatic electron nonlinear) waves.[20] The connections between EAWs and KEEN waves are currently being investigated.

6.3. Cyclotron Waves

A single particle of charge q and mass m in a magnetic field has a cyclotron frequency $f_c = qB/(2\pi m)$. In a plasma, the cyclotron mode frequency is shifted from f_c. For a single-species non-neutral plasma, cold fluid theory predicts that the cyclotron mode has frequencies:[21]

$$f = f_c + f_{E\times B}\left\{(m_\theta - 2) + \left[1 - \left(\frac{R_p}{R_w}\right)^{2m_\theta}\right]\right\}. \tag{6.31}$$

One can see that for $m_\theta = 1$ the cyclotron mode frequency is downshifted by one diocotron frequency. Figure 6.16 shows the measured cyclotron

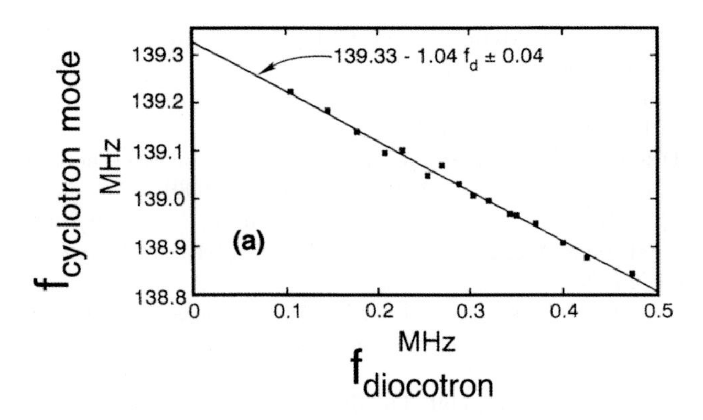

Fig. 6.16. Electron cyclotron mode frequency[21] plotted versus measured diocotron frequency. The solid line demonstrates the validity of Eq. (6.31) for single-species plasmas.

mode frequency of a pure electron plasma plotted against the diocotron frequency, demonstrating that for a single species plasma the lowest order cyclotron mode is downshifted by one diocotron frequency.

For multispecies non-neutral plasmas, cyclotron modes can be used to identify the composition of the plasma, but the exact frequency shifts are still a work in progress.

Acknowledgments

This work was supported by National Science Foundation Grant PHY-9093877 and Department of Energy Grant DE-SC0002451. The author thanks the organizers of the Les Houches Winter School for the opportunity to contribute to the workshop. The author also thanks Jo Ann Christina for typing and editing the manuscript, and Prof. C. F. Driscoll, Prof. D. H. E. Dubin and Prof. T. M. O'Neil for many years of enlightened guidance and theoretical support.

References

1. R. C. Davidson, *Physics of Nonneutral Plasmas* (Addison-Wesley, London, 1990).
2. J. S. deGrassie and J. H. Malmberg, *Phys. Fluids* **23**, 63–81 (1980).
3. T. M. O'Neil, *Phys. Today* **52**, 24–30 (1999).
4. J. S. deGrassie and J. H. Malmberg, *Phys. Rev. Lett.* **39**, 1077–1080 (1977).

5. W. D. White, J. H. Malmberg and C. F. Driscoll, Resistive-wall destabilization of diocotron waves, *Phys. Rev. Lett.* **49**, 1822–1825 (1982).
6. K. S. Fine, *Experiments with the* $l = 1$ *Diocotron Mode*, PhD Dissertation, UCSD Physics Department (1988).
7. C. F. Driscoll, Observation of an unstable $l = 1$ diocotron mode on a hollow electron column, *Phys. Rev. Lett.* **64**, 645–648 (1990).
8. J. R. Danielson, T. R. Weber and C. M. Surko, Next generation trap for positron storage, Non-Neutral Plasma Physics VII, *AIP Conf. Proc.* **1114**, 199–206 (2009).
9. K. S. Fine, C. F. Driscoll and J. H. Malmberg, Measurements of a nonlinear diocotron mode in pure electron plasmas, *Phys. Rev. Lett.* **63**, 2232–2235 (1989).
10. K. S. Fine and C. F. Driscoll, The finite length diocotron mode, *Phys. Plasmas* **5**, 601–607 (1998).
11. S. A. Prasad and T. M. O'Neil, Waves in a cold pure electron plasma of finite length, *Phys. Fluids* **26**, 665–672 (1983).
12. A. W. Trivelpiece and R. W. Gould, *J. Appl. Phys.* **30**, 1784–1793 (1959).
13. A. A. Kabantsev and C. F. Driscoll, "Correction to electron plasma mode frequency formula" (2006). Available at: nnp.ucsd.edu/pdf_files/correction_electron_freqs.PDF. (Accessed 14 June 2013)
14. F. Anderegg, E. M. Hollmann, and C. F. Driscoll, Rotating field confinement of pure electron plasmas using Trivelpiece–Gould modes, *Phys. Rev. Lett.* **78**, 4875–4878 (1998).
15. J. K. Jennings, R. L. Spencer and K. C. Hansen, Numerical calculation of axisymmetric electrostatic modes for cold finite-length non-neutral plasmas, *Phys. Plasmas* **2**, 2630–2639 (1995).
16. N. Shiga, F. Anderegg, D. H. E. Dubin, C. F. Driscoll and R. W. Gould, Thermally excited fluctuations as a pure electron plasma temperature diagnostic, *Phys. Plasmas* **13**, 022109:1–12 (2006).
17. F. Anderegg, N. Shiga, D. H. E. Dubin, C. F. Driscoll and R. Gould, Thermally excited Trivelpiece–Gould modes as a pure electron plasma temperature diagnostic, *Phys. Plasmas* **10**, 1556 (2003); F. Anderegg, N. Shiga, J. R. Danielson, D. H. E. Dubin, C. F. Driscoll and R. W. Gould, Thermally excited modes in a pure electron plasma, *Phys. Rev. Lett.* **90**, 115001:1–4 (2003).
18. J. P. Holloway and J. J. Dorning, Undamped plasma waves, *Phys. Rev. A* **44**, 3856–3868 (1991).
19. F. Valentini, T. M. O'Neil and D. H. E. Dubin, Excitation of nonlinear electron acoustic waves, *Phys. Plasmas* **13**, 052303:1–7 (2006).
20. B. Afeyan, K. Won, V. Savchenko, T. W. Johnston, A. Ghizzon and P. Bertrand, Kinetic electrostatic electron nonlinear (KEEN) waves and their interactions driven by the ponderomotive force of crossing laser beams. in Hammel, B. A., Meyerhofer, D. D., Meyer-ter-Vehn, J. and Azechi, H. (eds), *Proc. of the Inertial Fusion Sciences and Applications 2003*, pp. 213–217 (American Nuclear Society, Monterey, 2004).

21. R. W. Gould and M. A. LaPointe, Cyclotron resonance in a pure electron plasma, *Phys. Rev. Lett.* **67**, 3685–3688 (1991); R. W. Gould and M. A. LaPointe, Cyclotron resonance phenomena in a pure electron plasma, *Phys. Fluids B* **4**, 2038–2043 (1992); R. W. Gould, Theory of cyclotron resonance in a cylindrical nonneutral plasma, *Phys Plasmas* **2**, 1404–1411 (1995).

Chapter 7

Internal Transport in Non-neutral Plasma

François Anderegg
University of California, San Diego, Physics Dept. 0319, La Jolla CA 92093, USA
fanderegg@ucsd.edu

This chapter discusses transport processes across the magnetic field in a Penning–Malmberg trap geometry. The first section of the chapter explores test particle transport in elongated plasmas, shaped like "cigars" (three-dimensional system), and in short plasmas, shaped like "pancakes" (two-dimensional system). The second section covers heat transport and the final section discusses the transport of angular momentum (viscosity).

7.1. Types of Collisions

This chapter addresses one of the fundamental problems of the magnetic confinement of plasmas, namely how collisions affect particles, heat and angular momentum transport across the magnetic field. In the early days of plasma physics, transport was described in terms of velocity scattering collisions. These early efforts have come to be known as the "classical" theory of transport. Experiments, however, observe collisional transport that is much larger than classical theory predicts. More recent theories using long-range $E \times B$ drift collisions are in precise agreement with these experiments.

In this chapter, I will consider non-neutral plasmas in the usual regime, i.e. away from the maximum density limit known as the "Brillouin limit". More precisely, we will limit our discussion to plasmas satisfying $\lambda_D > r_c$ where the Debye length $\lambda_D \equiv (T/4\pi n e^2)^{1/2}$ and the cyclotron radius $r_c \equiv \mathrm{v}_\perp c/eB$; here we are using cgs units, and the temperature T as the unit of energy. The case of interest $\lambda_D > r_c$ corresponds to low density plasma in a large magnetic field.

7.1.1. *Classical velocity scattering collisions*

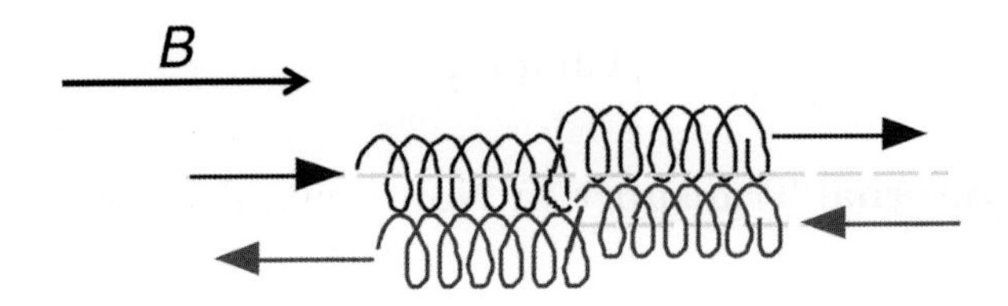

Fig. 7.1. Velocity scattering collision. The impact parameter $\rho < r_c$ is shown by the spacing of the two dashed lines.

Figure 7.1 illustrates a velocity scattering collision; the "blue" particle (coming from the left) collides with the "red" particle (coming from the right). The (perpendicular) impact parameter ρ is small enough for the particles to scatter their velocity vectors. This type of collision occurs for the following range of impact parameter $b < \rho < r_c$, where $b \equiv e^2/T$ is the distance of closest approach for two charged particles. After the collision, the distance separating the particle guiding centers has increased. We will estimate and discuss this increase in Section 7.2. These velocity scattering collisions are referred to as "classical" collisions, in the sense that they have been studied for a long time.

7.1.2. *Long-range $E \times B$ drift collisions*

An unusual type of long-range collisions dominates over classical collisions in plasmas where $\lambda_D > r_c$. Figure 7.2 illustrates a "blue" particle coming from the left and a "red" particle coming from the right. Since the impact parameter $\rho < \lambda_D$, the particles feel each other's electric field and $E \times B$ drift around each other. As above, careful evaluation of the collision will show that the distance separating their guiding center once again increases after the collision. It is worth noting that the collision left the velocity vector of each particle almost unchanged.

7.2. Test Particle Transport

In Subsections 7.2.1, 7.2.2 and 7.2.3 we will discuss three-dimensional (3D) systems and in Subsection 7.2.4 we will consider two-dimensional (2D) systems. In 2D systems, particles bounce rapidly along the plasma column

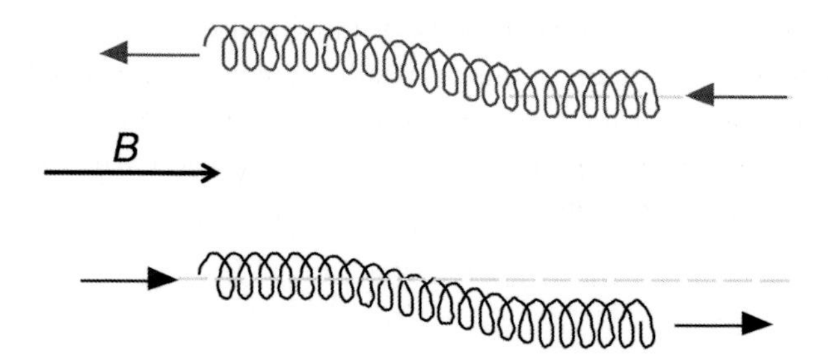

Fig. 7.2. Long-range $E \times B$ drift collisions; the impact $\rho < \lambda_D$ is shown by the spacing of the two dashed lines.

compared to their $E \times B$ rotation rates. This rapid bouncing effectively averages the cross-field transport along the magnetic field axis.

7.2.1. *Classical diffusion*

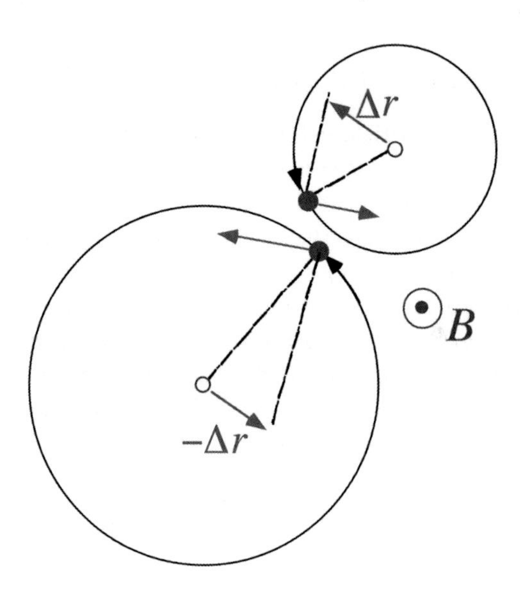

Fig. 7.3. Schematic of particle collisions in a plane perpendicular to the magnetic field B.

Figure 7.3 shows two particles colliding, and scattering their velocity vector $v_\perp$. The two new $v_\perp$ result in new positions of the guiding center of each particle. The displacement of the guiding center after the collision Δr is of the order of a cyclotron radius r_c, resulting in a diffusion coefficient $D \sim \nu r_c^2$, where ν is the collision rate of such collisions. The proper mathematical treatment was performed first by Longmire and Rosenbluth[1] and gives the classical diffusion coefficient:

$$D^{\text{clas}} = \frac{5}{4}\nu_{ii} r_c^2 , \tag{7.1}$$

where

$$\nu_{ii} = \frac{16}{15}\sqrt{\pi} n \overline{\mathrm{v}} b^2 \ln\left(\frac{r_c}{b}\right) , \tag{7.2}$$

and n is the particle density. Here the indice ii refers to ion–ion collisions, but it is worth noting that the same equation is valid for any type of "same particle" collision. For example, electron–electron collisions are simply faster than ion–ion collisions by a factor $(m_i/m_e)^{1/2}$ which comes from the particle thermal speed $\overline{\mathrm{v}} = (T/m)^{1/2}$. Physically, ν_{ii} is the rate at which a thermal particle loses half of its momentum to other particles.

7.2.2. *Long-range* $\boldsymbol{E \times B}$ *drift diffusion*

The electric field of a particle extends up to a Debye length λ_D before it is shielded by the other particles; therefore two particles within a Debye length "feel" each other's electric field and $E \times B$ drift at a velocity $\mathrm{v}_{E\times B} = (E \times B)/B^2 = -E\hat{r}/B$ where the electric field magnitude is of the order $E \sim q/\lambda_D^2$ and consequently $\mathrm{v}_{E\times B} \sim q/(\lambda_D^2 B)$. These two particles drift around for a time $\tau_d \sim \lambda_D/\overline{\mathrm{v}}$ resulting in a step size across the magnetic field $\Delta r = \mathrm{v}_{E\times B} \cdot t_d \sim q/(\lambda_D B \overline{\mathrm{v}})$, and these collisions occur at a rate $\sim n\overline{\mathrm{v}}\lambda_D^2$. Therefore the resulting diffusion coefficient is $D \sim n\overline{\mathrm{v}}\lambda_D^2 \cdot \Delta r^2 = n\overline{\mathrm{v}} b^2 r_c^2$; the step size is approximately the same as for velocity scattering collisions.

The proper mathematical analysis was performed by Dubin[2] and gives:

$$D^{E\times B} = 6\sqrt{\pi}\nu_c r_c^2 \ln\left(\frac{\lambda_D}{r_c}\right) \ln\left(\frac{\overline{\mathrm{v}}}{\Delta \mathrm{v}}\right) , \tag{7.3}$$

with

$$\nu_c = n\overline{\mathrm{v}} b^2. \tag{7.4}$$

Particles with small relative axial velocities $\Delta \mathrm{v}$ produce the largest amount of transport. The smallest $\Delta \mathrm{v}$ (longest interaction times) of relevance are limited by shear in the perpendicular drift velocity, or by the diffusion itself giving $\Delta v_m = (D_v\sqrt{\lambda_D r_c})^{1/3}$. It is worth noting that Eq. (7.3) is often written using ν_{ii}:

$$D^{E\times B} = \frac{45}{8}\nu_{ii} r_c^2 \frac{\ln(\lambda_D/r_c)}{\ln(r_c/b)} \ln\left(\frac{\bar{v}}{\Delta v_m}\right), \tag{7.5}$$

allowing easy and direct comparison with Eq. (7.1).

7.2.3. *Experimental measurements of test particle transport*

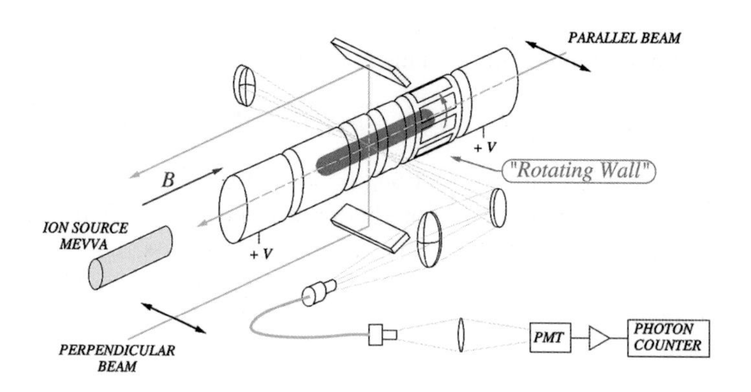

Fig. 7.4. Schematic of a Penning–Malmberg trap.

To test the theory results presented above, we conducted experiments in a Penning–Malmberg trap containing a pure magnesium ion plasma.[3] Figure 7.4 shows the cylindrical trap and confined Mg^+ ion plasma, with parallel and perpendicular laser beams. The electrodes are of radius $R_w = 2.86$cm and are contained in a vacuum chamber with $P \sim 10^{-9}$T. Magnesium ions are obtained from a brief discharge of a metal vapor vacuum arc source (MEVVA).[4] The ions are trapped by positive voltages applied to the end cylinder, whereas free electrons are ejected axially.

Basic radial confinement is provided by a uniform axial magnetic field $0.8 < B < 4$ tesla, which by itself would result in plasma loss time from neutral collisions and static field asymmetries of $\tau_L \lesssim 2000$s. The loss time is made essentially "infinite" (that is, larger than two weeks) by applying weak "rotating-wall" potentials ($\leq 1.\,V_{pp}$) to eight azimuthal wall sectors. This rotating field, varying as $e^{im_\theta\theta - i\omega t}$ with $m_\theta = 1$, adds angular momentum and energy to the plasma, balancing the background drag and energy losses.

The plasma density, temperature and rotating velocity profiles are obtained from laser-induced fluorescence measurements. The plasma has an axial length of $L_p \simeq 10$ cm, and individual ions bounce axially at a rate of $f_b = \overline{v}/2L_p \approx 3$ kHz for $T = 0.1$ eV. The ion plasma tends to cool to about 0.05 eV due to collisions with neutrals. We can increase the ion temperature either by ion cyclotron heating or by compressional heating applied to an end cylinder.

Since both transport rely on collisions, we will first measure the collision rate responsible for velocity space isotropization. We induce anisotropies between $T_\parallel$ and $T_\perp$, and then measure the relaxation of $T_\parallel(t)$ and $T_\perp(t)$ to a common final temperature T. This is modeled as

$$\frac{d}{dt}(T_\perp - T) = -\nu_{\perp 0}(T_\perp - T)\,, \tag{7.6}$$

where the relaxation rate $\nu_{\perp 0} = 1.5\nu_{ii} = 3\nu_{\perp\parallel}$. Figure 7.5 shows the measured $\nu_{\perp 0}$ and the Fokker–Planck prediction with no adjustable parameters. Thus there are no anomalous velocity scattering collisions.

The ion spin orientation is used to "tag" the test particles. The ground state of Mg^+ is a $3^2S_{1/2}$ state. In a magnetic field, the orientation can be

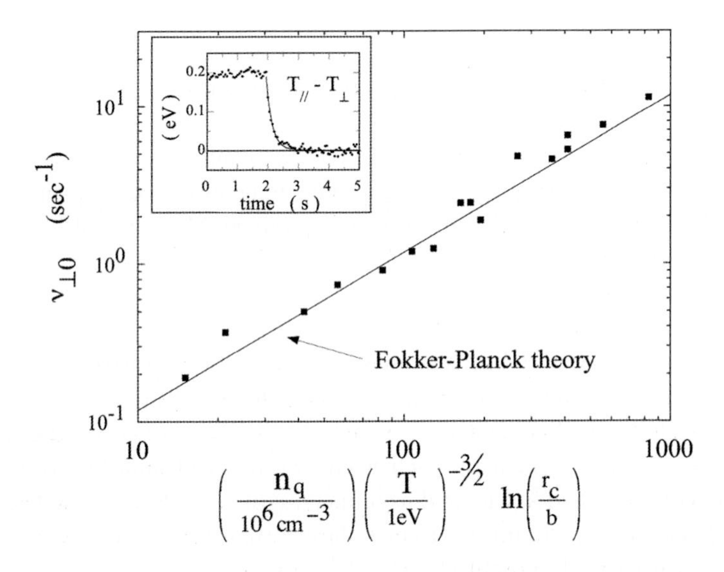

Fig. 7.5. Measured velocity space collision rate $\nu_{\perp 0}$. The insert shows a $T_\parallel - T_\perp$ measurement, with anisotropy maintained until $t = 2$ seconds, then exponential relaxation to $T_\parallel = T_\perp$.

$M_j = +1/2$ or -1/2; and the difference of energy between these two states is small (4.6×10^{-4}eV at 4 tesla). Nevertheless, measurements indicate that the ion spin state remains unchanged for much longer than the time required for the tagged ions to diffuse radially. Also, the transport processes are identical for ions with spin $M_j = +1/2$ or -1/2. The transport measurement is a three-step process,[5] as outlined in Figure 7.6.

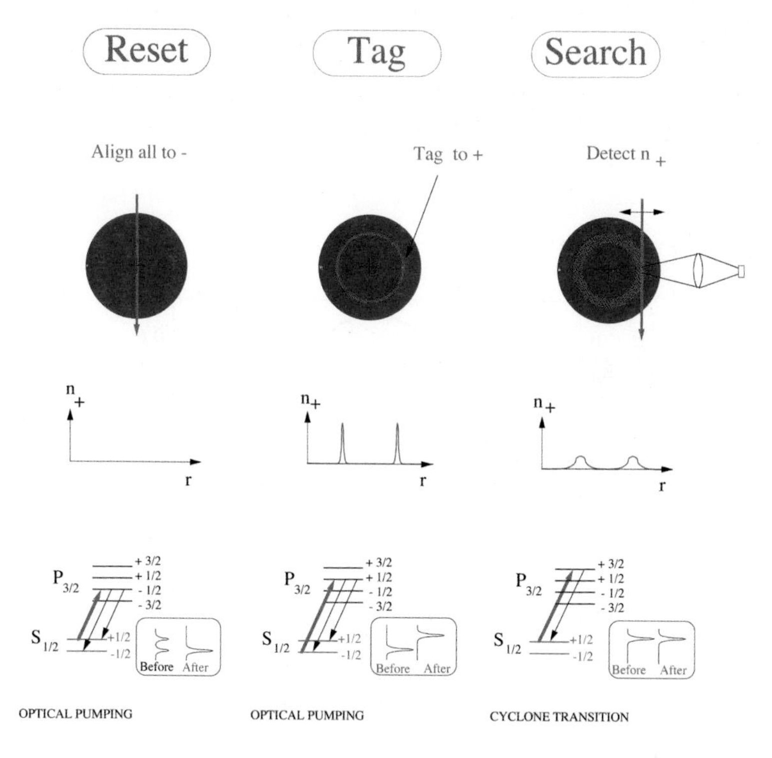

Fig. 7.6. The three steps used to measure test particle transport and relevant energy levels of magnesium singly ionized.

(a) "Reset." The plasma is completely (> 98%) polarized into the $S_{1/2}^{M_j=-1/2}$ state by direct optical pumping with an axial beam tuned to $S_{1/2}^{M_j=+1/2} \rightarrow P_{3/2}^{M_j=-1/2}$ transition. This beam is perpendicular to the magnetic field and passes through the center of the (drift-rotating) plasma.

(b) "Tag." The spin of particles at a chosen radial position r_t is reversed by direct optical pumping with an axial beam tuned to $S_{1/2}^{M_j=-1/2} \rightarrow$

$P_{3/2}^{M_j=+1/2}$. The tagging beam is left on for many rotations of the plasma and also for many bounce periods to ensure that most ions at the beam radius are pumped; typically 50 ms are required to locally tag more than 80% of ions into $S_{1/2}^{M_j=+1/2}$ state.

(c) "Search." The density evolution $n_t(r_s, t)$ of tagged particles at a chosen search position r_s is measured non-destructively with a perpendicular beam tuned to the peak of the "cyclone" transition $S_{1/2}^{M_j=+1/2} \rightarrow P_{3/2}^{M_j=+3/2}$ which decays only to $S_{1/2}^{M_j=+1/2}$. This search beam is weak and is on less than 10% of the time, in order to minimize "sideband" excitation of other transitions.

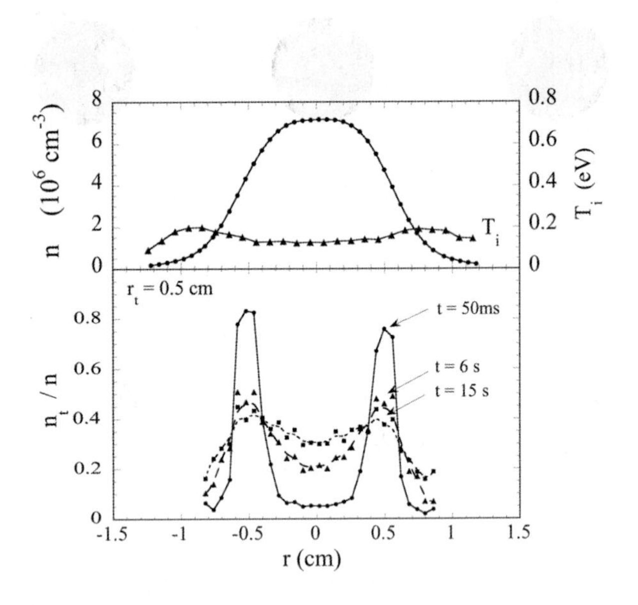

Fig. 7.7. Measured density $n(r)$ and temperature $T(r)$. Measured test particle density $n_t(r, t)/n(r)$ at three times showing radial diffusion towards n_t/n = constant.

Each "reset–tag–search" cycle determines $n_t(r_s, t)$ for $0.05 < t < 100$ s, and repeating this cycle for 31 different search positions gives the test particle evolution $n_t(r_s, t)$. Figure 7.7 shows the normalized test particle density $n_t(r)/n(r)$ for three times where the ions were tagged at $r_t = 0.5$cm. The test particles are transported radially toward a constant density n_t/n. The test particle flux Γ_t is obtained from these data as

$$\Gamma_t(r,t) = -\frac{1}{r}\int_0^r dx\, x \frac{\partial}{\partial t} n_t(x,t) + \int_0^r dx\, x \frac{2n_t(r,t) - n(x)}{\tau_s(x)}. \qquad (7.7)$$

The second term corrects for the slow spontaneous spin flip at a separately measured rate $\tau_s^{-1}(T) \sim 10^{-3}\ \text{sec}^{-1}$ to $10^{-2}\ \text{sec}^{-1}$. With this correction, the total number of test particles is conserved to within 10%. This flux is then compared to a model consisting of a local diffusion coefficient $D(r)$ and a convective velocity $V(r)$:

$$\Gamma_t(r,t) = -D(r)n(r)\frac{\partial}{\partial r}\frac{n_t(r,t)}{n(r)} + V(r)n_t(r,t). \qquad (7.8)$$

Figure 7.8 shows the measured normalized flux $\Gamma_t(t)/n_t(t)$ versus the measured normalized gradient $(n/n_t(t))(\partial/\partial r)(n_t(t)/n)$ for one radius $r_s = 0.318$ cm. The data points represent different times in the evolution. The linear fit to the data gives $D(r_s) = 3.3 \times 10^{-3}\,\text{cm}^2/\text{s}$ and $V(r_s) \simeq 0$ to experimental accuracy. The normalized flux is proportional to the normalized gradient over a range of 20, in accordance with Fick's law. Here, the diffusion coefficient measured for various densities, magnetic fields and temperatures is shown in Figure 7.9. The diffusion coefficients are normalized by $nB^{-2}\ln(\lambda_D/r_c)$ for comparison with classical and long-range $E \times B$ drift collisional theories. The data are about ten times larger than predicted by

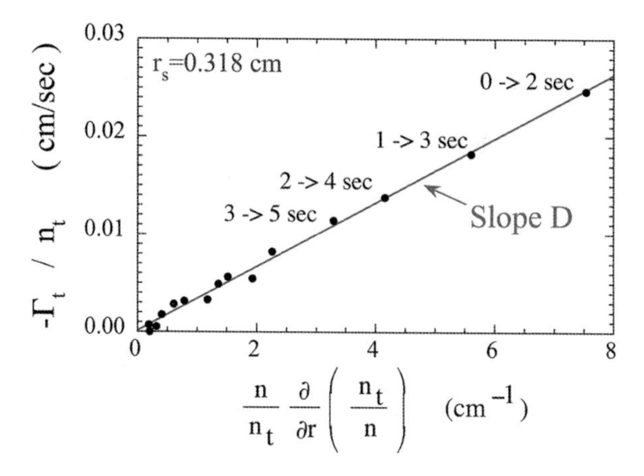

Fig. 7.8. Measured normalized test particle flux Γ_t/n_t versus normalized test particle gradient showing diffusive transport.

classical theory. $D^{\rm clas}$ is shown by solid black lines. Overall, Eq. (7.3) describing long-range $E \times B$ drift collisions is in quantitative agreement with the data.

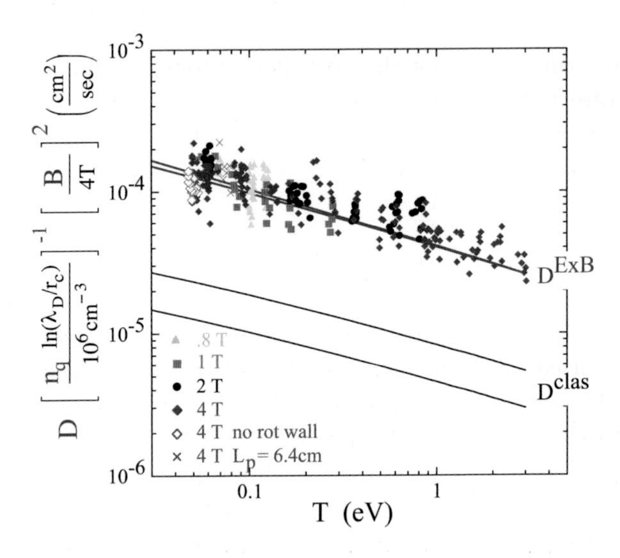

Fig. 7.9. Measured test particle diffusion coefficient D normalized to density $nB^{-2}\ln(\lambda_D/r_c)$ versus temperature. The black theory curve shows $D^{\rm clas}$ and the blue line shows the $E \times B$ drift collision.

7.2.4. *Test particle transport in 2D systems*

This diffusion due to long-range collisions is enhanced in the 2D regime, where individual particles in the finite length plasma bounce repeatedly in z before moving substantially in (r, θ). This suggests that pairs of particles may experience multiple correlated collisions, giving an enhanced transport step size Δr. This enhancement increases with N_b, the number of times a thermal particle bounces axially before shear in the θ rotation of the column separates the particle from its neighbors. Thus, rotational shear reduces the 2D transport enhancement.[6]

7.2.4.1. *Shear-free case*

First, let us consider a short plasma where the rotation is shear free (i.e., rotating like a rigid object). From the 2D perspective, each ion is a z-

average "rod" of charge which $E \times B$ drifts in (r, θ) due to the fields of all the other ions. The $E \times B$ drift dynamics of a collection of N charged rods is isomorphic to a 2D gas of N point vortices: each rod, with charge $q_L \equiv q/L_p$ per unit length, is equivalent to a point vortex with circulation

$$\gamma \equiv q_L(4\pi c/B) = 4\pi cq/(BL_p). \tag{7.9}$$

(Here, we take $\gamma > 0$ and $\omega_E > 0$ for positive ions by choosing $\mathbf{B} = -B\hat{z}$; for electrons, choosing $B = +B\hat{z}$ would similarly give $\gamma > 0$.)

Early work on the diffusion of 2D point vortices focused on the case of a quiescent, homogeneous shear-free gas.[7,8] When the vortices are distributed randomly, representing high-temperature thermal fluctuations, Taylor and McNamara predicted that the diffusion coefficient (for diffusion in one direction) has the following simple form:

$$D_{\mathrm{TM}} = \frac{\gamma}{8\pi}\sqrt{N}. \tag{7.10}$$

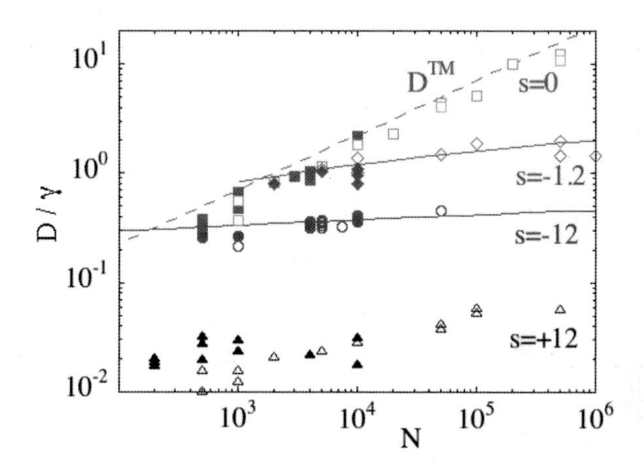

Fig. 7.10. Diffusion coefficient D from numerical simulations versus number of particles N, for shear rate s = 0, -1.2, -12., +12. Solid symbols are from MD simulations; open symbols are from PIC simulations. The dashed line is Eq. (7.10); solid lines are from Eq. (7.13).

The squares of Figure 7.10 display the diffusion rate obtained from more recent simulations by Dubin[9,10] with varying N compared to the predictions of Taylor–McNamara theory. The 2D diffusion coefficient in the absence of shear increases as the square root of the number of particles in the system.

7.2.4.2. *Effect of shear on 2D test particle transport*

The shear in the plasma rotation is defined as

$$S(r) = r\frac{\partial}{\partial r}\omega_E(r) \equiv r\omega'_E(r). \tag{7.11}$$

Note that two particles separated radially by a distance ρ will become separated azimuthally by a distance ρ in a time S^{-1}. The dimensionless scaled shear s is then defined as

$$s(r) \equiv \frac{r\omega'_E}{2\pi ceBn(r)} \sim \frac{S(r)}{\omega_E}, \tag{7.12}$$

where the last approximation is only valid when the rotation results from near-uniform $n(r)$.

Experimentally, the magnitude of the rotational shear can be controlled by varying the ion density and temperature, or by adjusting the character of the torques and drags on the column. This shear in the θ rotation has only a weak effect on diffusion due to long-range collisions in the 3D regime; but it is the controlling parameter in defining the 2D regime.

A rigorous Boltzmann calculation gives the radial diffusion coefficient D_{B} due to these small impact parameter collisions; and a quasilinear calculation based on the Kubo formula gives the diffusion coefficient D_{K} from multiple distant collisions. The total diffusion[9] is then

$$\begin{aligned} D_{\mathrm{2D}} = D_{\mathrm{B}} + D_{\mathrm{K}} &= \frac{\gamma}{2s}\frac{1}{\pi^2}\ln^2\Lambda_{\mathrm{B}} + \frac{\gamma}{2s}\ln(r/\rho_{\mathrm{min}}) \\ &= \frac{2\pi qc}{L_p Bs}\left[\frac{\ln^2\Lambda_{\mathrm{B}}}{\pi^2} + \ln(r/\rho_{\mathrm{min}})\right], \end{aligned} \tag{7.13}$$

with collision logarithms given by $\Lambda_{\mathrm{B}} \equiv 2.713\pi^2 S^2/n\gamma^2$ and $\rho_{\mathrm{min}} \approx (4D_k/S)^{1/2}$ for the regimes of interest here.

Of course, this estimate is valid only when it is less than the zero-shear result, i.e. when $D_{\mathrm{2D}} < D_{\mathrm{TM}}$. However, one can see that even a very small shear, of order $S/n\gamma \approx N^{-1/2}$, is required to make the zero-shear D_{TM} inapplicable. In other words, even small shears wipe out the large-scale Dawson–Okuda vortices required to give D_{TM}.

Dubin tested this theory using numerical simulations[9] of N identical point vortices, initially placed randomly inside a circular patch, with an applied uniform external shear rate S. To obtain the diffusion coefficient, as test particles he chose all vortices in the band of radii from $0.43R$ to $0.57R$, and followed their mean square change in radial position, $\langle\delta r^2(t)\rangle$. The diffusion coefficient was found from fitting $\langle\delta r^2(t)\rangle = 2Dt$.

Figure 7.10 displays the diffusion rates obtained from the simulations with varying N compared to the predictions of theory. The simulations with $s = 0$ agree closely with D_{TM} (the dashed line). Simulations with negative imposed shears of $s = -1.2$ and -12 show reduced diffusion, in quantitative agreement with D_{2D} (the solid lines).

However, the simulations with $s = +12$ show about ten times less diffusion, apparently due to trapping effects in collisions between vortices which are prograde with respect to the shear. In essence, the trapping size l is *infinite* for $s > 0$, and the effects of bound vortex pairs cannot be ignored; thus, the Boltzmann analysis is inapplicable. Furthermore, the method of integration along unperturbed orbits, which is essential to the Kubo analysis, also fails. Analysis of these trapped particle effects is a significant unsolved theory problem.

The experimental setup to measure 2D test particle transport is similar to the 2D case except that the plasma is short, so the ions bounce axially more rapidly than they drift-rotate. Figure 7.11 shows measured density, temperature velocities and rotational shear profiles. The shear that matters is the $E \times B$ velocity shear. The laser measures the particle velocity v_{tot} (also known as fluid velocity), and the diamagnetic drift v_{dia} is calculated from the measured density and temperature profile as $\text{v}_{\text{dia}}(r) = \nabla(n(r) \cdot T(r))$. The velocity of the guiding center is then $\text{v}_{E\times B} = \text{v}_{\text{tot}} - \text{v}_{\text{dia}}$. The shear $S(r)$ is calculated as $S(r) = r\partial\omega_{E\times B}/\partial r$, and the normalized shear $s(r) = S(r)/\omega_{E\times B}$.

Figure 7.12 shows the measured diffusion coefficient enhancement D/D_{3D} versus N_b. For $N_b < 1$, the dashed curve represents D_{3D} and the data points of Figure 7.9 would overlay the dashed line. For $N_b > 1$, the diffusion increases proportional to N_b, approaching the shear-free Taylor–McNamara limit as $N \to 10^4$. The theory line represents $D^{\text{2D}}_{\text{theory}} \simeq 0.5 D_{\text{3D}} f_b/|S| = 0.5 D_{\text{3D}} N_b$.

Figure 7.13 shows the same experimental data as Figure 7.12, but with the diffusion scaled as DL_pB and plotted versus normalized shear s. The diffusion approaches D_{TM} for $|s| < 10^{-3}$, decreases as $|s|^{-1}$ for $10^{-2} < |s| < 10^{-1}$ and equals D_{3D} for $|s| > 1$. Here the column lengths ranged over $0.7 < L_p < 10$cm and the temperature ranged over $0.05 < T < 3$ eV, with fixed $B = 3$ T.

It is worth noting that not all transport is diffusive, and that in the case of very small shear, a significant part of the transport may be convective, with convective cells as large as the system.

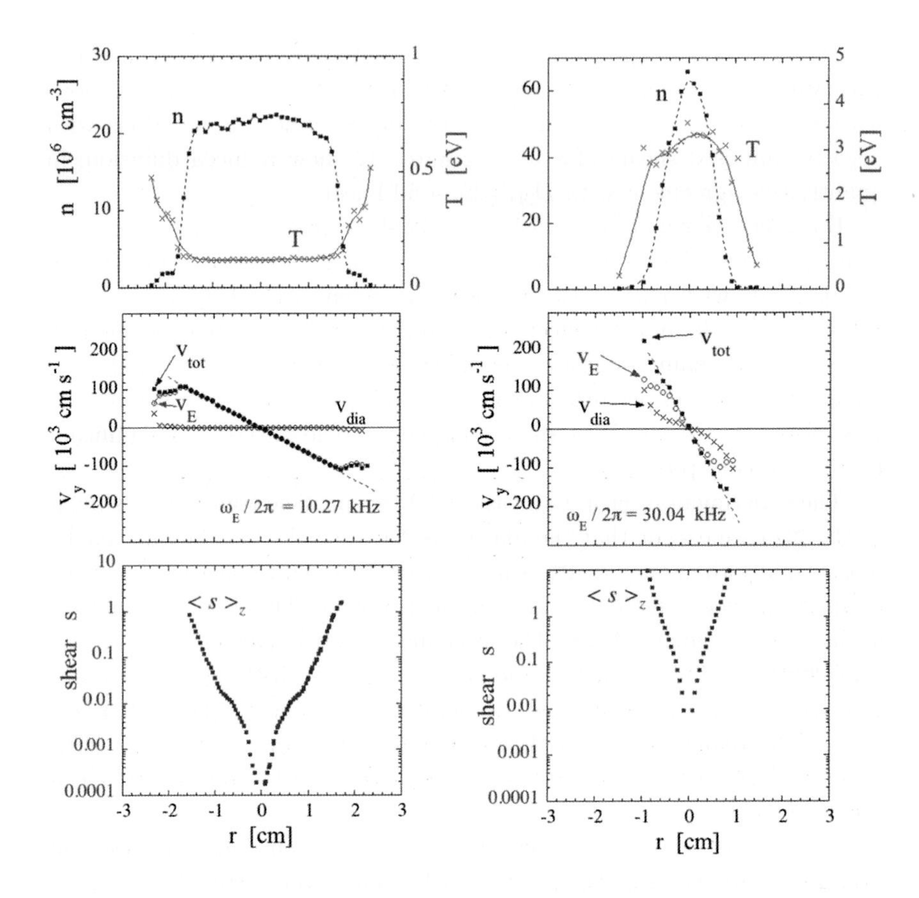

Fig. 7.11. Measured density, temperature, velocities and normalized shear $\langle s \rangle$ for the low shear case (left) and large shear case (right).

7.3. Heat Transport

Here we consider "collisional" heat transport across the magnetic field. The term "collisional" refers to transport driven by fluctuating fields from thermal motions of individual particles in our quiescent plasmas; in contrast, "turbulent" transport would be driven by nonthermal fluctuations such as unstable waves or broadband turbulence.

Let us consider the cross-field energy flux presumed to be of the form:

$$\Gamma_{\text{heat}} = -K \frac{\partial T}{\partial r}. \tag{7.14}$$

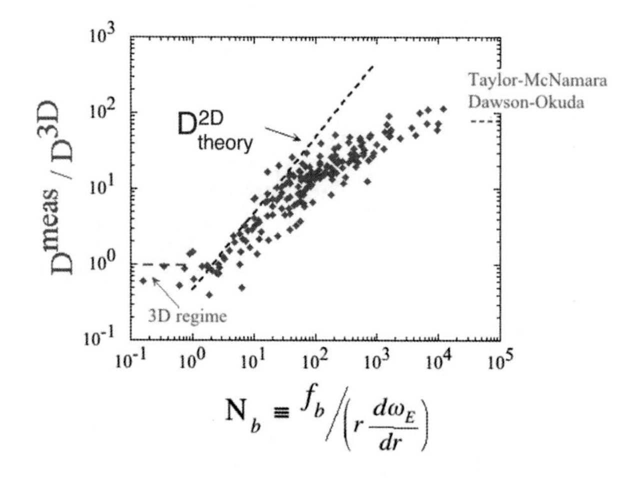

Fig. 7.12. Measured diffusion coefficient normalized by the 3D prediction plotted versus number of particle bounces N_b, showing enhanced diffusion in the 2D bounce-average regime.

The heat conductivity K is related to the thermal diffusivity χ by $K = \frac{5}{2}n\chi$, where the factor $\frac{5}{2}n$ corresponds to the specific heat at constant pressure. The thermal diffusivity can be thought of as a diffusion coefficient for energy, and it has the same units as the test particle diffusion coefficient (area/time).

7.3.1. *Classical heat transport*

In the classical (velocity scattering) collision description, heat and particles exhibit similar diffusion, $\chi^{\text{clas}} \propto D^{\text{clas}} \propto \nu r_c^2$; the proper mathematical treatment[11] gives

$$\chi^{\text{clas}} = \frac{16\sqrt{\pi}}{15}\nu_c r_c^2 \ln\left(\frac{r_c}{b}\right) = \nu_{ii} r_c^2. \tag{7.15}$$

Note that the classical thermal diffusivity scales as $\chi^{\text{clas}} \propto nT^{-1/2}B^{-2}$.

7.3.2. *Long-range E × B drift heat transport*

In contrast, long-range interactions can exchange (parallel) energy between particles separated by up to λ_D, suggesting that $\chi^{\text{LR}} \propto \nu_c\lambda_D^2$. The proper mathematical treatment[12] gives:

$$\chi^{\text{LR}} = 0.49\nu_c\lambda_D^2 \propto T^{-1/2}B^0. \tag{7.16}$$

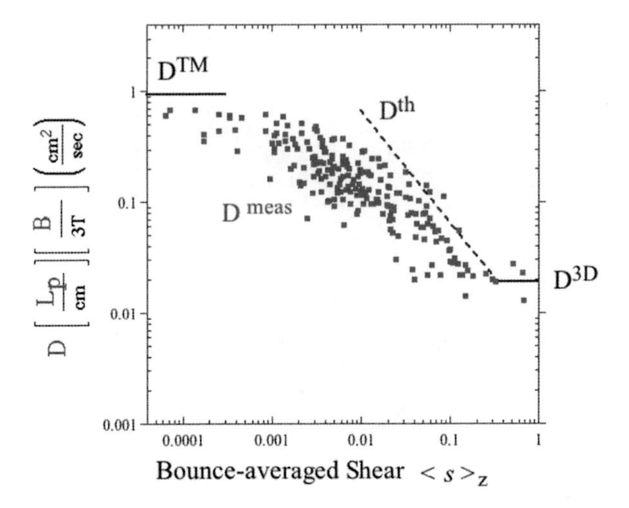

Fig. 7.13. Measured diffusion versus measured shear, showing shear-reduction of diffusion from the zero-shear Taylor–McNamara limit to the 3D regime. In the 2D regime, the diffusion scales as L_p^{-1} and B^{-1}.

It is important to note that the long-range thermal diffusivity is independent of the magnetic field, contrasting sharply with the "standard wisdom of plasma physics", namely that large magnetic fields produce better energy confinement.

7.3.3. *Experimental measurements of cross-magnetic field heat transport*

We measure heat transport across the magnetic field in the same Penning–Malmberg trap containing a pure magnesium ion plasma,[13,14] as shown in Figure 7.4. A laser beam parallel to the magnetic field cools (or heats) the plasma locally, creating a radial temperature gradient. A weak probe beam then measures the temperature profile $T(r, t)$. Figure 7.14 shows the steady-state density profile, and the temperature profile at $t = 0$ shows that the plasma was cooled on axis by the cooling beam. The cooling laser is then turned off and the temperature profile relaxes toward a uniform temperature within a time scale of a few seconds. Both the rate of change of temperature and its gradient are measured. The radial (cross-magnetic

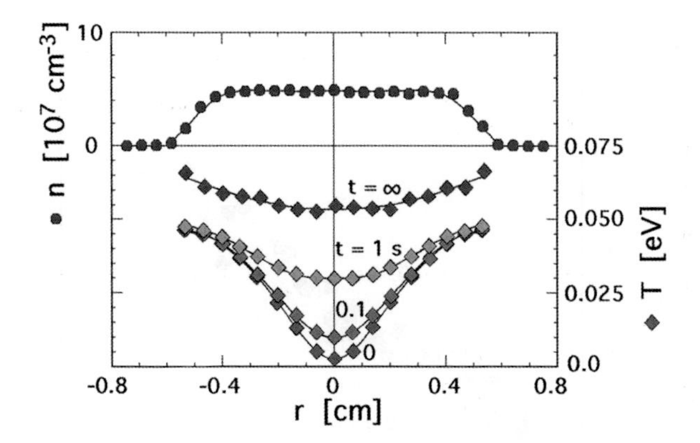

Fig. 7.14. Measured density profile and temperature profile starting from locally ($r = 0$) cooled initial conditions.

field) heat flux is calculated using measured quantities, as

$$\Gamma_q = -\frac{1}{r}\int_0^r r'dr'\left(\frac{3}{2}n\frac{\partial}{\partial t}T - \frac{\partial}{\partial t}q_{\text{ext}}\right). \tag{7.17}$$

Here, q_{ext} is the measured external heat input after the cooling laser is turned off, mainly due to collisions with residual background gas at room temperature; $\frac{\partial}{\partial t}q_{\text{ext}}$ is typically a small correction to the data. Figure 7.15 shows that the measured heat flux is proportional to the temperature gradient, as indicated by the unconstrained linear fit (the dashed line). The small non-zero intercept could represent a non-diffusive flux; but here it seems to be insignificant, arising from uncertainties in the data or imperfect corrections $\frac{\partial}{\partial t}q_{\text{ext}}$. Thus, Figure 7.15 demonstrates diffusive heat transport.

We obtain the thermal diffusivity χ from Figure 7.15 using

$$\chi(n, B, T) = -\frac{2}{5n}\frac{\Gamma_q}{\nabla T}. \tag{7.18}$$

Values of $\chi(n, B, T)$ were obtained for different equilibrium plasmas covering a range of 50 in density, 1000 in temperature, and 4 in magnetic field. Figure 7.16 shows χ versus T, with a single averaged value of χ for each evolution such as Figure 7.15. The averaging has little consequence since the range of n, T and χ in a single evolution is small. The dashed curve in Figure 7.16 shows the predicted classical thermal diffusivities χ^{clas} (Eq. (7.15)) for various densities and magnetic fields used. The solid line shows the predicted long-range thermal diffusivity χ^{LR} (Eq. (7.16)) which

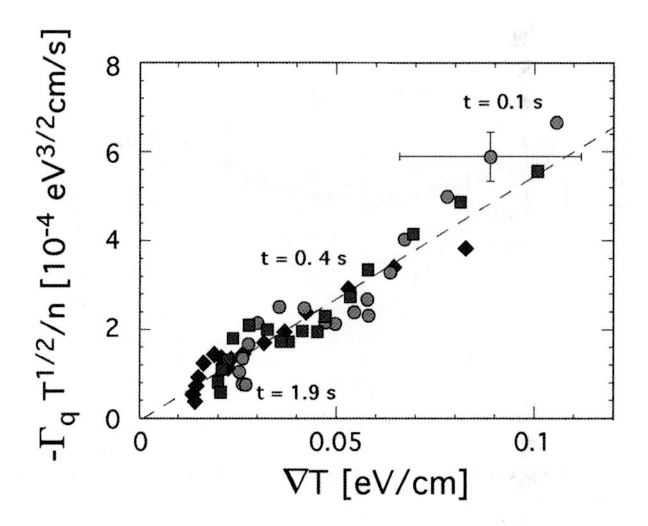

Fig. 7.15. Measured normalized radial heat flux versus temperature gradient, demonstrating diffusive heat transport.

depends only on temperature. The measured thermal diffusivities are up to 100 times larger than the classical prediction and are independent of B and n.

This long-range heat transport should occur in neutral plasmas also, predominantly in regimes of low density and strong magnetic field.

7.4. Transport of Angular Momentum

This section describes how a plasma confined in a Penning–Malmberg trap relaxes toward a state of thermal equilibrium; in particular, the rate at which a non-uniform density and non-uniform rotation profile relax toward a uniform density and uniform rotation. Viscosity acting on the rotational shear causes transport of particles (some inward, some outward) with a consequent relaxation toward a uniform rotation profile. The shear in the rotation creates an azimuthal drag force

$$F_\theta = \frac{1}{nr^2}\frac{\partial}{\partial r} r^3 \eta \frac{\partial \omega_r}{\partial r}, \tag{7.19}$$

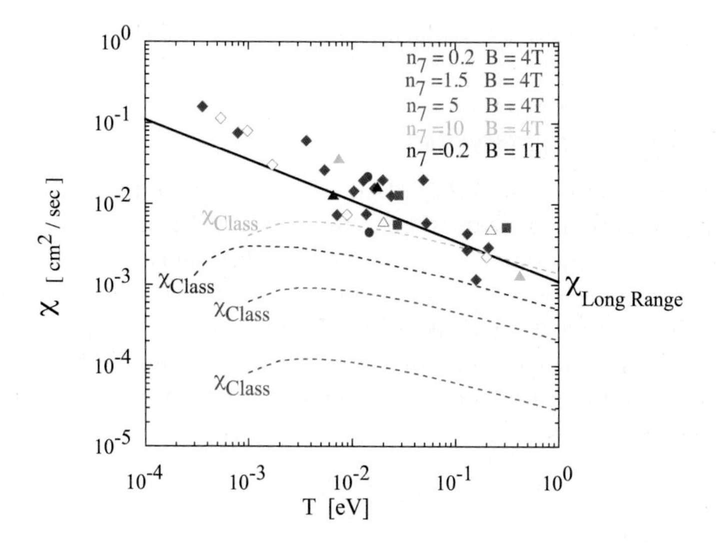

Fig. 7.16. Measured cross-magnetic field thermal diffusivity χ plotted as a function of temperature, demonstrating heat transport dominated by long-range collisions.

where η is the coefficient of shear viscosity and ω_r is the total rotation frequency,

$$\omega_r \equiv \underbrace{\frac{c}{rB}\frac{\partial \phi}{\partial r}}_{E\times B \text{ rotation}} + \underbrace{\frac{1}{mn\Omega_c r}\frac{\partial}{\partial r}(nT)}_{\text{diamagnetic drift}} . \tag{7.20}$$

The drag force F_θ causes an $F_\theta \times B$ drift, giving a radial flux Γ_r

$$\Gamma_r = F_\theta \frac{cn}{qB} = \frac{c}{qBr^2}\frac{\partial}{\partial r} r^3 \eta \frac{\partial \omega_r}{\partial r} . \tag{7.21}$$

The F_θ drag force vanishes when the plasma rotates uniformly (rigid rotor). The coefficient of shear viscosity η is related to the kinematic viscosity κ by $\kappa = \eta/nm$. The kinematic viscosity is a diffusion coefficient of angular momentum.

7.4.1. *Classical viscosity*

In classical theory, momentum is transferred across the magnetic field by a distance of order r_c, leading to a classical kinematic viscosity $\kappa^{\text{clas}} = \eta^{\text{clas}}/nm \propto \nu_c r_c^2$. The proper mathematical treatment[1] gives

$$\kappa^{\text{clas}} = \frac{2}{5}\sqrt{\pi}\nu_c r_c^2 \ln\left(\frac{r_c}{b}\right) \propto B^{-2}. \tag{7.22}$$

Like the classical diffusion coefficient, and the classical thermal diffusivity, classical viscosity scales as $\kappa^{\text{clas}} \propto nT^{-1/2}B^{-2}$.

7.4.2. *Long-range viscosity*

Similar to the heat transport, particles separated by up to a Debye length λ_D can exchange momentum. The distance over which angular momentum "steps" in a collision time is now of order λ_D rather than r_c, so the kinematic viscosity is independent of the magnetic field scaling as $\kappa^{\text{LR}} \propto \nu_c \lambda_D^2$. When guiding centers on different field lines interact, the long-range Coulomb force causes an exchange of momentum, and is responsible for the finite viscosity. Rigorous calculation of the viscosity due to long-range interaction has been performed,[15,16] giving:

$$\kappa^{\text{LR}} = \sqrt{\pi}\nu_c \lambda_D^2 \ln\left(\frac{\overline{\text{v}}}{\Delta \text{v}_m}\right) \propto T^{-1/2}B^0\,, \tag{7.23}$$

where $\Delta \text{v}_m = (D_\text{v}\lambda_D)^{1/3}$. The maximum interaction time is for particles with small relative axial velocity Δv, and the minimal effective Δv_m is determined by the axial velocity diffusion coefficient D_v. The long-range viscosity scales as $\kappa^{\text{LR}} \propto T^{-1/2}$ and is independent of B and of the plasma length.

7.4.3. *Viscosity in 2D system*

When the plasma temperature is sufficiently large that the frequency at which particles bounce along the magnetic field from one end of the plasma to the other $\omega_b \equiv \pi\,\overline{\text{v}}/L_p$ is larger than the fluid rotation frequency ω_r, the plasma is considered to be 2D. In this regime two particles on nearby field lines collide many times as they bounce back and forth along the magnetic field, and the effective number of bounces (limited by shear) is given by

$$N_b \equiv \frac{f_b}{r\omega_E'} = \frac{\overline{\text{v}}/2L_p}{r\omega_E'} \propto \frac{B}{L_p}. \tag{7.24}$$

These correlated multiple collisions lead to an increase in the viscous transport. A rigorous calculation[16] obtains an effective radial interaction distance of $d \equiv 2r_c \frac{\partial L_p}{\partial r} N_b$,

$$\begin{aligned} \kappa_{\text{2D}}^{\text{LR}} &= 16\pi^2 \nu_c d^2 N_b\, g\left(\frac{2d}{r}\right) \\ &\approx 64\pi^2 \nu_c r_c^2 \left(\frac{\partial L_p}{\partial r}\right)^2 N_b^3(0.1) \propto B^1 L_p^{-3}, \end{aligned} \tag{7.25}$$

where the numerical integral $g(x)$ has been approximated by 0.1.

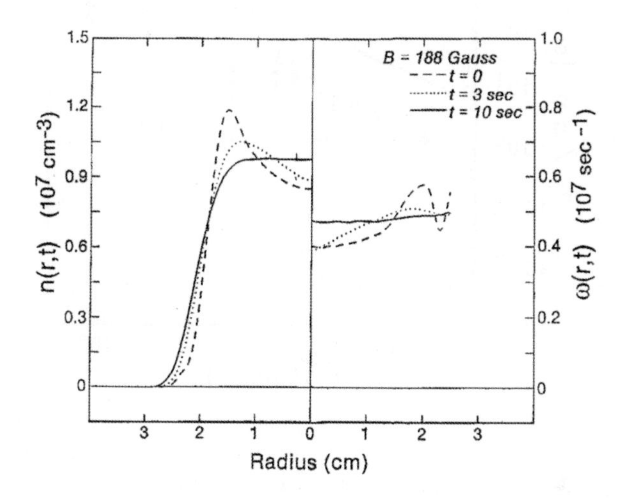

Fig. 7.17. Measured electron density profiles $n(r,t)$ and total rotation profiles at three different times, showing viscous relaxation to the rigid thermal equilibrium.

Experiments on pure electron plasmas[17,18] have measured the density profile relaxation towards equilibrium, as shown in Figure 7.17. From the measured integrated line charge $Q(r,t)$, the radial flux of particles is calculated:

$$N_b = \Gamma_r = \frac{1}{L_p r} \int_0^r dr' r' \frac{d}{dt} Q(r,t). \tag{7.26}$$

This particle flux is positive at some radii and negative at others. When a constant density (i.e., uniform rotation) is achieved, the F_θ force and the flux Γ_r vanish. The measured flux Γ_r is compared to a model of radial flux (Eq. (7.21)). All quantities in the model are measured and η is obtained.

Figure 7.18 shows $\kappa = \eta/nm$ measured plotted against the confining magnetic field B. The data come from two different apparati, each containing axially short plasmas. The solid lines are theory predictions of $\kappa = \eta/nm$. The measured kinematic viscosity is up to 10^8 times larger than the classical prediction, and scales as the 2D long-range theory. However, significant questions remain, both theoretically and experimentally, as to the parameters affecting this 10^8 times enhancement.

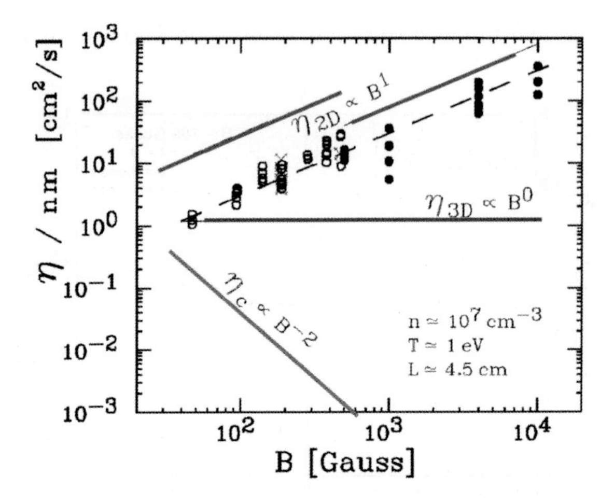

Fig. 7.18. Measured electron viscosity $\kappa = \eta/nm$ versus magnetic field B compared to the classical prediction of Eq. (7.22), to the 3D long-range Eq. (7.23) and 2D long-range Eq. (7.25).

7.5. Table of Transport Coefficients

Table 7.1 summarizes the transport coefficients discussed in this presentation. The first column is the test particle diffusion D, the second column is the thermal conductivity χ and the last column is the kinematic viscosity $\kappa = \eta/\text{nm}$. Each row is a different type of collision. The first row describes "classical" close range collisions with impact parameters ρ smaller than the cyclotron radius r_c. The second row shows the diffusion coefficient due to individual long-range $E \times B$ drift collisions with impact parameters $r_c < \rho < \lambda_D$. The last row presents the effect of 2D bounce-averaged collisions on test particle diffusion and viscosity. It is worth noting that according to theory there is no 2D enhancement of the thermal conductivity χ, but this is yet to be checked experimentally. All coefficients in this table are in cgs units, with dimensions of cm^2/sec.

Table 7.1. Summary of transport coefficients.

Regime and Mechanism	Test Particle Diffusion D	Thermal Conductivity $\chi = 2\kappa/5n$	Viscosity $\eta/(n\,m)$
$\rho < r_c$ close collisions	$\frac{4}{3}\sqrt{\pi}\nu_c r_c^2 \ln(\frac{r_c}{b})$ Longmire + Rosenbluth '56, Spitzer '55	$\frac{16}{15}\sqrt{\pi}\nu_c r_c^2 \ln(\frac{r_c}{b})$ Rosenbluth+Kaufmann '58	$\frac{2}{5}\sqrt{\pi}\nu_c r_c^2 \ln(\frac{r_c}{b})$ Longmire + Rosenbluth '56, Simon '55
$\rho > r_c$ long-range ExB collisions	$6\sqrt{\pi}\nu_c r_c^2 \ln\left(\frac{\lambda_D}{r_c}\right) \ln\left(\frac{\bar{v}}{\Delta v}\right)$ Anderegg et al. '97 Dubin '97; esitimate in Lifshitz + Pitaevskii	$0.49\ \nu_c \lambda_D^2$ +wave contribution Dubin+O'Neil '97; Hollmann et al '00 wave cont. estimate in Rosenbluth + Liu '76	$1.8\nu_c \lambda_D^2 \ln(\frac{\bar{v}}{v_{min}})$ +wave contribution O'Neil '85, Dubin '88 Driscoll '88
$\omega_b >> \omega_E$ 2D bounce-averaged collisions	$8\pi^2 \nu_c r_c^2 N_b \ln(\frac{r}{d})$ Dubin&Jin' 01 Anderegg et al '02	—	$16\pi^2 \nu_c d^2 N_b\, g(\frac{2d}{r})$ (monotonic rotation freq.) Dubin + O'Neil '98 Kriesel' 01

Acknowledgments

This work was supported by the National Science Foundation Grant PHY-9093877 and the Department of Energy Grant DE-SC0002451. The author thanks the organizers of the Les Houches Winter School for the opportunity to contribute to the workshop. The author also thanks Jo Ann Christina for typing and editing the manuscript, and Prof. C. F. Driscoll, Prof. D. H. E. Dubin and Prof. T. M. O'Neil for many years of enlightened guidance and theory support.

References

1. C. L. Longmire and M. N. Rosenbluth, Diffusion of charged particles across a magnetic field, *Phys. Rev.* **103**, 507–510 (1956); A. Simon, Diffusion of like particles across a magnetic field, *Phys. Rev.* **100**, 1557–1559 (1955); L. Spitzer, *Physics of Fully Ionized Gases.* (Interscience, New York, 1956).
2. D. H. E. Dubin, Test particle diffusion and the failure of integration along unperturbed orbits, *Phys. Rev. Lett.* **79**, 2678–2681 (1997).
3. F. Anderegg, X. P. Huang, C. F. Driscoll, E. M. Hollmann, T. M. O'Neil and D. H. E. Dubin, Test particle transport due to long range interactions, *Phys. Rev. Lett.* **78**, 2128–2131 (1997).

4. R. A. McGill, I. G. Brown and J. E. Galvin, Some novel design features of the LBL metal vapor vacuum arc ion sources, *Rev. Sci. Instrum.* **61**, 580–582 (1990).
5. F. Anderegg, X. P. Huang, E. M. Hollmann, C. F. Driscoll, T. M. O'Neil and D. H. E. Dubin, Test particle transport from long-range collisions, *Phys. Plasmas* **4**, 1552–1558 (1997).
6. C. F. Driscoll, F. Anderegg, D. H. E. Dubin, D.-Z. Jin, J. M. Kriesel, E. M. Hollmann and T. M. O'Neil, Shear reduction of collisional transport: Experiments and theory, *Phys. Plasmas* **9**, 1905–1914 (2002).
7. J. B. Taylor and B. McNamara, Plasma diffusion in two dimensions, *Phys. Fluids* **14**, 1492–1499 (1971).
8. J. M. Dawson, H. Okuda and R. N. Carlile, Numerical simulation of plasma diffusion across a magnetic field in two dimensions, *Phys. Rev. Lett.* **27**, 491–494 (1971); H. Okuda and J. M. Dawson, Theory and numerical simulation on plasma diffusion across a magnetic field *Phys. Fluids* **16**, 408–426 (1973).
9. D. H. E. Dubin and D. Z. Jin, Collisional diffusion in a 2-dimensional point vortex gas, *Phys. Lett. A* **284**, 112–117 (2001).
10. D. H. E. Dubin, Collisional diffusion in a two-dimensional point vortex gas or a two-dimensional plasma, *Phys. Plasmas* **10**, 1338–1350 (2003).
11. M. N. Rosenbluth and A. N. Kaufman, Plasma diffusion in a magnetic field, *Phys. Rev.* **109**, 1–5 (1958).
12. D. H. E. Dubin and T. M. O'Neil, *Phys. Rev. Lett.* **78**, 3868–3871 (1997).
13. E. M. Hollmann, F. Anderegg and C. F. Driscoll, Measurement of cross-magnetic-field heat transport due to long-range collisions, *Phys. Plasmas* **7**, 1767–1773 (2000).
14. E. M. Hollmann, F. Anderegg and C. F. Driscoll, Measurement of cross-magnetic-field heat transport in a pure ion plasma, *Phys. Rev. Lett.* **82**, 4839–4842 (1999).
15. T. M. O'Neil, A new theory of transport due to like-particle collisions, *Phys. Rev. Lett.* **55**, 943–946 (1985).
16. D. H. E. Dubin, Collisional transport in non-neutral plasmas, *Phys. Plasmas* **5**, 1688–1694 (1998).
17. C. F. Driscoll, J. H. Malmberg and K. S. Fine, Observation of transport to thermal equilibrium in pure electron plasmas, *Phys. Rev. Lett.* **60**, 1290–1293 (1988).
18. J. M. Kriesel and C. F. Driscoll, Measurements of viscosity in pure-electron plasmas, *Phys. Rev. Lett.* **87**, 135003:1–4 (2001).

Chapter 8

Antihydrogen Formation and Trapping

Niels Madsen

Department of Physics, College of Science, Swansea University, Swansea SA2 8PP, United Kingdom
n.madsen@swansea.ac.uk

Antihydrogen, the bound state of a positron and an antiproton, is the only neutral pure antimatter system available to date, and as such provides an excellent testbed for probing fundamental symmetries between matter and antimatter.

In this chapter we will concentrate on the physics issues that were addressed in order to achieve the first trapping of antihydrogen. Antihydrogen can be created by merging antiprotons and positrons in a Penning–Malmberg trap. However, traps for antihydrogen are at best about $\sim$50 μeV deep and, as no readily available cooling techniques exist, the antihydrogen must be formed trapped. Antiprotons are sourced from an accelerator and arrive with a typical energy of 5.3 MeV. The large numbers of positrons needed means that the self-potential of the positrons are of order 2–5 V. With such energetic ingredients a range of plasma control and diagnostic techniques must be brought to bear on the particles to succeed in making any antihydrogen cold enough to be trapped.

8.1. Introduction

Antihydrogen ($\bar{\mathrm{H}}$), the bound state of an antiproton ($\bar{\mathrm{p}}$) and a positron (e^+), is the simplest anti-atom and also the simplest neutral antimatter system. As such it holds the promise of precise comparisons between matter and antimatter using spectroscopic techniques.[1,2] Depending on the particular aspect of the (anti)atom investigated, one can study various aspects of the fundamental CPT-theorem which states that all physical laws must be unchanged under the simultaneous application of Charge conjugation, Parity inversion and Time reversal. Antihydrogen, due to its neutrality, furthermore allows for direct investigations of the gravitational interaction of antimatter.

The first low-energy $\bar{\mathrm{H}}$ suitable for scrutiny was made by the ATHENA collaboration working at the CERN Antiproton Decelerator (AD)[3] in 2002.[4] Following this, the off-spring ALPHA collaboration was started in 2005 with the intermediate goal of trapping $\bar{\mathrm{H}}$ and ultimate goal of performing precision comparisons of H and $\bar{\mathrm{H}}$. ALPHA succeeded in trapping $\bar{\mathrm{H}}$ in 2010[5] and further in holding onto it for 1000 s in 2011.[6] These results led to the first detection of a resonant quantum transition in an anti-atom in 2012,[7] thus kicking off the field of $\bar{\mathrm{H}}$ spectroscopy. This chapter will focus on the challenges that were faced and overcome in order to achieve all of these results.

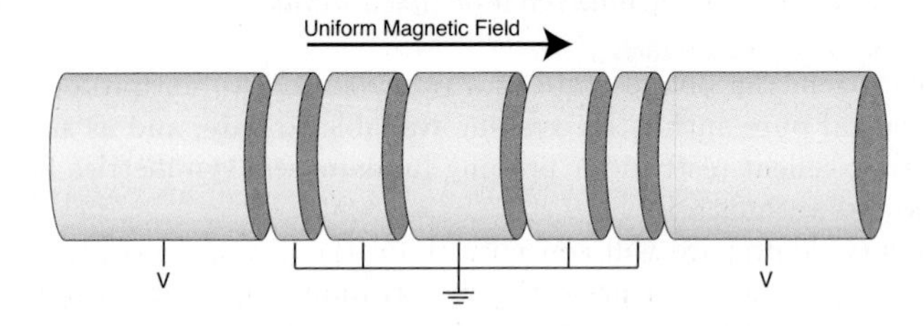

Fig. 8.1. Example of a Penning–Malmberg trap typical of those used in $\bar{\mathrm{H}}$ experiments (see text). In practice, since more electrodes are needed for all antiparticle manipulations to be possible, the $\bar{\mathrm{H}}$ experiments ALPHA, ASACUSA, ATHENA and ATRAP use of order fifty individually controllable co-axial electrodes.

Further active groups in the field of $\bar{\mathrm{H}}$ are the ATRAP collaboration, who also produce $\bar{\mathrm{H}}$[8] and have provided recent evidence for $\bar{\mathrm{H}}$ trapping,[9] the ASACUSA collaboration, who recently produced their first $\bar{\mathrm{H}}$,[10] the AEgIS collaboration,[11] whose setup has just been installed at CERN and the recently approved GBAR collaboration.[12]

8.2. Introduction to Antihydrogen Formation and Trapping

The most common method for forming $\bar{\mathrm{H}}$ is by merging plasmas of e^+s and $\bar{p}$s. These plasmas are typically contained in a cylindrical variation of the Penning trap called the Penning–Malmberg trap (Figure 8.1). In a Penning trap transverse confinement is secured by an axial magnetic field and axial confinement is achieved by axial electric fields. The classic Penning trap uses hyperbolic electrodes and is not well suited for injecting and extracting particles, a common practice in $\bar{\mathrm{H}}$ physics, where both $\bar{p}$s,

e^+s and electrons (e^-) are needed and are derived from various sources. Therefore, the Penning–Malmberg variant is used. Here the electric field is provided by cylindrical electrodes, co-axial with the magnetic field and individually controllable. This allows for a flexible system which can easily contain all three species of particles in various configurations. All charged particle traps in ALPHA are of this type and the majority of this chapter will assume that such traps are used. In ALPHA the axial magnetic field is provided by a large warm bore solenoidal magnet and is typically set to 1 T.

Antihydrogen may also be formed by collisions of $\bar{p}$s and positronium (Ps).[13,14] Ps is a bound state of a e^+ and an e^- and its two ground states of anti-parallel and parallel spins (para- and ortho-Ps) have annihilation lifetimes of 125 ps and 142 ns, respectively. The cross-section for $\bar{H}$ formation is greatly enhanced if the Ps is in an excited state.[15] This latter process was experimentally demonstrated by the ATRAP collaboration,[16] but it remains to be seen if it is advantageous for further experiments with $\bar{H}$, since the experiment showed very low formation rates. For $\bar{H}$ trapping experiments the potential benefit of this method is that one avoids the problem of the self-field (space charge) of the e^+ and it could therefore be easier to make cold, trappable $\bar{H}$. The AEgIS collaboration, which aims to measure the gravitational pull on a beam of $\bar{H}$, is planning to use a method similar to ATRAP's to make a beam of $\bar{H}$.[11] The difference here is that the Ps will be made by shooting e^+ onto a porous structure. A related process, that requires two Ps atoms, has been proposed for making $\bar{H}^+$ ions for the GBAR experiment.[12]

In most of these experiments the electrodes at the centre of the system are cooled to cryogenic temperatures in the range 1–15 K depending on the experiment. The cooling serves two purposes. The main purpose is to assist in the formation of cold $\bar{H}$ by helping to cool the constituent particles ($\bar{p}$s and e^+s). Positrons and e^-s will cool as the strong axial magnetic field in the traps causes them to emit cyclotron radiation and approach thermal equilibrium with the surroundings. Antiprotons will not emit an appreciable amount of radiation but can be cooled by merging them with e^-s at some stage in the process. An additional advantage of a cryogenic system is that the whole chamber acts as a cryo-pump, thereby keeping the pressure low, often in the 10^{-14} mbar range or lower. This is a prerequisite for most $\bar{H}$ experiments, as the particle manipulations necessary may last from minutes to hours.

Neutral atoms may be trapped by acting on their intrinsic magnetic moment, that arises mainly due to the e^- spin (e^+ spin for anti-atoms). The potential energy of an atom with magnetic moment $\bar{\mu}$ in a magnetic field $\bar{B}$ is given by

$$U = -\bar{\mu} \cdot \bar{B}. \tag{8.1}$$

Trapped atoms typically move slowly relative to the changes of magnetic field direction such that their spin will tend to be aligned with the magnetic field. Depending on the quantum state of the atom, the spins can be aligned both in parallel and anti-parallel to the magnetic field. Atoms with spins aligned in parallel to the magnetic field will have a lower internal energy in higher fields, and *vice versa* for atoms with anti-parallel magnetic moments. The former are called high-field seekers and the latter low-field seekers as they will tend towards high and low field strength locations respectively. One may create a trap for such atoms by generating a three-dimensional maximum or minimum of magnetic field strength. As it is only possible to produce a minimum in three dimensions one may only create traps for the low-field seekers. A magnetic minimum can be created by introducing two axially separated co-axial coils and a co-axial transverse multipole, such as the Ioffe–Pritchard configuration shown in Figure 8.2.

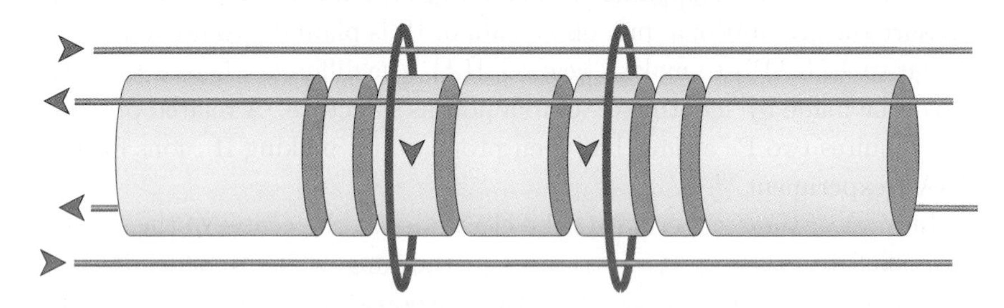

Fig. 8.2. Combined Penning–Malmberg and Ioffe–Pritchard trap. The coil geometry causes a three-dimensional minimum in magnetic field strength at the centre of the trap. The classical setup uses a transverse quadrupole generated by four wires and energised as shown.

When trapping atoms it is often possible to either use laser cooling or collisional cooling with a cryogenic fluid to cool the atoms enough that they may be trapped. The main disadvantage of these methods is that only a small fraction of the available atoms may be trapped, a distinct disadvantage for trapping the already scarce $\bar{\mathrm{H}}$. Furthermore, $\bar{\mathrm{H}}$ does not survive collisions with matter, so in practice there are no methods to do cooling-assisted trapping and $\bar{\mathrm{H}}$ must therefore be synthesised trapped. This means that the processes for $\bar{\mathrm{H}}$ formation must be performed inside the magnetic minimum trap and it must be ensured that the $\bar{\mathrm{H}}$ atoms are sufficiently cold to be trapped. The potential well depth for ground state (anti)hydrogen in a magnetic trap is 0.67 K/T, which is small compared to almost all other energy scales in the experiments.

8.3. Antiproton Catching and Pre-cooling

Antiprotons are delivered by the CERN AD[3]. The $\bar{\mathrm{p}}$s are created by colliding 26 GeV/c protons from the CERN PS with an iridium target.[17] The AD collects the antiprotons from this energetic collision at a mean momentum of 3.57 GeV/c and decelerates them through a number of stages to 100 MeV/c, at which point they are bunched and ejected to the experiments. A typical bunch of antiprotons is 200 ns long and contains 3×10^7 antiprotons. At 100 MeV/c the antiprotons have a kinetic energy of 5.3 MeV. In order to capture these, further deceleration is necessary. The simplest, but also least efficient, way to do this is to allow them to pass through a degrading foil where collisions with the nuclei and e^-s will decelerate them further. The passage through the degrading foil causes the antiproton energy spread to increase and typically about half the antiprotons annihilate in the foil. The foil thickness is tuned to secure a maximum of antiprotons emerging in the energy range of the capture setup and typically consists of $\sim$0.2 mm aluminium-equivalent material.

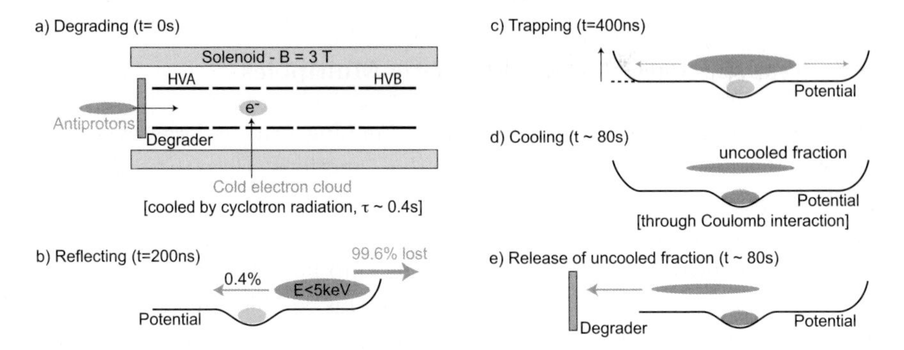

Fig. 8.3. Antiproton capture and pre-cooling (see text for explanation). On (a) are indicated the two high voltage electrodes, HVA and HVB, used for the capture process.

Antiprotons are captured as illustrated on Figure 8.3. Before the antiprotons arrive e^-s have been pre-loaded into the centre of the trap. Various ways of loading these are used by different experiments. In ALPHA the e^-s are sourced from an e^- gun that is located on a removable assembly outside the main solenoidal magnet. A blocking electrode (HVB in Figure 8.3.a) is excited to a high voltage (HV) of typically -5 kV. When the antiprotons arrive, antiprotons with low enough kinetic energy will deflect off of this potential, and in the mean time a HV electrode closer to the degrading foil (HVA) will be quickly switched to the same potential as HVB and thus block their escape from the trap.[18] For this process to be of maximum efficiency the degrading foil must be as close to HVA as possible, in order to allow capture of the low-energy tail, and the incoming antiproton pulse length should be less than the round trip time of a maximum energy $\bar{p}$ in the trap, which for ALPHA is around 400 ns. In this way up to about 1% of the $\bar{p}$s may be captured, though other factors typically limit the number to about half that.

Once captured the $\bar{p}$s will collide with the pre-loaded e^-s and the mixture will cool towards thermal equilibrium with the surroundings.[19] With 20 million e^-s, ALPHA typically cools all $\bar{p}$s that overlap with the e^-s in about 80 s. The cooling efficiency, that is the fraction of caught $\bar{p}$s successfully cooled, depends on the spatial overlap of the e^- plasma and the $\bar{p}$s.[20] Tuning the transverse position of the incoming $\bar{p}$ beam for optimum cooling efficiency thus allows the beam to be centred radially in the capture region. By also tuning the size of the e^- plasma, efficiencies of up to 80% have been achieved. Typically around 40000 cold $\bar{p}$s remain for further experimentation at the end of the process.

8.4. Trapped Particles and Magnetic Multipoles

As is well known, particles confined in a Penning trap undergo cyclotron (ω_c), magnetron (ω_m) and axial bounce (ω_z) motions. Here the order typically is $\omega_c \gg \omega_z \gg \omega_m$. In thermal equilibrium a plasma will undergo a rigid rotation at a modified magnetron frequency given by the total transverse electric field from the plasma and the trap potentials.[21]

When magnetic gradients are introduced, this equilibrium situation will be perturbed. A simple example is to examine how the previously parallel field lines distort when a transverse multipole is introduced. Figure 8.4 illustrates how field lines, and thus low-energy particles, originating from a circle in the centre of the figure, behave on moving along the axis. This distortion gives rise to a number of deleterious effects of which the simplest is a dynamic aperture. Assuming that particles stay on field lines then as

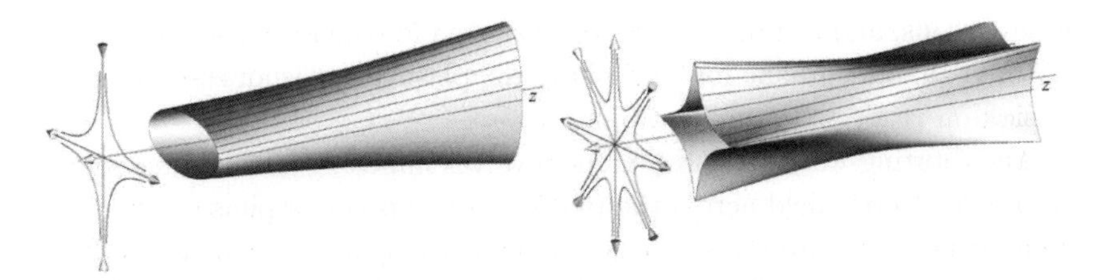

Fig. 8.4. Magnetic field lines for a transverse quadrupole (left) and octupole (right), superimposed on a uniform axial field. For each configuration, the vectors on the left represent the directions of the axially-invariant multipole and axial field components. The surface is created by following the field lines from a centred circular locus, and the lines shown within the surfaces are field lines.

either the multipole field strength or the length the axial well is increased, one end of the axial bounce will move further and further outwards radially. Eventually, the exact parameters depending on the initial radial position of the particle, the particle will collide with the walls of the apparatus and be lost. This can be described by a critical radius, that is given by[22]

$$r_c = R_w \exp\left(-\frac{1}{2}\frac{B_w}{B_z}\frac{L}{R_w}\right) \quad \text{and} \quad r_c = \frac{R_w}{\sqrt{1+\frac{B_w}{B_z}\frac{L}{R_w}}}, \tag{8.2}$$

for a quadruple and an octupole, respectively. Here, R_w is the trap wall radius, B_w is the multipole field strength at the wall, B_z is the axial field strength and L is the full axial length of the trap.

This effect, as well as a range of other deleterious effects caused by multipolar fields superimposed on Penning–Malmberg traps[23,24] means that it is best to avoid axially long and/or radially large plasmas.[25] Also clear from these issues is that it is better for charged particle confinement for the atom trap to have a higher-order multipole for a given field strength at the wall (i.e. trap depth), as higher-order fields drop faster in strength as the radius decreases. This was the prime motivation for ALPHA choosing an octupole in its design of its magnetic minimum trap.[26]

8.5. The Rotating-wall Technique

As we saw above, there are good reasons to work with radially small plasmas in order to avoid issues caused by the strong multipole fields inherent to $\bar{\text{H}}$ trapping. Furthermore, it was already evident in the first experiments on cold $\bar{\text{H}}$ in ATHENA, that tailoring the size of the plasmas such that all $\bar{\text{p}}$s would enter the e^+ plasma makes more efficient use of the $\bar{\text{p}}$s.[27] For these reasons, and to tailor the e^- plasma used for the initial cooling of the captured $\bar{p}$s, a technique for controlling the plasma size is an essential part of work on $\bar{H}$ formation and trapping.

The rotating-wall (RW) technique involves imposing a transverse rotating dipole electric field across the particles by applying a phase-shifted radiofrequency signal to the sectors of an azimuthally segmented electrode.[28] The electrode is typically split into four, six or eight sectors. Through this rotating field, angular momentum may be transferred between the field and the plasma, and it is thus possible to control the rotation frequency of the plasma, and subsequently its size. A number of different regimes have been identified, where the overall distinction is whether the rotating wall acts on single particles[29] or on the plasmas as a whole.[20] When acting on the plasma, which is the case most relevant to $\bar{H}$, it may act through resonances with oscillatory modes of the plasma,[30] or in the most commonly used it may act in the strong-drive regime, where the rigid rotation frequency of the plasma locks to the external drive.[31] Common to all of these methods is that the plasma should only be partially located in the segmented section of the electrode stack to avoid RW induction of a bulk diocotron motion rather than coupling to the internal motion of the plasma.

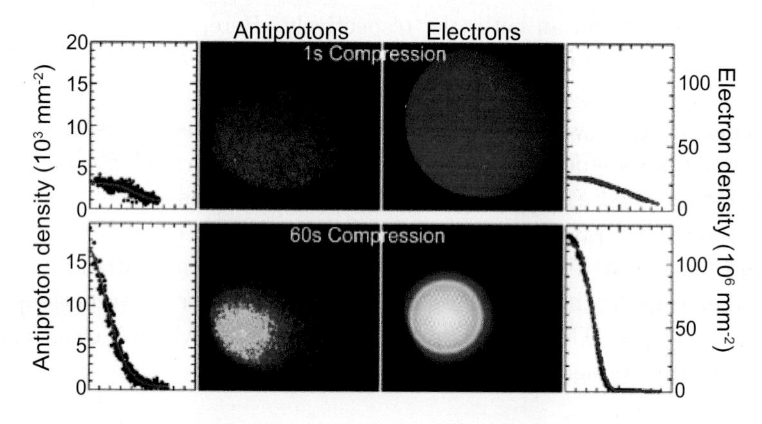

Fig. 8.5. Antiproton and e^- images showing the effect of rotating-wall compression and the resulting radial profiles. The solid red lines are Gaussian-like fits to the radial profiles (i.e. $\exp(-|r/r_0|^k)$, where k≈2).

A final feature is that compression leads to heating of the plasma, and that successful compression requires some form of cooling. For e^+s and e^-s this cooling is readily available by means of the cyclotron radiation or though the buffer gas as used for e^+ accumulation. However, $\bar{p}$s are typically cooled through collisions with e^-s. By tuning the rotating wall to act on e^-s, one may compress $\bar{p}$s through collisions with the e^-s.[20] Using the latter technique, by first optimising the compression of e^-s, ALPHA routinely compresses $\bar{p}$ plasmas to sizes of $\sim$0.1 mm in a 3 T field. Figure 8.5 shows an example of the compression of the $\bar{p}$s in an e^- plasma.

8.6. Antiproton Preparation

To prepare $\bar{p}$s for $\bar{H}$ formation and trapping a typical sequence consists of the following steps:

(1) Electrons are pre-loaded into the trap and their size is tuned to gain optimum cooling efficiency whilst remaining reasonably small. A typical value is a radius of 1 mm, which is roughly the size of the incoming $\bar{p}$ beam.
(2) Antiprotons are caught by switching the HV system.
(3) Antiprotons cool on the pre-loaded e^-s for about 80 s.
(4) The HV is turned off, by first turning off HVA, which leaves $\bar{p}$s that have not been cooled to escape and annihilate on the degrader/Faraday cup (Figure 8.3 (e) and Figure 8.6). These are counted as a way of monitoring the cooling efficiency.
(5) The $\bar{p}/e^-$ mixture is compressed using the RW, which has been optimised on e^-s, and then on the mixture.
(6) About 90% of the e^-s are ejected by briefly opening the side of the well where the mixture is contained. The $\bar{p}$s are slow, so they remain trapped, but this e^- "kick" heats them.
(7) The $\bar{p}/e^-$ mixture is re-compressed using the RW. The RW tends to compress towards a constant density (rotation frequency), so compression of the less numerous e^- plasma results in a smaller e^- plasma and, therefore, ultimately a smaller $\bar{p}$ plasma.
(8) As transport tends to heat the $\bar{p}$s further, they are now transported with e^-s to the mixing section of the apparatus (Figure 8.6), such that minimal movements are required after they have been freed from e^-s.
(9) A final e^- kick-out is carried out. As there are now fewer particles, the space-charge potential is smaller, and so the necessary wells are shallower, which means that we can apply lower voltage pulses to let the e^-s escape, which in turn reduces the heating of the $\bar{p}$s. At this point the $\bar{p}$ plasma typically consists of 40000 $\bar{p}$s at around 400 K (heated by the ejection process) with a radius of 0.3 mm.

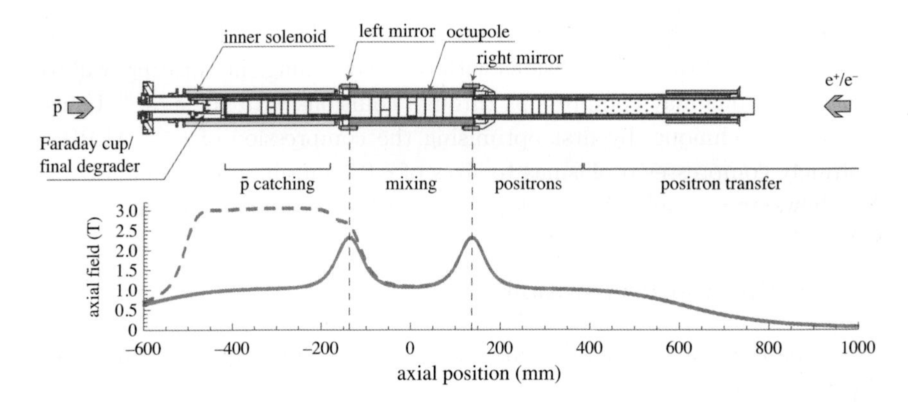

Fig. 8.6. The upper figure shows a vertical cut-away of the ALPHA experimental setup designed to trap $\bar{\mathrm{H}}$. The trap electrodes are shown along the centre of the apparatus. The bottom graph shows the axial magnetic field along the axis of the apparatus. A large external solenoid (not shown) supplies a 1 T axial field throughout. The inner solenoid on the left of the apparatus can raise the general 1 T bias axial field formed by up to 2 T in order to increase the efficiency of trapping $\bar{p}$s (which enter from the left).[32] Also visible on the graph is the axial magnetic minimum generated at the zero position (solid line) by powering the mirror coils of the neutral trap.

While 400 K is not the limit of the above technique it is clearly a far cry from the ∼0.5 K depth of the neutral trap and it seems unlikely that the rough manipulations involved in ejecting the e^-s, which also have a significant space-charge potential, could allow $\bar{p}$s to reach the very low temperatures needed for $\bar{\mathrm{H}}$ trapping. As we shall see in the following section, the final temperature of the $\bar{p}$s is not yet likely to be a limit in current trapping experiments, but some development on making them as cold as the surroundings has nonetheless been accomplished. ALPHA were the first to report cryogenic $\bar{p}$s, which were achieved through evaporative cooling[33] and will be described in more detail in the following section. A separate approach, demonstrated for $\bar{p}$s by the ATRAP collaboration, is to use adiabatic cooling, which makes use of the E/f adiabatic invariant of particles in a harmonic oscillator, in order to cool the plasma while expanding it axially.[34] Here E is the particle's energy and f the oscillation frequency. This technique is potentially very powerful but has not yet been used by ALPHA and therefore will not be described further.

The temperature of all particle species discussed here is measured by gently lowering the confining potential while counting the number of escaping particles and thus measuring the exponential tail of the Boltzmann

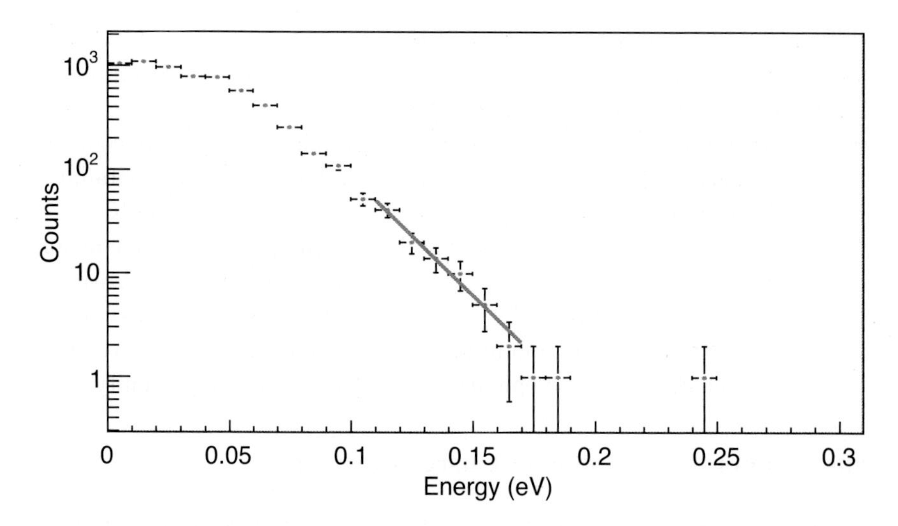

Fig. 8.7. Antiprotons lost as a function of trap depth. Annihilation events are counted while the trap depth is lowered over 100 ms. The detector has a 25±5 % absolute efficiency for detecting a $\bar{p}$ annihilation. The graph shows the number of escaped particles at each point in this process. The temperature is extracted by fitting the release of the first ~ two hundred escaping particles, and the fit yields 219±33 K. The counting uncertainty is purely statistical, and the extracted temperature has not been corrected for adiabatic cooling nor space-charge (see Section 8.6). The horizontal uncertainty represents the energy resolution of the measurement.

distribution.[35] In order for this to work, it is important that very low numbers of particles can be counted, as the space-charge field of the particle clouds as well as the changes in the cloud during ejection will otherwise mask the temperature information. Figure 8.7 shows a temperature measurement of a typical ensemble of $\bar{p}$s used for the trapping experiments described here. Particle-in-cell simulations of our ejections show that the various effects, such as space charge and adiabatic cooling, lead us to overestimate temperatures by up to 40%. The temperatures given here are the raw data, not compensated for any such effects.

8.7. Positron Preparation

The e^+s are accumulated and transferred using well-established techniques.[36,37] Positrons are accumulated in a Surko-type buffer-gas based e^+ accumulator using neon-ice moderated e^+s from a ^{22}Na radioactive source. After accumulation the e^+s are transferred ballistically in a single pass to the main ALPHA apparatus, where they are captured much like the $\bar{p}$s and then cool through cyclotron radiation. Finally they are compressed using the RW technique and positioned in the centre of the trap system for merging with the $\bar{p}$s. Using these techniques as well as evaporative cooling, ALPHA typically obtains e^+ plasmas with a radius of 0.8 mm containing 1×10^6 e^+s at a density of 5×10^7 cm^{-3} and a temperature of $\sim$40 K.

The e^+ temperature may be measured using the same technique as for the $\bar{p}$s. The low temperatures in these experiments require the sensitivity to be of the order of one particle. With the $\bar{p}$s this is relatively easy using scintillating panels and photomultiplier tubes. However, for e^+, ALPHA uses a multi-channel plate to enhance the sensitivity to close to unity.[38]

8.8. Evaporative Cooling of Charged Particles

Evaporative cooling is a well-established discipline in atom traps and relies on the simple observation that if one removes particles from a sample that are more energetic than the average particle of the sample, the average kinetic energy drops. By applying this technique and ejecting particles slower than the equilibration rate, one can ensure that one always ejects particles that are more energetic than the average, and thus that the temperature of the sample drops with the total number of particles. The trick is thus to tune this evaporation such that the fraction of particles at low temperatures actually increases while the total number decreases.

ALPHA demonstrated cooling of $\bar{p}$s to cryogenic temperatures using this technique.[33] Figure 8.8 shows the measured temperature as a function of well depth during a demonstration of the method. Compared to evaporative cooling of atoms, one particular feature stands out. Evaporative cooling of charged particles in a Penning trap lets the particles evaporate axially from the radial centre only. This means that no angular momentum is carried away in the loss process. Conservation of total canonical angular momentum[21] then implies that the plasma will expand radially as $N_0/N = \langle r^2\rangle/\langle r_0^2\rangle$ when the angular momentum is redistributed among fewer particles. Here N_0 and r_0 are, respectively the initial number and radius of the $\bar{p}$s and N and r are, respectively, the number and radius after evaporative cooling. This effect limits the range of usefulness of the technique as the radial size of the plasma must be kept small to avoid the deleterious effects discussed in previous sections.

In the ALPHA experiment limited evaporative cooling is used on both $\bar{p}$s and e^+s. The $\bar{p}$s are typically cooled to a temperature of $\sim$100 K, a

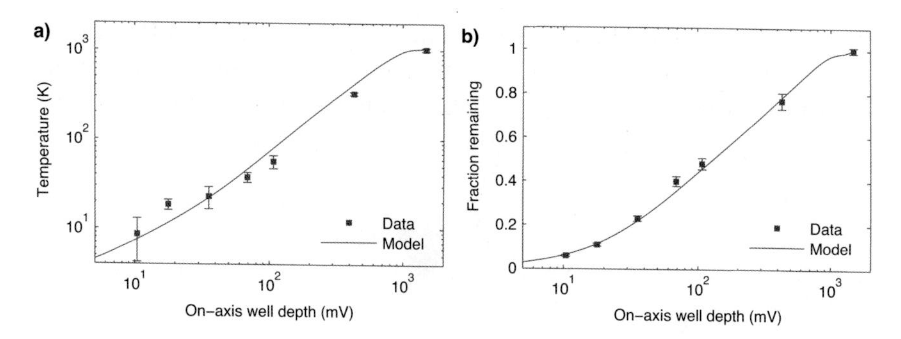

Fig. 8.8. Evaporative cooling of $\bar{\text{p}}$s. (a) Temperature vs the on-axis well depth. (b) The fraction of $\bar{\text{p}}$s remaining after evaporative cooling vs on-axis well depth. The initial number of $\bar{\text{p}}$s was approximately 45000 for an on-axis well depth of (1484±14) mV.[33]

radius of 0.4 mm, which leaves 15000 $\bar{\text{p}}$s. As we shall see, and as has been verified experimentally, the $\bar{\text{p}}$ temperature does not seem to play a crucial role in the current mode of experimentation, so the main purpose of the evaporative cooling has been to give stable conditions for the auto-resonant injection system (see below), which does require cold particles. The e^+s are strongly influenced by the octupole field of the atom trap as well as other unidentified heating sources. The e^+s are therefore evaporatively cooled directly before the $\bar{\text{H}}$ synthesis is initiated. The temperatures are typically of the order of 40 K, for a e^+ plasma with a radius of 0.8 mm and containing 1 million e^+s.

8.9. Merging Antiprotons and Positrons

After preparing cold $\bar{\text{p}}$s and e^+s, the next phase in the experiment is to merge the plasmas to form $\bar{\text{H}}$. Experiments by ATHENA determined early on that the traditional $\bar{\text{H}}$ synthesis technique that launched eV $\bar{\text{p}}$s into a cold e^+ plasma, generated $\bar{\text{H}}$ that was too energetic to be trapped.[39] ALPHA first attempted to solve this issue by merging the plasmas by slowly moving the potentials holding the $\bar{\text{p}}$s and e^+s in neighbouring wells.[40] This led to some new loss mechanisms in the neutral trap magnetic fields.[41] However, the technique did not result in trapped $\bar{\text{H}}$, so a novel technique using autoresonance[42,43] to excite the $\bar{\text{p}}$s collectively in their well until they enter the e^+ plasma was introduced.[44] This technique makes use of the fact that an anharmonic oscillator will lock to an external drive under certain conditions and, by carefully chirping the drive frequency, energy can be transferred to the oscillator in a deterministic fashion. To trap $\bar{\mathrm{H}}$, ALPHA prepares the $\bar{\mathrm{p}}$s and e^+s in neighbouring wells, energises the magnetic trap, evaporatively cools the e^+s and finally merges the two by autoresonantly exciting the $\bar{\mathrm{p}}$s into contact with the e^+s (see Figure 8.9).

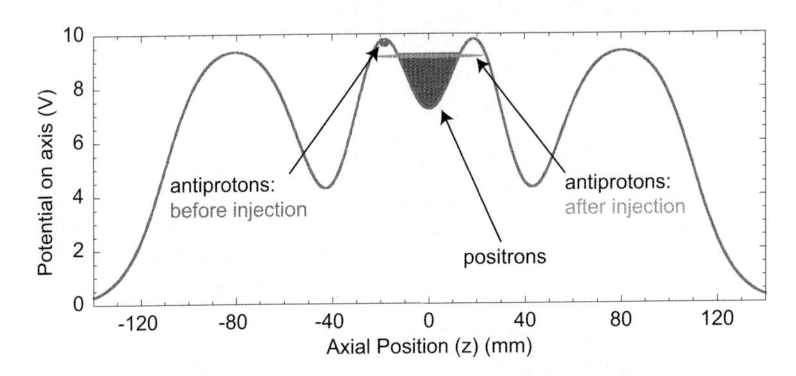

Fig. 8.9. The electrical potentials used for merging cold $\bar{\mathrm{p}}$s and e^+ to make cold $\bar{\mathrm{H}}$. See text in Section 8.9 for details.

8.10. Trapped Antihydrogen and its Detection

To trap $\bar{\mathrm{H}}$ it is synthesised with the neutral trap energised following the recipe given above. After autoresonant excitation the merged species are held for one second before all charged particles are ejected from the trap. Following this the superconducting neutral trap is de-energised rapidly with an e-folding time of 9 ms by means of dissipating its stored energy into external resistors. The trap depth is reduced to less than 1% of the full value in about 30 ms.[26] We look for signs of trapped $\bar{\mathrm{H}}$ in this time period.

When searching for trapped $\bar{\mathrm{H}}$ two potentially false signals contribute. The first is from the cosmic rays passing through the apparatus. To deal with this background, ALPHA uses a unique annihilation detector constructed with three layers of double-sided silicon strip detectors in a barrel geometry around the core of the apparatus[45] (see Figure 8.10). The detector allows the reconstruction of the tracks of the charged pions that appear from $\bar{\mathrm{p}}$ annihilations, and by combining these tracks an annihilation vertex can be found, localising the exact position in space where the $\bar{\mathrm{p}}$ annihilated. Cosmic rays can be distinguished by their very different tracking geometry. Using these techniques we obtain a detection efficiency of $\bar{\mathrm{p}}$ annihilations of 57±6% with a background rate of 47±2 mHz. The effective background for

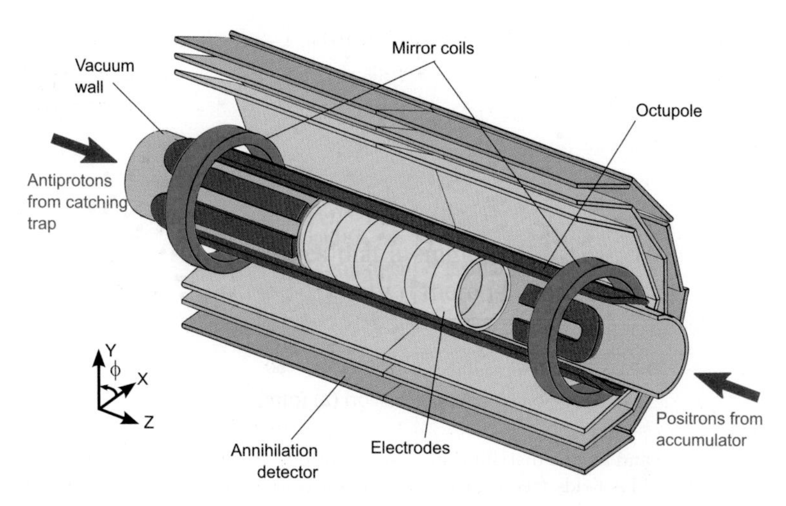

Fig. 8.10. Cutaway drawing of the $\bar{\mathrm{H}}$ synthesis and trapping region. The superconducting atom-trap magnets, the annihilation detector, and some of the Penning–Malmberg trap electrodes are shown. An external solenoid (not shown) provides the 1 T magnetic field for the Penning–Malmberg trap.

detection of trapped $\bar{\mathrm{H}}$ is further reduced by only searching for annihilation signals in the 30 ms window of de-energising the neutral trapping fields. The background fake rate is thus $1.4 \pm 0.1 \times 10^{-3}$ counts per experiment or one event in 700 experiments, thus effectively eliminating this issue.[6,46]

However, the annihilation detector only registers $\bar{\mathrm{p}}$ annihilations. In the ATHENA experiment, a similar detector was used, but with the additional capacity of detecting e^+ annihilations.[4] With the added material for the neutral trap magnets, e^+ detection was not an option for ALPHA, and instead an extra layer of silicon was added to allow the track curvature to be measured.[45] Thus the second type of potential false positives stems from bare trapped $\bar{\mathrm{p}}$.

Before de-energising the neutral trap, charged particles are cleared from the trap by a series of electric pulses. Charged particles, in particular $\bar{\mathrm{p}}$s, may be trapped on their motional magnetic moment, so-called mirror-trapping. Trapping depends on the relative difference between their transverse (to the magnetic field) and longitudinal velocities. Clearing electric fields applied allow us to eject $\bar{\mathrm{p}}$s with transverse energy less than 20 eV. As most of the manipulating potentials leading to $\bar{\mathrm{H}}$ formation are lower than this it is unlikely that any $\bar{\mathrm{p}}$s remain after clearing.[47] However, experimental errors and unforeseen circumstances could lead to a population

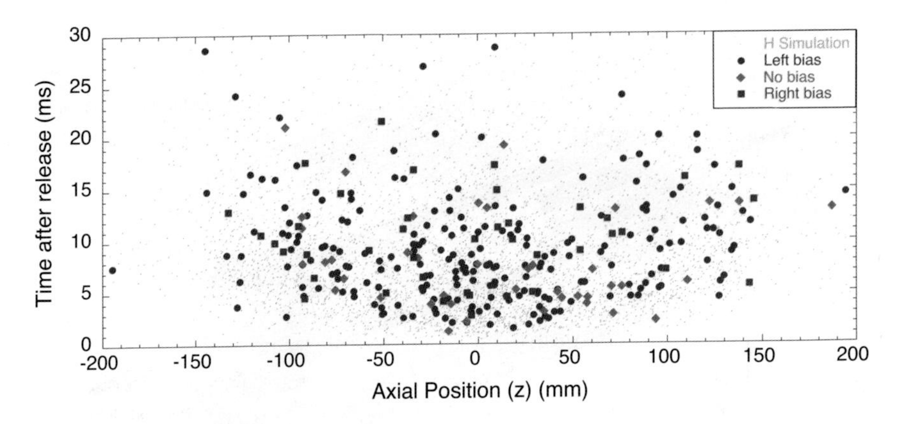

Fig. 8.11. Time and axial z distribution of annihilations on release of $\bar{\mathrm{H}}$ from the neutral trap for different bias fields (see legend) and comparison with simulation (grey dots).

of energetic $\bar{\mathrm{p}}$s in the trap whose signal could be confused for trapped $\bar{\mathrm{H}}$. To avoid these issues ALPHA applied three controls for fake trapped $\bar{\mathrm{H}}$ signals. The first was to carry out the full $\bar{\mathrm{H}}$ trapping process with all fields and manipulations but while heating the e^+ plasma to ~1000 K and thus suppressing the $\bar{\mathrm{H}}$ formation, and therefore any true $\bar{\mathrm{H}}$ trapping signal. This technique was pioneered by ATHENA in the first observation of cold $\bar{\mathrm{H}}$.[4] The second control consisted of applying axial bias fields across the atom trap volume in order to cause any mirror-trapped $\bar{\mathrm{p}}$s still present to be preferentially located either on the left or right side of the trapping volume, thus making them clearly distinguishable from neutral $\bar{\mathrm{H}}$ which escape symmetrically. The third and final control was to simulate the release of both $\bar{\mathrm{p}}$s and $\bar{\mathrm{H}}$ atoms from the trapping volume. Due to their different dynamics and energies their behaviour during de-energisation can easily be distinguished with the ALPHA annihilation detector.[47] Figure 8.11 shows 309 annihilation events recorded in 2010, most of which were taken with a wait time after $\bar{\mathrm{H}}$ synthesis of less than one second. For most of the data presented here, a static electric bias field of 500 Vm^{-1} was applied during the confinement and shutdown stages to deflect bare $\bar{\mathrm{p}}$s that may have been mirror-trapped. The bias field ensured that annihilation events could only be produced by neutral $\bar{\mathrm{H}}$. We observed significant signal at holding times of up to 1000 s, sufficient to envisage the first measurements on ground state $\bar{\mathrm{H}}$.[6]

8.11. Conclusions and Outlook

In this review we have presented a range of techniques used to manipulate trapped charged particles and plasmas in order to achieve $\bar{\text{H}}$ trapping. The current experimental performance is about one trapped $\bar{\text{H}}$ per experiment. This comes as no surprise, as there is an energy difference of 11 orders of magnitude from the 5.3 MeV $\bar{\text{p}}$s delivered by the AD to the 50 μeV depth of the magnetic trap for $\bar{\text{H}}$. The confinement of $\bar{\text{H}}$, and holding it for long enough for it to be in the ground state, is a milestone in $\bar{\text{H}}$ physics and opens up a range of possibilities for precision measurements. The 1s–2s ground state transition in atomic hydrogen has been measured to the unprecedented precision of 4.2×10^{-15} in a beam,[48] and to 1.2×10^{-12} in a trap.[49] This is, however, not the only transition of interest, and proposals exist to measure the hyperfine splitting in $\bar{\text{H}}$ using a beam. However, nothing prevents this from being done in a trap, and thus gaining an increased precision on the $\bar{\text{p}}$ magnetic moment, which is currently only known to a precision of 4.4×10^{-6}.[50] A rate of about one trapped atom per experiment will cause obvious limitations, but as ALPHA's recent demonstration of the first quantum transition in $\bar{\text{H}}$ showed, it is feasible to make measurements on very few $\bar{\text{H}}$ atoms when one makes use of the unique detection techniques that have been developed, and are indeed possible, when working with antimatter.[7] Trapped $\bar{\text{H}}$ also opens the door to testing the weak equivalence principle on $\bar{\text{H}}$ by "simply" letting it fall in the gravitational field. The current apparatus is not specifically well suited to such a measurement, but it is relatively easy to envisage changes that would allow such an experiment and, again, the position sensitive detection possible with $\bar{\text{p}}$s will be an important asset as demonstrated in a recent first look at the gravitational influence on antihydrogen.[51]

Eventually one atom per experiment will place overly stringent requirements on the stability of all the components and thus limit the measurement resolution achievable. As a result ALPHA is upgrading the experimental setup to improve both the temperature in the atom trapping region, the electronic noise, the homogeneity of the magnetic field and the residual gas pressure; all issues that are believed to influence the rate of trappable $\bar{\text{H}}$ with the current techniques. At CERN, further developments are under way which should eventually lead to higher trapping rates. CERN has commenced the design and construction of a new decelerator facility called ELENA[52] that will decelerate the 5.3 MeV $\bar{\text{p}}$s to 100 keV with an expected efficiency of $\sim$100%. Even when including the unavoidable losses in connection with degrading such a beam to trappable energies, this will still mean an increase in $\bar{p}$ capture per delivery of about two orders of magnitude greater than current rates. The ELENA facility is intended to go online in 2017 and will be a huge boost to $\bar{p}$ and $\bar{H}$ physics.

References

1. N. Madsen, Cold antihydrogen: a new frontier in fundamental physics, *Phil. Trans. Roy. Soc. A* **368**, 3671–3682 (2010).
2. G. M. Shore, Strong equivalence, Lorentz and CPT violation, anti-hydrogen spectroscopy and gamma-ray burst polarimetry, *Nucl. Phys. B* **717**, 86–118 (2005).
3. S. Maury, The Antiproton Decelerator: AD, *Hyp. Int.* **109**, 43–52 (1997).
4. M. Amoretti *et al.* (ATHENA), Production and detection of cold antihydrogen atoms, *Nature* **419**, 456–459 (2002).
5. G. B. Andresen *et al.* (ALPHA), Trapped antihydrogen, *Nature* **468**, 673–676 (2010).
6. G. B. Andresen *et al.* (ALPHA), Confinement of antihydrogen for 1,000 seconds, *Nature Physics* **7**, 558–564 (2011).
7. C. Amole *et al.* (ALPHA), Resonant quantum transitions in trapped antihydrogen atoms, *Nature* **483**, 439–443 (2012).
8. G. Gabrielse *et al.* (ATRAP), Driven Production of Cold Antihydrogen and the First Measured Distribution of Antihydrogen States, *Phys. Rev. Lett.* **89**, 233401 (2002).
9. G. Gabrielse *et al.* (ATRAP), Trapped Antihydrogen in Its Ground State, *Phys. Rev. Lett.* **108**, 113002 (2012).
10. Y. Enomoto *et al.* (ASACUSA), Synthesis of Cold Antihydrogen in a Cusp Trap, *Phys. Rev. Lett.* **105**, 243401 (2010).
11. M. Doser *et al.* (AEgIS), AEGIS: An experiment to measure the gravitational interaction between matter and antimatter, *J. Phys.: Conf. Ser.* **199**, 012009 (2010).
12. P. Perez and Y. Sacquin (GBAR), The GBAR experiment: gravitational behaviour of antihydrogen at rest , *Class. Quantum Grav.* **29**, 184008 (2012).
13. J.W. Humberston, M. Charlton, F.M. Jacobsen and B.I. Deutch, On antihydrogen formation in collisions of antiprotons with positronium, *J. Phys. B* **20**, L25–29 (1987).
14. B. I. Deutch, F. M. Jacobsen, L. H. Andersen, P. Hvelplund, H. Knudsen, M. H. Holzscheiter, M. Charlton and G. Laricchia, Antihydrogen Production by Positronium-Antiproton Collisions in an Ion Trap, *Phys. Scr.* **T22**, 248–255 (1988).
15. M. Charlton, Antihydrogen production in collisions of antiprotons with excited states of positronium, *Phys. Lett. A* **143**, 143–146 (1990).
16. C. H. Storry *et al.* (ATRAP), First Laser-Controlled Antihydrogen Production, *Phys. Rev. Lett.* **93**, 263401 (2004).
17. D. Möhl, Production of low-energy antiprotons, *Hyp. Int.* **109**, 33–41 (1997).

18. G. Gabrielse *et al.* (TRAP Collaboration), *Phys. Rev. Lett.* **57**, 2504 (1986).
19. G. Gabrielse *et al.* (TRAP), Cooling and slowing of trapped antiprotons below 100 meV, *Phys. Rev. Lett.* **63**, 1360–1363 (1989)
20. G. B. Andresen *et al.* (ALPHA), Compression of Antiproton Clouds for Antihydrogen Trapping, *Phys. Rev. Lett.* **100**, 203401 (2008).
21. T. M. O'Neil, A confinement theorem for nonneutral plasmas, *Phys. Fluids* **23**, 2216–2218 (1980).
22. J. Fajans, N. Madsen and F. Robicheaux, Critical loss radius in a Penning trap subject to multipole fields, *Phys. Plas.* **15**, 032108 (2008)
23. J. Fajans *et al.*, Effects of Extreme Magnetic Quadrupole Fields on Penning Traps and the Consequences for Antihydrogen Trapping, *Phys. Rev. lett.* **95**, 155001 (2005)
24. G. B. Andresen *et al.* (ALPHA), A novel antiproton radial diagnostic based on octupole induced ballistic loss, *Phys. Plas.* **15**, 032107 (2008)
25. G. B. Andresen *et al.* (ALPHA), Antimatter Plasmas in a Multipole Trap for Antihydrogen, *Phys. Rev. Lett.* **98**, 23402 (2007)
26. W. Bertsche *et al.* (ALPHA), A magnetic trap for antihydrogen confinement, *Nucl. Inst. & Meth. A* **566**, 746–756 (2006).
27. M. Amoretti *et al.* (ATHENA), Dynamics of antiproton cooling in a positron plasma during antihydrogen formation, *Phys. Lett. B* **590**, 133–142 (2004)
28. X. -P. Huang *et al.*, Steady-State Confinement of Non-neutral Plasmas by Rotating Electric Fields, *Phys. Rev. Lett.* **78**, 875–878 (1997)
29. C. A. Isaac *et al.*, Compression of Positron Clouds in the Independent Particle Regime, *Phys. Rev. Lett.* **107**, 033201 (2011).
30. F. Anderegg, E. M. Hollmann and C. F. Driscoll, Rotating Field Confinement of Pure Electron Plasmas Using Trivelpiece-Gould Modes, *Phys. Rev. Lett.* **81**, 4875–4878 (1998).
31. J. R. Danielson and C. M. Surko, Torque-Balanced High-Density Steady States of Single-Component Plasmas, *Phys. Rev. Lett.* **94**, 035001 (2005).
32. G. B. Andresen *et al.* (ALPHA), Production of antihydrogen at reduced magnetic field for anti-atom trapping, *J. Phys. B* **41**, 011001 (2008).
33. G. B. Andresen *et al.* (ALPHA), Evaporative Cooling of Antiprotons to Cryogenic Temperatures, *Phys. Rev. Lett.* **105**, 013003 (2010).
34. G. Gabrielse *et al.* (ATRAP), Adiabatic Cooling of Antiprotons, *Phys. Rev. Lett.* **106**, 073002 (2011).
35. D. L. Eggleston *et al.*, Parallel energy analyzer for pure electron plasma devices, *Phys. Fluids. B* **4**, 3432–3439 (1992).
36. C. M. Surko and R. G. Greaves, Emerging science and technology of antimatter plasmas and trap-based beams, *Phys. Plasmas* **11**, 2333–2348 (2004).
37. L. V. Jørgensen *et al.* (ATHENA), New Source of Dense, Cryogenic Positron Plasmas, *Phys. Rev. Lett.* **95**, 025002 (2005).
38. G. B. Andresen *et al.* (ALPHA), Antiproton, positron, and electron imaging with a microchannel plate/phosphor detector, *Rev. Sci. Inst.* **80**, 123701 (2009).
39. N. Madsen *et al.* (ATHENA), Spatial Distribution of Cold Antihydrogen Formation, *Phys. Rev. Lett.* **94**, 033403 (2005).

40. G. B. Andresen *et al.* (ALPHA), Antihydrogen formation dynamics in a multipolar neutral anti-atom trap, *Phys. Lett. B* **685**, 141–145 (2010).
41. G. B. Andresen *et al.* (ALPHA), Magnetic multipole induced zero-rotation frequency bounce-resonant loss in a Penning-Malmberg trap used for antihydrogen trapping, *Phys. Plas.* **16**, 100702 (2009).
42. B. Meerson and L. Friedland, Strong autoresonance excitation of Rydberg atoms: The Rydberg accelerator, *Phys. Rev. A* **41**, 5233–5236 (1990).
43. J. Fajans and L. Friedland, Autoresonant (nonstationary) excitation of pendulums, Plutinos, plasmas, and other nonlinear oscillators, *Am. Jour. Phys.* **69**, 1096–1102 (2001).
44. G. B. Andresen *et al.* (ALPHA), Autoresonant Excitation of Antiproton Plasmas, *Phys. Rev. Lett.* **106**, 025002 (2011).
45. G. B. Andresen *et al.* (ALPHA), Antihydrogen annihilation reconstruction with the ALPHA silicon detector, *Nucl. Inst. & Meth. A* **684**, 73–81 (2012).
46. G. B. Andresen *et al.* (ALPHA), Search for trapped antihydrogen, *Phys. Lett. B* **695**, 95–104 (2011).
47. G. B. Andresen *et al.* (ALPHA), Discriminating between antihydrogen and mirror-trapped antiprotons in a minimum-B trap, *New J. Phys.* **14**, 015010 (2012).
48. C. G. Parthey *et al.*, Improved Measurement of the Hydrogen $1S - 2S$ Transition Frequency, *Phys. Rev. Lett.* **107**, 203001 (2011).
49. C. L. Cesar *et al.*, Two-Photon Spectroscopy of Trapped Atomic Hydrogen, *Phys. Rev. Lett.* **77**, 255–258 (1996).
50. J. DiSciacca *et al.* (ATRAP), *Phys. Rev. Lett.* **110**, 130801 (2013).
51. ALPHA Collaboration and A. E. Charman, *Nature Communications* **4**, 1785 (2013).
52. T. Eriksson *et al.*, ELENA - a preliminary cost and feasibility study, CERN-AD-2007-079 OP, unpublished (2007). http://cdsweb.cern.ch/record/1072485/files/ab-2007-079.pdf.

Chapter 9

Quantum Information Processing with Trapped Ions

Christian F. Roos
Institute for Quantum Optics and Quantum Information,
Innsbruck, Technikerstraße 21a, A-6020, Austria
christian.roos@uibk.ac.at

Trapped ions constitute a well-isolated small quantum system that offers low decoherence rates and excellent opportunities for quantum control and measurement by laser-induced manipulation of the ions. These properties make trapped ions an attractive system for experimental investigations of quantum information processing. In this chapter, the basics of storing, manipulating and measuring quantum information encoded in a string of trapped ions will be discussed.

9.1. Introduction

Quantum information science has its origins in studies attempting to generalize theories of classical information processing to the quantum domain. In the 1980s and the 1990s, it was shown that by making explicit use of the laws of quantum physics, certain information processing parts could be carried out more efficiently than by classical information processing.[1] In quantum computing, bits are replaced by qubits, i.e. two-level quantum systems storing the most basic unit of information in superposition states $|\psi\rangle = \alpha|\uparrow\rangle + \beta|\downarrow\rangle$. Logic gates are realized as (entangling) unitary operations acting on a small number of quantum bits. Processing of quantum information can be effected by a sequence of quantum gates acting on a register of quantum bits initially prepared in a suitable pure quantum state in combination with measurements providing information about the state of the quantum register. While quantum computation might be the best-known branch of quantum information processing (QIP), the term also encompasses many other fields, such as quantum communication, the study of properties of entangled states, the enhancement of measurements by protocols processing quantum information and quantum simulation.

While the study of QIP started as a purely theoretical endeavour, it developed into a major research field when proposals began discussing actual QIP in experiments. The first practical proposal for QIP was made for trapped ions in 1995 and was soon followed by other proposals involving systems like ultracold neutral atoms, photons, quantum dots or superconducting devices. Any system considered for QIP needs to fulfil a set of basic requirements. These include the initialization of the system to a pure quantum state, the ability of coherent control over the quantum state, long coherence time and the possibility of carrying out quantum measurements providing information about the system's quantum state. All of these requirements are met to an excellent degree in experiments with trapped ions.

The development of radiofrequency and Penning traps has greatly advanced our possibilities for carrying out quantum physics experiments with charged particles under controlled conditions. Moreover, the availability of lasers for exciting atomic transitions has provided physicists with a versatile tool that can be used for trapping and cooling, quantum state manipulation and state detection.

Due to the long range of the Coulomb force, it is easy to store single ions in radiofrequency traps for days or even weeks. An ion can be localized to a region as small as 10 nm by laser cooling it to the ground state of the confining potential. If the ion trap is mounted in an ultrahigh vacuum environment with a pressure of 10^{-10} mbar or below, then the ion will be perturbed only very infrequently by collisions with the remaining residual gas atoms. In addition, if care is taken to control electric and magnetic stray fields, then the ion becomes a quantum object with coherence times in the range of milliseconds to minutes.[2]

The development of these experimental techniques was originally driven by the goal of using the ion for precision spectroscopy and in particular for the development of frequency standards. For this purpose, an ion needs to be prepared in a well-defined quantum state, coherently probed for as long as possible, and its quantum state read out with high efficiency. Here, quantum state detection refers to the ability to discriminate between the ion being in either one of two long-lived states. An efficient state read-out based on fluorescence detection is possible if one of two states can be excited to a third short-lived state which decays back to the original state.[3–5]

The very same properties that make a trapped ion suitable for precision spectroscopy also make it an interesting object for experiments related to quantum optics, QIP and fundamental tests of quantum physics. For QIP, each ion encodes one quantum bit (qubit) of information in a superposition of two long-lived internal states that can be initialized, manipulated and read out by laser light. The key point is to find a means of realizing entangling interactions between the quantum bits that allows the system to be scaled up – at least in principle – to a large number of quantum bits. The solution, proposed by Ignacio Cirac and Peter Zoller in 1995,[6] is to use a laser–ion interaction that couples the qubit states to the vibrational modes of the ion string. Due to the Coulomb interaction between the ions, the vibrational modes are shared between all ions in the string. This proposal and further ideas[7,8] inspired by it have been successfully implemented in the last couple of years and have led to the demonstration of different two-qubit quantum gates[9–12] and the realization of a few simple quantum algorithms. In other experiments, the same interactions have been used for creating entangled states and studying their entanglement.

9.2. Storing Quantum Information in Trapped Ions

In linear traps, the trapping frequency along the axial direction is typically much weaker than along the radial directions, i. e. $\omega_z \ll \omega_x, \omega_y$. For a collection of N ions, the equilibrium configuration minimizing their energy will be a linear ion string along the axial direction (with uneven ion spacings) as long as the anisotropy of the confining potential is sufficiently large. The typical distance between neighboring ions d is of the order of a few micrometers for axial frequencies of about 1 MHz.

The magnitude of d is interesting for several reasons. First of all, d is considerably bigger than optical wavelengths which makes it possible to address a single ion within an ion string by a strongly focussed laser beam[13] and to spatially resolve the fluorescence emitted by the ions on a charge-coupled device (CCD) camera. Secondly, the inter-ion distance d is much bigger than the Bohr radius. For this reason, state-dependent ion–ion interactions are extremely weak and the energy of an ion barely depends on the state of a neighboring ion. Therefore, trapped ions are well suited for storing quantum information in their internal states. However, the realization of quantum gates involving several ions is very challenging.

The storage of a quantum bit of information in a single ion requires the use of long-lived atomic states. Otherwise, the information will be quickly lost by spontaneous decay. As shown in Figure 9.1 (a),(b), there are different possibilities for encoding a qubit: it can be encoded as a *hyperfine qubit* in a pair of two hyperfine ground states if the ion has non-zero nuclear spin or as a *Zeeman qubit* in a superposition of two Zeeman ground states if the total angular momentum of the ground state is different from zero. Alternatively, a ground state and a metastable state can be used for encoding the qubit as an *optical qubit*, the name referring to the energy difference between the two states of the qubit that corresponds to that of a photon in the visible or near-infrared range of wavelengths.

The availability of states suitable for encoding a qubit is a first requirement that limits the choice of ion species. Further constraints arise from the requirements of having a qubit that can be efficiently detected and conveniently manipulated by laser light. For these reasons, experiments have focussed on singly charged alkali-earth ions and some further ion species. Ions that have been used for encoding hyperfine qubits are $^{9}\mathrm{Be}^+$, $^{25}\mathrm{Mg}^+$, $^{43}\mathrm{Ca}^+$, $^{111}\mathrm{Cd}^+$, $^{137}\mathrm{Ba}^+$ and $^{171}\mathrm{Yb}^+$. Optical qubits have been encoded in $^{40}\mathrm{Ca}^+$ and $^{88}\mathrm{Sr}^+$.

Hyperfine qubits have qubit states with transition frequencies in the microwave range. Therefore, single-qubit operations can in principle be realized by coupling the states with microwave fields. However, due to their long wavelengths, this approach is unable to address a single qubit within an ion string unless the fields are made spatially inhomogenous to achieve addressing in the frequency domain.

For this reason, the method of choice has been to use lasers for qubit manipulation. In the case of hyperfine qubits, two-photon Raman transitions are used for coupling the qubit states by using a laser field that off-resonantly couples the qubit states to a short-lived excited state.[14] Optical qubit states are directly coupled in a single-photon dipole-forbidden transition using a laser with ultra-stable frequency.[15] In the following sections, procedures for qubit preparation, manipulation and measurement will be discussed.

9.3. Preparation, Manipulation and Detection of an Optical Qubit

Most experiments processing quantum information using optical qubits have been carried out with $^{40}\mathrm{Ca}^+$ ions (see Figure 9.1 (c)) and this will allow us to illustrate the procedures necessary for preparing, manipulating and detecting the qubit. In $^{40}\mathrm{Ca}^+$, we make the identifications $|\downarrow\rangle = |S_{1/2}, m\rangle$ and $|\uparrow\rangle = |D_{5/2}, m'\rangle$ where $|\uparrow\rangle$ ($|\downarrow\rangle$) is an eigenstate of the operator σ_z with eigenvalue $+1$ (-1) and m, m' denote different Zeeman states of the

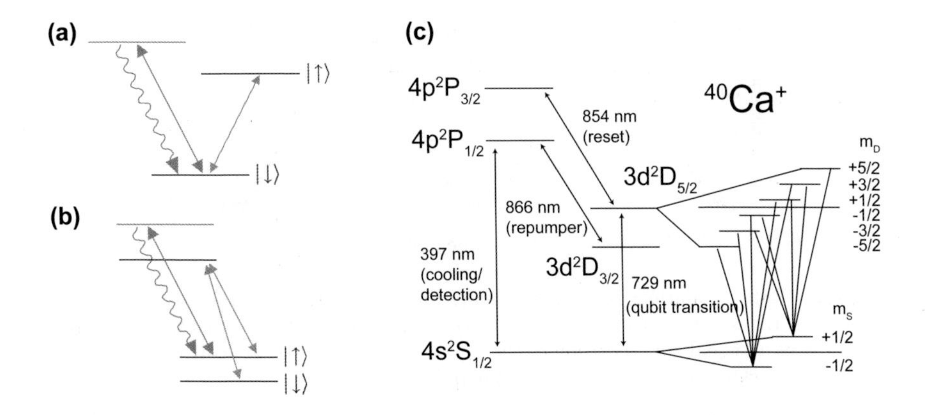

Fig. 9.1. Quantum information may be encoded in an optical qubit (a) or in a hyperfine qubit (b). The qubit is manipulated by laser pulses exciting either a single-photon or a Raman transition. (c) Energy levels and transitions used in the experiments with $^{40}Ca^+$. The right-hand side shows the Zeeman structure of the $S_{1/2} \leftrightarrow D_{5/2}$ quadrupole transition. An optical qubit can be encoded in a suitable pair of Zeeman states.

$S_{1/2}$ ground and $D_{5/2}$ metastable state. The metastable $D_{5/2}$ state has a lifetime of about 1.15 s, which is long compared with the typical duration of coherent manipulation and state detection.

Laser cooling allows the preparation of the vibrational state of a single trapped ion or an ion crystal in a pure quantum state. For QIP, the ion's internal state also needs to be initialized in a pure state which is achieved by optical pumping. To illustrate this process, we consider the problem of initializing a $^{40}Ca^+$ ion in the $S_{1/2}, m = 1/2$ state. The metastable state $D_{3/2}$ and $D_{5/2}$ are easily pumped to the $S_{1/2}$ by coupling them to the $P_{1/2}$ or $P_{3/2}$ state with lasers running at 866 and 854 nm, respectively. To initialize the ion in the $m = 1/2$ Zeeman ground state, an ion placed in a weak magnetic field that defines the quantization axis can be excited with σ_+-circularly polarized light on the $S_{1/2} \leftrightarrow P_{1/2}$ transition. In this way, the polarization of the laser beam makes sure that only the $m = -1/2$ state couples to the excited state. In practice, about 99% of the population can be transferred to the qubit state by this method.

The qubit can be coherently manipulated by exciting it with a monochromatic travelling-wave laser beam. The Hamiltonian describing

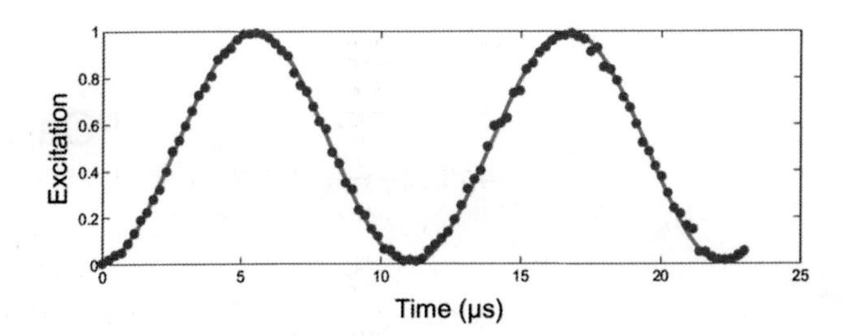

Fig. 9.2. Rabi oscillations of an ion initially prepared in the $|\downarrow\rangle$ state and resonantly excited on the qubit transition. A $\pi/2$ pulse, corresponding to an excitation of about 3 μs creates an equal superposition of the $|\downarrow\rangle$ and $|\uparrow\rangle$ state. π pulse, corresponding to a 6 μs pulse, exchanges the population of the qubit states.

the laser–ion interaction is given by

$$H = \frac{\hbar\Omega}{2}(\sigma_+ e^{-i\Delta t} e^{i(k\hat{x}+\phi)} + \sigma_- e^{i\Delta t} e^{-i(k\hat{x}+\phi)}) + \hbar\nu(a^\dagger a + \frac{1}{2}), \tag{9.1}$$

where $\Delta = \omega_L - \omega_0$ is the detuning of the laser frequency from the qubit transition frequency, ϕ the phase of the laser field, k its wavenumber, $\hat{x}$ the operator describing the ion's position and Ω the Rabi frequency characterizing the coupling strength. The final term represents the harmonic confinement of the trap and does not matter for the moment. For a resonant laser beam, $\Delta = 0$, the equation reduces to

$$H_{carr} = \frac{\hbar\Omega}{2}(\sigma_+ e^{i\phi} + \sigma_- e^{-i\phi}) \tag{9.2}$$

after carrying out the Lamb–Dicke approximation and dropping rapidly time-varying terms. This interaction will only act on the qubit state without affecting the ion's motional state. By applying it for a duration τ, qubit state transformations,

$$U(\theta, \phi) = \exp(-\frac{i}{\hbar} H_{carr}\tau), \tag{9.3}$$

are carried out that correspond to rotations of the Bloch sphere around a rotation axis located in the equatorial plane of the Bloch sphere where the rotation angle θ is given by $\theta = \Omega\tau$. In particular, rotations around the x- and y-axis are realized by setting the laser phase ϕ either to 0 or to $\pi/2$.

Additionally, rotations around the z-axis can be accomplished by illuminating the ion with a far-detuned laser field coupling either to the quadrupole or to dipole transitions that induce a differential Stark shift of the qubit states. In comparison to the rotations induced by the resonant interaction (9.1), interactions based on light-shifts have the advantage of being independent of the path-length fluctuations of the optical beam path which shift the phase of the laser light. Also, as the interaction depends quadratically on the coupling strength Ω, it is easier to focus to a single ion.

The optical qubit can be conveniently detected by coupling the $S_{1/2}$ state to the short-lived excited state $P_{1/2}$ by a laser with a wavelength close to 397 nm. As the $P_{1/2}$ state decays after about 8 ns back to the $S_{1/2}$ state, tens of millions of photons can be scattered on this transition by near-resonant laser light provided that a second laser at 866 nm is used to prevent optical pumping to the $D_{3/2}$ state. By suitably setting the frequencies of these lasers, the ion can be Doppler cooled at the same time. If the qubit is originally in the $|\downarrow\rangle$ state, the ion will scatter laser light that is detected by either a photomultiplier tube or a CCD camera. On the other hand, if the qubit is in the $|\uparrow\rangle$ state, the ion does not couple to the laser fields and will not emit fluorescence photons. In this case, the only light that reaches the detectors is due to scattering of stray light from the trap electrodes.

The detection process constitutes a projective measurement that projects the qubit state onto either $|\downarrow\rangle$ or $|\uparrow\rangle$. Detection errors can arise if the qubit changes its state during the detection interval or if photons scattered off trap electrodes are mistaken for fluorescence photons. In a recent experiment, the Oxford ion-trapping group demonstrated qubit detection errors as small as 10^{-4}.[16]

9.4. Entangling Quantum Gates

The big challenge in trapped-ion QIP is the generation of entanglement between different qubits that are spatially separated by the ions' mutual Coulomb repulsion. As the distance between ions is much bigger than the Bohr radius, no state-dependent ion–ion interactions exist in the absence of additional external electromagnetic fields. Fortunately, the very same force leads to a coupling of the ion motion which in turn can be used for mediating entanglement between the qubits. For this, laser fields are used that lead to correlated changes of the state of the qubits and the motional degrees of the ion crystal and thus entangle the two subsystems. The laser–ion interactions can be chosen in such a way that ultimately an entangling operation has been carried out between two or more qubits and no entanglement persists between the qubits and the vibrational degrees of freedom.

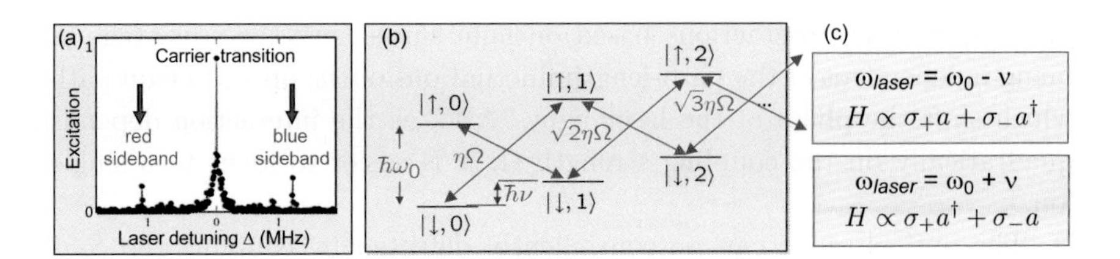

Fig. 9.3. Coherent excitation on vibrational sidebands (a) is used for coupling the qubit to the motional state. By exciting the red (blue) sideband, a qubit state change is accompanied by a decrease (increase) of the vibrational quantum number (b), thus realizing an (anti-) Jaynes–Cummings Hamiltonian (c).

Another possible way of creating entanglement between ion qubits is the use of qubit-state-dependent external potentials in addition to the trap potential which would enable entanglement generation by microwave fields.[17] A third strategy, developed to entangle ions over macroscopic distances, consists of entangling ion qubits with photons that are subsequently projected onto an entangled state by a measurement.[18] Currently, however, it is not very efficient as the entanglement generation process is non-deterministic and succeeds only in about 1 of 10^8 trials.

All current quantum gate realizations that deterministically entangle ions rely on a transient coupling of the internal states to the vibrational modes of the ion string. For coupling internal and vibrational states by laser light, we can once again turn to Eq. (9.1)'s description of the basic interaction between a single ion and a laser beam. We express the ion's position operator $\hat{x}$ by the phonon creation and annihilation operators a and $a^\dagger$ and write $k\hat{x} = \eta(a+a^\dagger)$ by making use of the definition of the Lamb–Dicke parameter η. Processes involving a change of vibrational quanta can be made resonant by setting the detuning of the laser frequency to either $\Delta = -\nu$ or to $\Delta = +\nu$. In the former case where the light field is red-detuned with respect to the ion's transition frequency by an amount equal to the ion's oscillation frequency, the ion can be excited from the lower to the upper electronic state while simultaneously reducing the vibrational quantum number by one (see Figure 9.3). In the Lamb–Dicke regime, the interaction is described by

$$H_{RSB} = i\hbar\frac{\eta\Omega}{2}(\sigma_+ a e^{i\phi} - \sigma_- a^\dagger e^{-i\phi}). \tag{9.4}$$

This type of interaction, which is often called red sideband excitation or lower vibrational sideband excitation, is used for resolved sideband cooling.

In a similar way, the ion can be excited on the blue or upper vibrational sideband by choosing $\Delta = +\nu$. Then, the Hamiltonian describing the interaction is given by

$$H_{BSB} = i\hbar\frac{\eta\Omega}{2}(\sigma_+ a^\dagger e^{i\phi} - \sigma_- a e^{-i\phi}). \tag{9.5}$$

The generalization of (9.4) and (9.5) to the case of a string of trapped ions is straightforward:[19] in this case, a and $a^\dagger$ denote annihilation and creation operators, respectively, of one of the vibrational modes of the string that is brought into resonance by making the laser detuning equal to the vibrational mode frequency. σ_+ and σ_- are the internal state operators of the ion with which the laser is interacting.

The laser-based gate operations can be broadly classified according to the manner in which the lasers interact with the ions:

(1) Quantum gates induced by a laser beam which interacts with a single ion at a time as originally proposed in the seminal paper by I. Cirac and P. Zoller.[6] In these gates, a single ion is entangled with a vibrational mode of the ion string and the entanglement is subsequently transferred from the vibrational mode to the internal state of a second ion.
(2) Quantum gates induced by a bichromatic laser which collectively interacts with two or more ions. Here, a vibrational mode becomes transiently entangled with the qubits before getting disentangled at the end of the gate operation, resulting in an effective interaction between the qubits capable of entangling them.[7,8,20,21]

Even though both classes of gates are applicable to hyperfine qubits as well as optical qubits, for a long time experiments with optical qubits have relied on the former[11] and experiments with hyperfine qubits on the latter type of interaction.[10] In any case, the main goal consists of demonstrating fast operations creating entanglement with high fidelity.

9.4.1. *Cirac–Zoller-type gate interactions*

An interaction on the red or the blue sideband leads to entanglement between the ion's internal and vibrational states. For experiments aiming at entangling internal states of several ions, the goal consists in finding pulse sequences that transiently entangle internal and vibrational states in such a way that in the end the internal states are entangled, with no entanglement remaining between the internal states and the motion. The simplest example of such a pulse sequence is the following: initially, two ions are prepared in the state $|\downarrow\rangle_2|\uparrow\rangle_1|0\rangle$ by optical pumping, sideband cooling and a carrier π-pulse to flip the state of ion 1. Next, a $\pi/2$ pulse is applied to ion 2 on the blue sideband which induces the mapping

$$|\downarrow\rangle_2|\uparrow\rangle_1|0\rangle \rightarrow |\downarrow\rangle_2|\uparrow\rangle_1|0\rangle + |\uparrow\rangle_2|\uparrow\rangle_1|1\rangle,$$

and entangles ion 2 with the motion (here, for simplicity, the normalization factor was dropped). In the last step, the entanglement is transferred from the vibrational mode to ion 1 by a π-pulse on the blue sideband, this time applied to ion 1, in order to map the state to

$$|\downarrow\rangle_2|\uparrow\rangle_1|0\rangle + |\uparrow\rangle_2|\uparrow\rangle_1|1\rangle \rightarrow |\downarrow\rangle_2|\uparrow\rangle_1|0\rangle + |\uparrow\rangle_2|\downarrow\rangle_1|0\rangle.$$

While the sequence given above maps a product state onto a maximally entangled state, it does not constitute a quantum gate operation since it performs the mapping only for a specific input state while it transforms all other input states into two-qubit states that are entangled with the vibrational state.

However, slightly more complicated pulse sequences using the same basic interactions can indeed realize a universal two-qubit gate as was first shown by I. Cirac and P. Zoller. In their proposal for a quantum controlled-NOT gate operation,[6] the state $\psi = \alpha|\uparrow\rangle + \beta|\downarrow\rangle$ of the control qubit is swapped onto a vibrational mode initially prepared in the ground state by a π-pulse on the blue sideband. After this pulse, the vibrational mode is in a superposition $\psi = \alpha|0\rangle + \beta|1\rangle$ of the two lowest vibrational states so that the problem is reduced to carrying out a controlled-NOT gate between the mode and the target qubit.[22] This operation can be further broken down into a controlled-phase gate enclosed in a Ramsey experiment consisting of two $\pi/2$-carrier pulses on the target qubit. For the conditional phase gate, in reference[6] the use of a 2π-pulse on the red sideband of a transition connecting one of the target qubit states to an auxiliary level was suggested. The gate operation is completed by mapping the quantum information from the vibrational mode back to the control qubit again using a π-pulse on the blue motional sideband.

The experimental realization of a controlled-NOT gate operation is important for trapped-ion quantum computing as the gate allows for realizing arbitrary unitary operations when combined with the single-qubit operations described in Section 9.3. Since its first demonstration in 2003,[11] the gate fidelity has been raised from the initial 70% to about 90%,[23] and the gate has been a building block in experiments demonstrating elementary quantum protocols.

9.4.2. Quantum gates based on bichromatic light fields

An alternative route towards the realization of a universal two-qubit gate consists in coupling both ions simultaneously to the vibrational mode. Using this approach, conditional phase gates $U = \exp(i\frac{\pi}{4}\sigma_z^{(1)} \otimes \sigma_z^{(2)})$ as well as Mølmer–Sørensen gates $U = \exp(i\frac{\pi}{4}\sigma_x^{(1)} \otimes \sigma_x^{(2)})$ that induce correlated spin flips are realizable by a bichromatic light field interacting with both ions simultaneously. Interestingly, there is a unified description for both types of gates even though at a first glance the physical mechanisms at work seem to be fairly different.[24] This description, which is based on the driven quantum harmonic oscillator, will be reviewed in the following section.

The Hamiltonian $\tilde{H} = \hbar\nu a^\dagger a + \hbar\Omega i(a^\dagger e^{i\omega t} - ae^{-i\omega t})$ describes a harmonic oscillator oscillating at frequency ν and driven by a periodic force with frequency ω and coupling strength Ω. Going into an interaction picture defined by $H_0 = \hbar\nu a^\dagger a$ yields the Hamiltonian

$$H = \hbar\Omega i(a^\dagger e^{i\delta t} - ae^{-i\delta t}), \tag{9.6}$$

where $\delta = \omega - \nu$. Under the action of the driving force, an oscillator that is initially in a coherent state remains in a coherent state. For a force that is slightly detuned from resonance, the coherent state maps out a circle in phase space and returns to the initial state after a period $\tau = 2\pi/\delta$. This operation multiplies the oscillator state by a phase factor whose magnitude is given by the ratio of the strength of the force and the detuning.[10,20] After time t, the propagator is $U(t) = \hat{D}(\alpha(t))\exp(i\Phi(t))$ with $\alpha(t) = i\left(\frac{\Omega}{\delta}\right)(1 - e^{i\delta t})$ and $\Phi = \left(\frac{\Omega}{\delta}\right)^2(\delta t - \sin\delta t)$. Here, $\hat{D}(\alpha) = e^{\alpha a^\dagger - \alpha^* a}$ denotes the displacement operator in the motional state space. After a time $\tau_N = 2\pi N/|\delta|$, $N = 1, 2, \ldots$, the motional state returns to its initial state in phase space with its phase changed by an amount $\Phi(\tau_N) = 2\pi N\left(\frac{\Omega}{\delta}\right)^2 \mathrm{sign}(\delta)$.

By making this phase change dependent on the internal states of a pair of ions, an entangling gate operation can be achieved. For

$$H = \hbar\Omega i(a^\dagger e^{i\delta t} - ae^{-i\delta t})\mathcal{O}, \tag{9.7}$$

where $\mathcal{O}$ is an operator acting on the qubit states, the propagator U is replaced by

$$U(t) = \hat{D}(\alpha(t)\mathcal{O})\exp(i\Phi(t)\mathcal{O}^2). \tag{9.8}$$

Choosing the interaction time τ such that $\alpha(\tau) = 0$ thus realizes a propagator that depends nonlinearly on $\mathcal{O}$ and does not alter the vibrational state.

9.4.3. *Conditional phase gates*

Conditional phase gates are realized by setting the operator $\mathcal{O}$ in Eq. (9.7) to $\mathcal{O} = \sigma_z^{(1)} + \sigma_z^{(2)}$ and choosing a driving force $\gamma(t) = \eta\Omega e^{i\delta t}$ with constant amplitude. With this choice, for $\tau = 2\pi/\delta$ the propagator becomes equal to

$$U_\gamma(\tau) = \exp\left(i4\pi\left(\frac{\eta\Omega}{\delta}\right)^2 \mathrm{sign}(\delta)\sigma_z^{(1)} \otimes \sigma_z^{(2)}\right), \qquad (9.9)$$

up to an unimportant global phase. By setting the coupling strength equal to $\eta\Omega = \delta/4$, a conditional phase gate is realized.

This type of gate was implemented for the first time[10] by Leibfried *et al.* who worked with two $^9\mathrm{Be}^+$ hyperfine qubits held in a trap with axial oscillation frequency ν. The required gate Hamiltonian was realized by a spatially and temporally varying differential ac-Stark shift. For this, the ions were placed in a standing wave created by two counter-propagating Raman beams with a frequency difference $\delta\omega = \sqrt{3}\nu + \delta$ that induced a differential Stark shift between the qubit states.[a] By choosing the ion distance d such that $|\Delta\vec{k}|d = 2\pi M$ with integer M and $\Delta\vec{k}$ the Raman k-vector of the transition, the ions were coupled to the stretch mode by a spatial light shift gradient only if they were in different qubit states. In this way, a state-dependent displacement force was created based on the strong field gradients created by the Raman light field.

9.4.4. *Mølmer–Sørensen gates*

In contrast to $\sigma_z \otimes \sigma_z$ gates that do not change the internal states of the ions, the $\sigma_\phi \otimes \sigma_\phi$ gate operations first investigated by Sørensen, Mølmer[7] and others[8] rely on collective spin flips $|\downarrow\downarrow\rangle \leftrightarrow |\uparrow\uparrow\rangle$, $|\downarrow\uparrow\rangle \leftrightarrow |\downarrow\uparrow\rangle$ by processes coupling to the lower and upper motional sidebands as illustrated in Figure 9.4. It can be shown that the Hamiltonian governing the action of the gate is described by setting $\mathcal{O} = \sigma_\phi^{(1)} + \sigma_\phi^{(2)}$ where $\sigma_\phi = \cos\phi\,\sigma_x + \sin\phi\,\sigma_y$. For a properly chosen coupling strength, the gate operation maps the product state basis $\{|\uparrow\uparrow\rangle, |\uparrow\downarrow\rangle, |\downarrow\uparrow\rangle, |\downarrow\downarrow\rangle\}$ onto a basis of entangled states.

In order to see that the simultaneous application of a red and a blue sideband excitation does indeed produce the desired interaction, one needs

[a]In the actual experiment, the magnitude of the Stark shift was not the same for both qubit states and therefore care had to be taken to cancel additional single-qubit phase shifts.

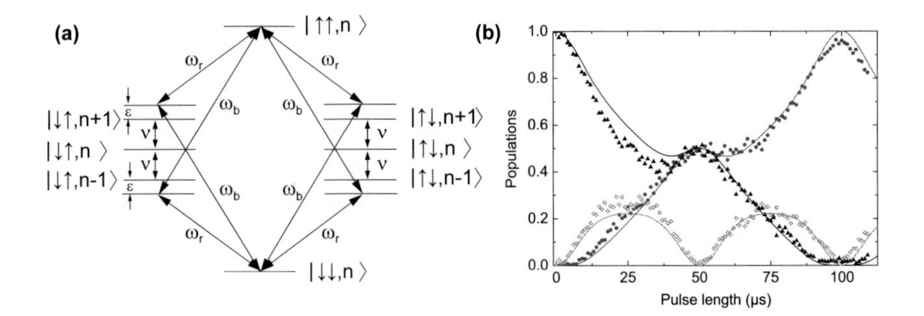

Fig. 9.4. (a) Mølmer–Sørensen gate. A bichromatic laser field with frequencies ω_b, ω_r satisfying $2\omega_0 = \omega_b + \omega_r$ is tuned close to the upper and lower motional sideband of the qubit transition. The field couples the qubit states $|\downarrow\downarrow\rangle \leftrightarrow |\uparrow\uparrow\rangle$ via the four interfering paths shown in the figure. Similar processes couple the states $|\uparrow\downarrow\rangle \leftrightarrow |\downarrow\uparrow\rangle$ with the same strength provided that the Rabi frequencies of the light fields ω_b, ω_r are equal. (b) Dynamics of the populations $p_{\downarrow\downarrow}$ (triangles), $p_{\uparrow\uparrow}$ (filled circles) and $p_{\downarrow\uparrow} + p_{\uparrow\downarrow}$ (open circles) induced by the bichromatic laser field. After $\tau = 50\mu s$, the $|\downarrow\downarrow\rangle$ has been transformed into the entangled state $|\downarrow\downarrow\rangle + i|\uparrow\uparrow\rangle$.

to sum up the Hamiltonians (9.4) and (9.5). Because of the chosen detunings, the phases in the Hamiltonians become time-dependent with $\phi_r = \epsilon t$ in (9.4) and $\phi_b = -\epsilon t$ in (9.5). Therefore, for a single ion we have

$$\begin{aligned} H_{RSB} + H_{BSB} &= i\hbar\eta\Omega(\sigma_+ a e^{i\epsilon t} - \sigma_- a^\dagger e^{-i\epsilon t}) + i\hbar\eta\Omega(\sigma_+ a^\dagger e^{-i\epsilon t} - \sigma_- a e^{i\epsilon t}) \\ &= -i\hbar\eta\Omega\sigma_y(a^\dagger e^{-i\epsilon t} + a e^{i\epsilon t}), \end{aligned} \tag{9.10}$$

where for notational convenience we chose a different convention for defining the Rabi frequency Ω. If both ions are simultaneously illuminated by the bichromatic light field, a Hamiltonian is obtained that corresponds to setting $\mathcal{O} = \sigma_y^{(1)} + \sigma_y^{(2)}$ in (9.7) and which can be used to implement the gate operation $U = \exp(i\frac{\pi}{4}\sigma_y^{(1)} \otimes \sigma_y^{(2)})$. More generally, arbitrary $\sigma_\phi \otimes \sigma_\phi$ gates are generated by properly setting the global phase of the bichromatic light field. Figure 9.4 shows the population evolution under the action of a Mølmer–Sørensen gate in an experiment[12] demonstrating the creation of Bell states with a fidelity above 99%.

9.5. Quantum State Tomography

All information about physical systems is inferred from measurements. In quantum physics, sometimes a measurement on only a single quantum system is performed, the outcome of which is one of the eigenvalues of the observable that is measured. More often, however, the expectation value $\langle A\rangle = \mathrm{Tr}(\rho A)$ of an observable A is measured using an ensemble of systems all prepared in the same quantum state ρ where Tr denotes the trace operation. Here, the measurement is either carried out in parallel on many quantum systems of the same type or it is carried out sequentially on an individual quantum system that is repeatedly prepared and measured to gather the required information.

In the field of quantum information, quantum measurements are a key element. Among others, measurement of observables called *entanglement witnesses* can provide information about whether a quantum state is entangled. There are, however, also questions that cannot be answered by the measurement of a single observable because the quantity to be measured cannot be written as a linear function of the density operator. A simple example of this kind is the task of determining the purity of a single-qubit state. Another example is the characterization of the entanglement of a two-qubit entangled state by means of an entanglement measure.

In these cases, one way to accomplish these tasks is to devise a strategy for determining the whole density matrix that describes the quantum state. In this way, the complete characterization of the quantum state by means of its density matrix provides the answer to any question one might want to pose. This approach has been called quantum state tomography. The following discussion of quantum state tomography will be restricted to the case of finite-dimensional Hilbert spaces. The assumption that the description of the quantum state to be measured requires only a finite number of dimensions is well satisfied in the experiments described in this chapter. Formally, the basic principle of quantum state tomography can be summarized as follows:

In a d-dimensional quantum system, it is possible to expand the density matrix ρ using a basis of d^2 hermitian matrices A_i that are mutually orthogonal. Mathematically, this condition is expressed as

$$\rho = \sum_{i=1}^{d^2} \lambda_i A_i, \tag{9.11}$$

with $\mathrm{Tr}(A_i A_j) = \mu_i \delta_{ij}$ where $\mu_i \neq 0$. Since the matrices A_i are assumed to be hermitian, they are observables with mean value

$$\langle A_j\rangle = \mathrm{Tr}(\rho A_j) = \sum_{i=1}^{d^2} \lambda_i \mathrm{Tr}(A_i A_j) = \lambda_i \mu_i, \tag{9.12}$$

and the coefficients that completely characterize the state are given by

$$\lambda_i = \mu_i^{-1}\langle A_i \rangle. \tag{9.13}$$

Therefore, a measurement of the set of observables $\mathcal{S} = \{A_i\}$ is sufficient for reconstructing the density operator ρ. The most basic example of a measurement of this kind is the reconstruction of the density matrix of a spin-1/2 system or any other two-level quantum system. In this case, a possible set of observables is given by $\mathcal{S} = \{\mathbb{I}, \sigma_x, \sigma_y, \sigma_z\}$ where $\mathbb{I}$ is the identity and σ_j are the Pauli matrices. Because of $\mathrm{Tr}(\sigma_i \sigma_j) = 2\delta_{ij}$ and $\mathrm{Tr}(\sigma_i) = 0$, the observables are orthogonal and the density matrix can be written as

$$\rho = \frac{1}{2}(\mathbb{I} + n_x\sigma_x + n_y\sigma_y + n_z\sigma_z) \text{ with } n_k = \langle \sigma_k \rangle. \tag{9.14}$$

A measurement of the density matrix is accomplished by measuring the components n_j of the Bloch vector $\vec{n} = (n_x, n_y, n_z)$.[b]

The generalization of this procedure to the reconstruction of an N-qubit system is straightforward. In this case, it is sufficient to measure observables $A_{\vec{j}}$ that are tensor products of Pauli matrices, i. e. $A_{\vec{j}} = A_{j_1} \otimes A_{j_2} \otimes \ldots$ with $A_{j_k} \in \{I, \sigma_x, \sigma_y, \sigma_z\}$.

For a qubit encoded in a trapped ion, a number of measurements need to be carried out. The observable σ_z is determined by a fluorescence measurement as described in section 9.3. The other two observables cannot be directly measured. Instead, for measuring σ_x, we apply the single-qubit rotation (9.3) that maps the eigenvector corresponding to the eigenvalue $+1$ of σ_x onto the $+1$-eigenvector of σ_z first before measuring σ_z on the transformed state. A similar procedure is carried out for measurements of σ_y. In the case of an N-qubit system encoded in a string of ions, this procedure requires the possibility of addressing single ions and also measuring the fluorescence of a single ion. To measure a spin correlation like $\sigma_x^{(1)} \otimes \sigma_z^{(2)}$, first the Bloch sphere of qubit 1 is rotated around the y-axis by a $\pi/2$ pulse. Then, a spatially resolved measurement of the fluorescence of both ions is performed in order to determine $\sigma_z^{(1)} \otimes \sigma_z^{(2)}$. This requires correlating the measurement results (either $+1$ or -1) found for ion 1 and ion 2 by multiplying them with each other. This procedure is easily generalized to N qubits where it requires measurements in 3^N different bases that are reduced to measurements of $\sigma_z^{(j_1)}$, $\sigma_z^{(j_1)} \otimes \sigma_z^{(j_2)}$, $\sigma_z^{(j_1)} \otimes \sigma_z^{(j_2)} \otimes \sigma_z^{(j_3)}$ where $j_m \in \{1, 2, \ldots N\}$ and so on. Figure 9.5 shows the experimentally reconstructed density matrix of a two-ion Bell state.

[b]There is no need to measure the observable $\mathbb{I}$ since $\langle \mathbb{I} \rangle = 1$ for any quantum state.

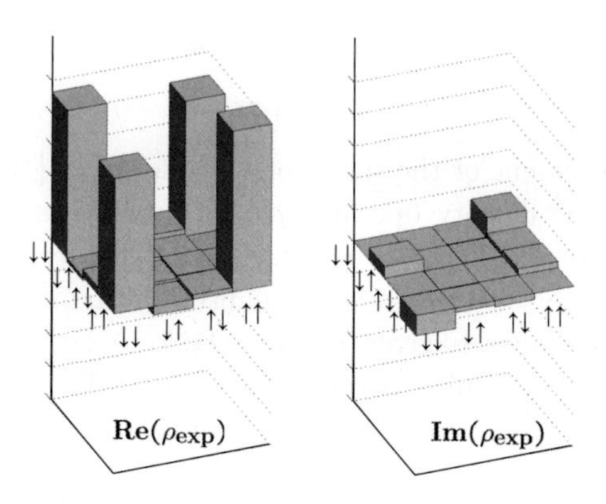

Fig. 9.5. Density matrix reconstruction of the two-ion Bell state $|\psi\rangle = \frac{1}{\sqrt{2}}(|\downarrow\rangle|\downarrow\rangle + |\uparrow\rangle|\uparrow\rangle)$. On the left, the real part of the experimentally reconstructed density matrix ρ is shown. As expected, the populations of the $\downarrow\downarrow$ and $\uparrow\uparrow$ have populations close to 0.5. In addition, the existence of off-diagonal elements close to 0.5 demonstrates the coherence between the $\downarrow\downarrow$ and $\uparrow\uparrow$ parts and proves that the state is indeed entangled and not just a mixture of $\downarrow\downarrow$ and $\uparrow\uparrow$.

Although the general strategy for measuring the density matrix ρ is very straightforward, its practical implementation becomes non-trivial once the inevitable condition that in any experiment only a finite number of copies of ρ is available for carrying out the measurements is accounted for. For this reason, it is impossible to measure the expectation values $\langle A_i \rangle$ of Eq. (9.13) to arbitrary precision. Any value entering this equation can therefore only be an estimate $\tilde{\lambda}_i$ of the observable's expectation value λ_i that will converge to the true value only in the limit of an infinite number of measurements.

This minor difference has important consequences, as are illustrated in Figure 9.6 for the case of a two-level system. For simplicity, the Bloch vector describing the state of the system is assumed to have a vanishing y-component. Any valid Bloch vector has to lie within the unit circle. However, errors in the measurement of $\langle \sigma_x \rangle$ and $\langle \sigma_z \rangle$ can give rise to a state reconstruction with a vector whose tip lies outside the "Bloch" circle. Then, the corresponding density has one eigenvalue bigger than one and one eigenvalue smaller than zero; thus, it is no longer positive semidefinite and does not describe a valid physical state. This problem occurs particularly for states of high purity that lie close to the boundary of the Bloch sphere. However, the likelihood of encountering this problem increases the higher the dimension of the state space becomes. In practice, these problems can

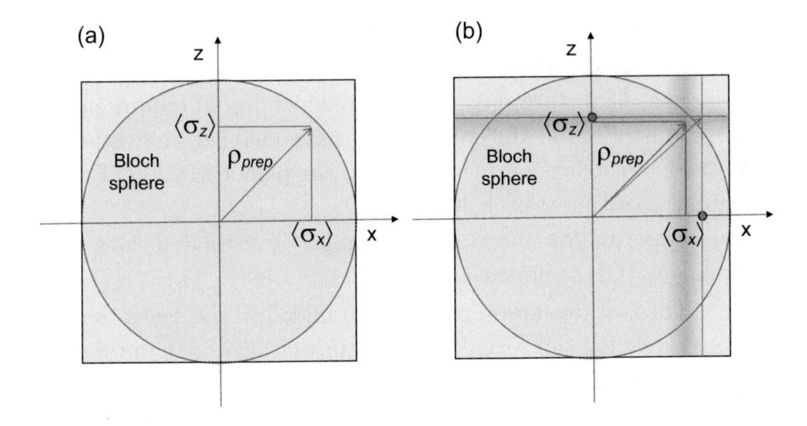

Fig. 9.6. Quantum state tomography of a two-level system. The Bloch vector describing the system is indicated by an arrow. Errors in the determination of Bloch vector components λ_x and λ_z can give rise to a reconstructed vector whose tip (indicated by the "+" sign) lies outside the unit circle containing the physically allowed states.

be overcome by state reconstruction schemes based on maximum-likelihood or Bayesian estimation that go beyond the simple linear reconstruction scheme sketched in Eqs (9.11)–(9.13). By these schemes, it is possible to reconstruct physically valid density matrices such as the one shown in Figure 9.5 from the same set of measurements as is used for the linear reconstruction technique.

While the principles of quantum state tomography were already considered 50 years ago, its extension to the complete characterization of quantum processes[1] started only about 15 years ago. In quantum process tomography, the characterization of a physical transformation of a state by means of unitary or decohering processes and measurements is achieved by preparing a set of input states that are subjected to the process in question and reconstructing the quantum state at the output. Process tomography is one possible way of characterizing experimental realizations of entangling gate operations.[23]

9.6. Elementary Quantum Protocols and Quantum Simulation

In current experiments, entangling gate operations have been applied to ion strings of up to 16 ions. Experiments investigating high-fidelity operations have shown that single-qubit gates can be performed with error rates as low a one in 10^5 whereas for two-qubit gates error rates as low as 1% have been observed. While these achievements are still far from the requirements of large-scale quantum computation, which would require millions of gates acting on thousands of qubits, they have enabled interesting entanglement studies in strings of a few ions and the demonstration of a number of elementary quantum protocols.

Various experiments have, for example, investigated Greenberger–Horne–Zeilinger (GHZ) states $\psi = \frac{1}{\sqrt{2}}(|\uparrow\uparrow \ldots \uparrow\rangle + |\downarrow\downarrow \ldots \downarrow\rangle)$ of up to 14 ions[9,25]. These states are maximally entangled in the sense that a measurement of single ion can reveal the full information about the state of the other ions. This property makes GHZ states fragile with respect to interactions with a decohering environment. It does not, however, signify that any measurement will completely destroy the quantum character of the state. If, for example, one ion of the three-qubit state $\psi = \frac{1}{\sqrt{2}}(|\uparrow\uparrow\uparrow\rangle + |\downarrow\downarrow\downarrow\rangle)$ is measured in the σ_z-basis $\{|\uparrow\rangle, |\downarrow\rangle\}$, the other two will be projected into either the product state $|\downarrow\downarrow\rangle$ or $|\uparrow\uparrow\rangle$. If, on the other hand, a measurement onto the eigenstates of σ_x $(|\uparrow\rangle \pm |\downarrow\rangle)$ is carried out, the remaining ions are projected into one of the Bell states $|\uparrow\downarrow\rangle - |\downarrow\uparrow\rangle$ as demonstrated in reference[26].

Elementary quantum algorithms with trapped ions include demonstrations of quantum teleportation, quantum error correction, entanglement purification and entanglement swapping[27–31]. More details about some of these experiments can be found in references.[32,33] Efforts to advance experiments along these lines focus on reducing error rates in the coherent operations, speeding them up and handling larger numbers of ions by using microfabricated ion traps.

Quantum simulations are another research direction that has recently come into the focus of trapped-ion experiments. Quantum simulation aims at simulating a quantum system of interest with a controllable laboratory system described by the same mathematical model. In this way, it might be possible to simulate quantum systems that cannot be efficiently simulated on a classical computer[34] or easily accessed experimentally. For this, the laboratory system needs to be very well understood in terms of the Hamiltonian describing it. To turn it into a useful quantum simulator, it should allow for parameter tunability and for the measurement of observables which provide important insights into the physics of the system to be simulated. There are two types of quantum simulators currently discussed in the literature. Digital quantum simulators[35] try to translate the unitaries describing the system dynamics into quantum circuits consisting of elementary gate operations. In this approach, a universal quantum computer could be used for efficiently simulating all quantum systems with local interactions. The second class, the so-called analog quantum simulators, builds on the principle of engineering a system which has the exact Hamiltonian of interest. The main motivation behind these approaches is to find solutions of problems in quantum many-body physics that cannot be efficiently simulated on classical computers. First experiments demonstrating basic elements of analog and digital quantum simulators with trapped ions have been carried out in the last couple of years (for an overview, see references[36,37]).

One class of Hamiltonians that seem to be particularly well suited for being simulated by trapped ions, are models describing systems of interacting spins. Experiments simulating the Ising Hamiltonian with a transverse field, $H = \sum_{i,j} J_{ij}\sigma_z^i\sigma_z^j + B\sum_i \sigma_x^i$ started with two to three ions[38,39] demonstrating adiabatic state changes and spin frustration. Since then, experiments involving up 16 ions[40] have been carried out in order to show the onset of a quantum phase transition. In all these experiments, each ion-qubit encodes a spin-1/2 system which interacts with the other spins via laser-mediated effective spin–spin interactions similar to those used for the entangling quantum gates discussed previously. The action of the transverse field can be implemented in a fairly straight-forward manner by further interactions coupling the two states of the qubits.

If the present experiments could be scaled up to a few tens of ions, trapped-ion experiments might enter the regime where numerical simulation aiming at predicting experimental outcomes becomes infeasible due to the large dimension of the Hilbert space required to describe the quantum dynamics. If these conditions are met, trapped-ion quantum experiments would no longer be just demonstration experiments but also tools for predicting quantum behavior in a new regime.

References

1. M. A. Nielsen and I. L. Chuang, *Quantum computation and quantum information* (Cambridge University Press, Cambridge, 2000).
2. J. J. Bollinger, D. J. Heinzen, W. M. Itano, S. L. Gilbert and D. J. Wineland, A 303 MHz frequency standard based on trapped Be^+ ions, *IEEE Trans. Instr. Meas.* **40**, 126–128 (1991).
3. W. Nagourney, J. Sandberg and H. Dehmelt, Shelved optical electron amplifier: Observation of quantum jumps, *Phys. Rev. Lett.* **56**, 2797–2799 (1986).

4. J. C. Bergquist, R. G. Hulet, W. M. Itano and D. J. Wineland, Observation of quantum jumps in a single atom, *Phys. Rev. Lett.* **57**, 1699–1702 (1986).
5. T. Sauter, W. Neuhauser, R. Blatt and P. E. Toschek, Observation of quantum jumps, *Phys. Rev. Lett.* **57**, 1696–1698 (1986).
6. J. I. Cirac and P. Zoller, Quantum computations with cold trapped ions, *Phys. Rev. Lett.* **74**, 4091–4094 (1995).
7. A. Sørensen and K. Mølmer, Quantum computation with ions in thermal motion, *Phys. Rev. Lett.* **82**, 1971–1974 (1999).
8. E. Solano, R. L. de Matos Filho and N. Zagury, Deterministic Bell states and measurement of the motional state of two trapped ions, *Phys. Rev. A* **59**, R2539–R2543 (1999).
9. C. A. Sackett, D. Kielpinski, B. E. King, C. Langer, V. Meyer, C. J. Myatt, M. Rowe, Q. A. Turchette, W. M. Itano, D. J. Wineland and C. Monroe, Experimental entanglement of four particles, *Nature* **404**, 256–259 (2000).
10. D. Leibfried, B. DeMarco, V. Meyer, D. Lucas, M. Barrett, J. Britton, W. M. Itano, B. Jelenković, C. Langer, T. Rosenband and D. J. Wineland, Experimental demonstration of a robust, high-fidelity geometric two ion-qubit phase gate, *Nature* **422**, 412–415 (2003).
11. F. Schmidt-Kaler, H. Häffner, M. Riebe, S. Gulde, G. P. T. Lancaster, T. Deuschle, C. Becher, C. F. Roos, J. Eschner and R. Blatt, Realization of the Cirac-Zoller controlled-NOT quantum gate, *Nature* **422**, 408–411 (2003).
12. J. Benhelm, G. Kirchmair, C. F. Roos and R. Blatt, Towards fault-tolerant quantum computing with trapped ions, *Nat. Phys.* **4**, 463–466 (2008).
13. H. C. Nägerl, D. Leibfried, H. Rohde, G. Thalhammer, J. Eschner, F. Schmidt-Kaler and R. Blatt, Laser addressing of individual ions in a linear ion trap, *Phys. Rev. A* **60**, 145–148 (1999).
14. C. Monroe, D. M. Meekhof, B. E. King, S. R. Jefferts, W. M. Itano, D. J. Wineland and P. L. Gould, Resolved-sideband Raman cooling of a bound atom to the 3D zero-point energy, *Phys. Rev. Lett.* **75**, 4011–4014 (1995).
15. C. Roos, T. Zeiger, H. Rohde, H. Nägerl, J. Eschner, D. Leibfried, F. Schmidt-Kaler and R. Blatt, Quantum state engineering on an optical transition and decoherence in a Paul trap, *Phys. Rev. Lett.* **83**, 4713–4716 (1999).
16. A. H. Myerson, D. J. Szwer, S. C. Webster, D. T. C. Allcock, M. J. Curtis, G. Imreh, J. A. Sherman, D. N. Stacey, A. M. Steane and D. M. Lucas, High-fidelity readout of trapped-ion qubits, *Phys. Rev. Lett.* **100**, 200502 (2008).
17. F. Mintert and C. Wunderlich, Ion-trap quantum logic using long-wavelength radiation, *Phys. Rev. Lett.* **87**, 257904 (2001).
18. D. L. Moehring, P. Maunz, S. Olmschenk, K. C. Younge, D. N. Matsukevich, L.-M. Duan and C. Monroe, Entanglement of single-atom quantum bits at a distance, *Nature* **449**, 68–71 (2007).
19. D. F. V. James, Quantum dynamics of cold trapped ions with application to quantum computation, *Appl. Phys. B* **66**, 181–190 (1998).
20. A. Sørensen and K. Mølmer, Entanglement and quantum computation with ions in thermal motion, *Phys. Rev. A* **62**, 022311 (2000).
21. G. J. Milburn, S. Schneider and D. F. James, Ion trap quantum computing

with warm ions, *Fortschr. Phys.* **48**, 801–810 (2000).
22. C. Monroe, D. M. Meekhof, B. E. King, W. M. Itano and D. J. Wineland, Demonstration of a fundamental quantum logic gate, *Phys. Rev. Lett.* **75**, 4714–4717 (1995).
23. M. Riebe, K. Kim, P. Schindler, T. Monz, P. O. Schmidt, T. K. Körber, W. Hänsel, H. Häffner, C. F. Roos and R. Blatt, Process tomography of ion trap quantum gates, *Phys. Rev. Lett.* **97**, 220407 (2006).
24. P. J. Lee, K. A. Brickman, L. Deslauriers, P. C. Haljan, L. Duan and C. Monroe, Phase control of trapped ion quantum gates, *J. Opt. B* **7**, 371–383 (2005).
25. T. Monz, P. Schindler, J. T. Barreiro, M. Chwalla, D. Nigg, W. A. Coish, M. Harlander, W. Hänsel, M. Hennrich and R. Blatt, 14-qubit entanglement: creation and coherence, *Phys. Rev. Lett.* **106**, 130506 (2011).
26. C. F. Roos, M. Riebe, H. Häffner, W. Hänsel, J. Benhelm, G. P. T. Lancaster, C. Becher, F. Schmidt-Kaler and R. Blatt, Control and measurement of three-qubit entangled states, *Science* **304**, 1478–1480 (2004).
27. M. Riebe, H. Häffner, C. F. Roos, W. Hänsel, J. Benhelm, G. P. T. Lancaster, T. W. Körber, C. Becher, F. Schmidt-Kaler, D. F. V. James and R. Blatt, Deterministic quantum teleportation with atoms, *Nature* **429**, 734–737 (2004).
28. M. D. Barrett, J. Chiaverini, T. Schaetz, J. Britton, W. M. Itano, J. D. Jost, E. Knill, C. Langer, D. Leibfried, R. Ozeri and D. J. Wineland, Deterministic quantum teleportation of atomic qubits, *Nature* **429**, 737–739 (2004).
29. J. Chiaverini, J. Britton, D. Leibfried, E. Knill, M. D. Barrett, R. B. Blakestad, W. M. Itano, J. D. Jost, C. Langer, R. Ozeri, T. Schaetz and D. J. Wineland, Implementation of the semiclassical quantum Fourier transform in a scalable system, *Science* **308**, 997–1000 (2005).
30. R. Reichle, D. Leibfried, E. Knill, J. Britton, R. B. Blakestad, J. D. Jost, C. Langer, R. Ozeri, S. Seidelin and D. J. Wineland, Experimental purification of two-atom entanglement, *Nature* **443**, 838–841 (2006).
31. M. Riebe, T. Monz, A. S. Villar, P. Schindler, M. Chwalla, M. Hennrich and R. Blatt, Deterministic entanglement swapping with an ion trap quantum computer, *Nat. Phys.* **4**, 839–842 (2008).
32. H. Häffner, C. F. Roos and R. Blatt, Quantum computing with trapped ions, *Physics Reports* **469**, 155–203 (2008).
33. R. Blatt and D. Wineland, Entangled states of trapped atomic ions, *Nature* **453**, 1008–1015 (2008).
34. R. Feynman, Simulating physics with computers, *Int. J. Theoret. Phys.* **21**, 467–488 (1982).
35. S. Lloyd, Universal quantum simulators, *Science* **273**, 1073–1078 (1996).
36. M. Johanning, A. F. Varón and C. Wunderlich, Quantum simulations with cold trapped ions, *J. Phys. B.* **42**, 154009 (2009).
37. R. Blatt and C. F. Roos, Quantum simulations with trapped ions, *Nat. Phys.* **8**, 277–284, (2012).
38. H. Friedenauer, H. Schmitz, J. Glueckert, D. Porras and T. Schaetz, Simulating a quantum magnet with trapped ions, *Nat. Phys.* **4**, 757–761 (2008).
39. K. Kim, M. Chang, R. Islam, S. Korenblit, L. Duan and C. Monroe, Entan-

glement and tunable spin-spin couplings between trapped ions using multiple transverse modes, *Phys. Rev. Lett.* **103**, 120502 (2009).

40. R. Islam, E. E. Edwards, K. Kim, S. Korenblit, C. Noh, H. Carmichael, G.-D.Lin, L.-M. Duan, C.-C. J. Wang, J. K. Freericks and C. Monroe, Onset of a quantum phase transition with a trapped ion quantum simulator, *Nat. Comm.* **2**, 377 (2011).

Chapter 10

Optical Atomic Clocks in Ion Traps

Helen S. Margolis
National Physical Laboratory,
Teddington, Middlesex TW11 0LW, United Kingdom
helen.margolis@npl.co.uk

Narrow optical transitions in single laser-cooled trapped ions are highly reproducible in frequency and therefore make ideal references for accurate atomic clocks. This chapter provides an introduction to trapped ion optical clocks, covering their principles of operation and describing the different systems being studied at present. The performance of state-of-the-art trapped ion optical clocks is discussed and the various contributions to their frequency uncertainty budgets are outlined. Finally the further developments likely to occur over the next few years are considered.

10.1. Introduction

Since 1967, the definition of the "second" in the International System of Units (SI) has been based on the ground-state hyperfine transition in the ^{133}Cs atom at 9 192 631 770 Hz. This definition can be realized with astounding accuracy, with the best caesium fountain primary frequency standards[1,2] having reached uncertainties below 3 parts in 10^{16}. However this performance is now being challenged by a new generation of atomic clocks operating at optical, rather than microwave, frequencies.

In selecting the best atomic transition to use as the reference frequency for an atomic clock, several factors must be considered. The first is the fractional frequency instability that can theoretically be achieved, which is inversely proportional to the quality factor Q of the reference transition (the ratio of its frequency ν_0 to its linewidth $\Delta\nu_0$). For a transition in the optical region of the spectrum, ν_0 can be nearly five orders of magnitude higher than for the caesium primary standard, whilst linewidth-limiting processes are similar. The potential gain in moving to optical atomic clocks is therefore very clear, although the improvement in stability that will be achieved in practice also depends on the signal-to-noise ratio with which the atomic absorption signal is observed. This scales theoretically as $N^{1/2}$, where N is the number of atoms detected. In selecting a reference transition, it is also important to consider its sensitivity to environmental perturbations such as electric and magnetic fields, since this will determine the reproducibility and accuracy of the clock.

Two different types of optical atomic clock are currently being investigated.[3] One approach, which offers the prospect of very low short-term instability, is based on an array of laser-cooled atoms trapped in an optical lattice.[4] The other approach, which is the focus of this chapter, is to use a single laser-cooled ion confined in a radiofrequency (RF) trap. Since the ion trap provides a very well-controlled environment, this second approach may have advantages in terms of the ultimate accuracy achievable.

10.2. Principles of Operation

As illustrated in Figure 10.1, there are three main parts to any trapped ion optical atomic clock. These are an ultra-stable probe laser (often referred to as the clock laser), a suitably narrow reference transition in the trapped ion to which the clock laser is stabilized and a femtosecond optical frequency comb which enables the stability of the reference to be transferred to other frequencies (both optical and microwave).

10.2.1. *Trapping, cooling and probing a single ion*

Trapped ion optical atomic clocks use RF or Paul-type ion traps[5] to confine an ion close to the minimum of a dynamic pseudopotential well created by a time-varying quadrupolar potential. In most clocks built to date, only a single ion is used, and so the potential need only be quadrupolar close to the centre of the trap. For this reason, hyperbolic electrodes are unnecessary and simpler electrode structures with better optical access are preferred, such as Paul–Straubel (ring) traps or endcap traps.[6] In some cases linear Paul traps are used.[7]

A single ion confined in the centre of an RF ion trap, where the electric field is zero, is a close approximation to the spectroscopic ideal of an absorber at rest in a perturbation-free environment.[8] Using laser-cooling techniques, the ion can be confined in the Lamb–Dicke regime where the

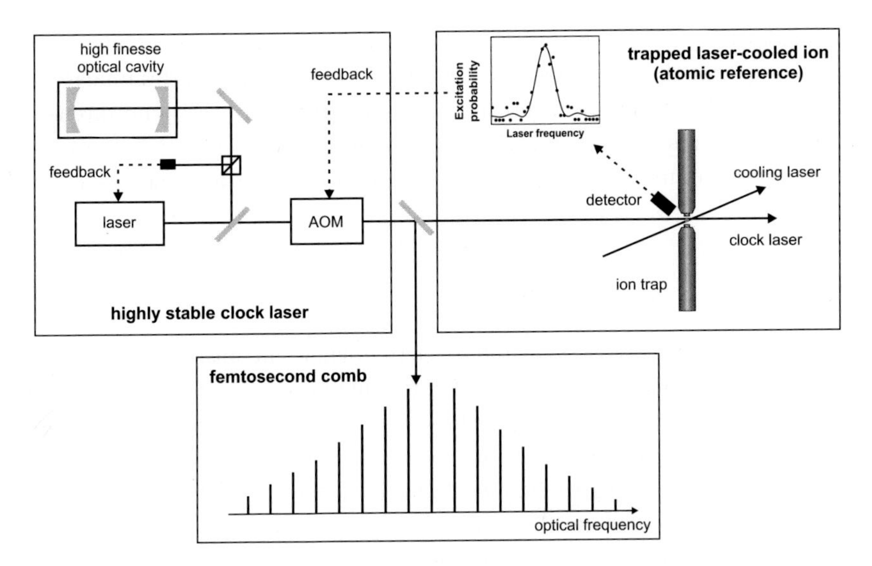

Fig. 10.1. The major components of a trapped ion optical atomic clock. An ultra-stable laser is locked to a narrow optical transition in the trapped ion, using an acousto-optic modulator (AOM) to bridge the frequency difference between the reference transition and the nearest resonance of the optical cavity. The femtosecond comb is used to transfer the frequency stability of the reference to other spectral regions.

spatial extent of the ion's wavepacket is smaller than the wavelength of the reference transition for the optical clock. In this regime, the absorption spectrum of the clock transition is free from the first-order Doppler effect and consists of an unshifted narrow carrier with weak sidebands at the characteristic frequencies of the ion's motion within the trap.[9] With the amplitude of the ion's micromotion minimized, for example using RF photon correlation techniques,[10] second-order Doppler shifts and electric field perturbations can also be reduced to very low levels. Lastly, since the ion trap is operated under ultrahigh vacuum conditions, collisional effects are negligible.

Most ions used for optical atomic clocks have both a strong allowed transition that is used for laser cooling and a narrow transition to a metastable state that is used as the frequency reference. Since the reference transition is narrow, it is necessarily very weak, but can be detected with close to 100% efficiency using the electron shelving technique.[11] This involves observing quantum jumps in the fluorescence from the cooling transition whenever the ion is driven into the metastable state by the clock laser. The resonance profile of the clock transition can be determined by measuring the quantum jump probability as a function of the clock laser frequency. The clock laser can then be stabilized to the reference transition by repeatedly stepping its frequency between the two estimated half-maximum intensity points of the resonance, and using the quantum jump rate imbalance between these points to derive an error signal for correction of the clock laser frequency.

10.2.2. *Clock laser stabilization*

The atomic transitions used as reference frequencies in optical atomic clocks typically have natural linewidths of a few hertz or less. To avoid degrading the achievable stability, the clock laser used to probe the transition must therefore have similar or narrower linewidth. Reduction of the free-running laser linewidth to this level can be achieved by locking the laser to a high finesse ultra-stable optical cavity using the Pound–Drever–Hall technique.[12] Such a cavity is constructed by optically contacting two high reflectivity concave mirrors onto each end of a spacer made of ultralow-expansion (ULE) glass. Typical cavities have an optical finesse of around 200 000 and are not usually tunable, so an AOM is used to shift the laser frequency to coincide with that of the atomic reference transition.

The frequency stability of the locked laser is determined by the stability of the optical cavity resonance frequency, and so isolation of the cavity from environmental perturbations is vital. The cavity is housed in an evacuated enclosure and controlled at a temperature close to the point at which its coefficient of thermal expansion is zero. The cavity housing and the optics for the Pound–Drever–Hall lock are normally mounted on a vibration-isolation platform and surrounded by an acoustic isolation enclosure. To achieve the best possible levels of frequency stability, much attention in recent years has been focused on designing cavities and their support structures to reduce their inherent sensitivity to vibrations.[13–16] The result is that the frequency stability of state-of-the-art optical cavities is limited by dimensional changes originating from thermal noise in the cavity mirror substrates and their coatings.[17,18]

For many years the benchmark for laser stability was set by a ULE-cavity-stabilized dye laser at NIST (National Institute of Standards and Technology), with a linewidth of 0.16 Hz for averaging times up to 20 s and a fractional frequency instability of 3×10^{-16} for an averaging time of 1 s.[19] However this performance has recently been surpassed by that of a diode laser stabilized to a cavity with a reduced thermal noise floor, achieved through a judicious choice of mirror substrate material, beam radius and cavity length.[20] The thermal noise floor can also be reduced by cooling the cavity to cryogenic temperatures, although alternative materials must be used to avoid unacceptably large coefficients of thermal expansion. Very recently, laser instability at the 1×10^{-16} level for averaging times from 0.1–1 s was demonstrated using a cavity constructed from mono-crystalline silicon and operated at a temperature of 124 K where the thermal expansion coefficient of silicon is zero.[21] This represents the lowest fractional frequency instability of any oscillator, either optical or microwave, reported to date.

10.2.3. *Femtosecond optical frequency combs*

To determine the absolute frequency of the reference transition in a trapped ion optical atomic clock relative to the caesium fountain primary frequency standard, a femtosecond optical frequency comb is used.[22,23] Such a comb is based on a mode-locked femtosecond laser that emits a periodic train of ultrashort pulses at a repetition rate f_{rep}. In the frequency domain, the output spectrum of such a laser is a comb of equally spaced modes at frequencies $f_n = nf_{\text{rep}} + f_0$, where n is an integer and the offset f_0 of the comb modes from integer harmonics of f_{rep} originates from dispersion within the laser cavity.

To use a comb to measure an optical frequency, three radiofrequencies must be determined: f_{rep}, f_0 and the beat frequency f_{b} between the optical frequency and the nearest comb mode. The pulse repetition rate f_{rep} can be determined directly from the beat signal between pairs of adjacent comb modes or between pairs of comb modes that are separated by a harmonic of the repetition rate, while f_0 is usually determined using the f:$2f$ self-referencing scheme.[24]

The accuracy with which absolute frequency measurements can be made is fundamentally limited by the accuracy of the microwave reference used for the optical frequency comb. For the most accurate measurements this should be traceable back to a local caesium fountain primary frequency standard. However optical frequency combs can also be used to measure optical frequency ratios rather than absolute frequencies. In this way optical clocks operating at different frequencies can be compared directly without the limitations imposed by the uncertainty in the current realization of the SI second.[25] This is extremely useful in enabling their systematic frequency uncertainties to be evaluated in a rapid and precise manner.

Table 10.1. Trapped ion optical frequency standards currently being studied, showing in each case the spectroscopic designation of the clock transition, the cooling and clock laser wavelengths $\lambda_{\rm cooling}$ and $\lambda_{\rm clock}$, and the natural linewidth $\Delta\nu_{\rm clock}$ of the clock transitions.

Ion	Clock transition	$\lambda_{\rm cooling}$ / nm	$\lambda_{\rm clock}$ / nm	$\Delta\nu_{\rm clock}$ / Hz
Alkali-like / quasi-alkali-like ions:				
^{40}Ca$^+$	^{2}S$_{1/2}$–^{2}D$_{5/2}$	397	729	0.14
^{88}Sr$^+$	^{2}S$_{1/2}$–^{2}D$_{5/2}$	422	674	0.4
^{171}Yb$^+$	^{2}S$_{1/2}$–^{2}D$_{3/2}$	370	436	3.1
^{171}Yb$^+$	^{2}S$_{1/2}$–^{2}F$_{7/2}$	370	467	$\sim 10^{-9}$
^{199}Hg$^+$	^{2}S$_{1/2}$–^{2}D$_{5/2}$	194	282	1.8
Alkaline-earth-like ions:				
^{27}Al$^+$	^{1}S$_0$–^{3}P$_0$	280 (^{25}Mg$^+$) or 313 (^{9}Be$^+$)	267	0.008
^{115}In$^+$	^{1}S$_0$–^{3}P$_0$	231	237	0.8

10.3. Systems Studied and State-of-the-art Performance

Narrow transitions suitable for optical atomic clocks have been identified in a number of different ions, and can be classified into two groups (Table 10.1). In the first group there are standards based on ions with alkali-like or quasi-alkali-like atomic structure. Of these, the ions with which most progress has been made towards the development of highly stable and accurate frequency standards are ^{40}Ca$^+$, ^{88}Sr$^+$, ^{171}Yb$^+$ and ^{199}Hg$^+$. All these have electric quadrupole clock transitions from the ^{2}S$_{1/2}$ ground state to low-lying metastable ^{2}D states, with natural linewidths in the range 0.14–3.1 Hz. However the ^{171}Yb$^+$ ion also has a ^{2}F$_{7/2}$ state lying below the ^{2}D states. This can only decay to the ground state via an electric octupole transition and as a result has an exceptionally long lifetime of roughly six years.[26] Higher laser power is required to drive this transition compared to the electric quadrupole transitions, causing an ac Stark shift which must be accurately controlled and quantified (see Section 10.4), but the advantage is that higher frequency stability may ultimately be achievable with a sufficiently narrow probe laser.

In the second group there are standards based on ions which have an atomic structure similar to that of the alkaline-earth elements, with two valence electrons. The reference transition in these systems is the strongly spin-forbidden $ns^2\,{}^1$S$_0$–$nsnp\,{}^3$P$_0$ transition, which is weakly allowed in atoms with non-zero nuclear spin due to hyperfine-interaction-induced level mixing. Not only are the natural linewidths of these transitions favourable, being 0.8 Hz for $^{115}In^+$ and 8 mHz for $^{27}Al^+$, but also certain systematic frequency shifts are very low. In particular there is no electric quadrupole shift of the clock transition frequency and the blackbody Stark shift is relatively small (see Section 10.4). However, the experimental implementation of these standards is quite challenging because deep UV laser sources are required for cooling and probing the ion. The $ns^2\,^1S_0$–$nsnp\,^1P_1$ resonance transitions are at inaccessible wavelengths of 159 nm and 167 nm for $^{115}In^+$ and $^{27}Al^+$, respectively. The $^{115}In^+$ ion is instead cooled on the $5s^2\,^1S_0$–$5s5p\,^3P_1$ intercombination line at 231 nm. However, because this is relatively weak, high power levels and bichromatic sideband cooling techniques must be used.[27] In $^{27}Al^+$, the corresponding transition is even weaker and cannot be used for laser cooling. Instead, the $^{27}Al^+$ ion is trapped along with an auxiliary ion that has a more convenient cooling wavelength, such as $^9Be^+$ or $^{25}Mg^+$. The Coulomb interaction between the two ions couples their motional states, leading to sympathetic cooling of the $^{27}Al^+$ ion and allowing the quantum state of the $^{27}Al^+$ clock ion to be transferred to the cooling ion for read-out.[28]

In spite of these technical challenges, the $^{27}Al^+$ optical clock is currently leading the field in terms of the performance levels achieved. With a probe time of 300 ms, a Fourier-transform-limited clock transition linewidth of 2.7 Hz has been observed,[29] corresponding to a Q-factor of 4.2×10^{14}. This is the highest achieved for any trapped ion optical clock. The $^{27}Al^+$ standard is also presently leading in terms of demonstrated stability and reproducibility. Results from a direct frequency comparison of two different $^{27}Al^+$ clocks at NIST, one using $^9Be^+$ as the auxiliary cooling ion and the other using $^{25}Mg^+$, show a combined fractional frequency instability of $2.8 \times 10^{-15}\tau^{-1/2}$ for averaging times up to a few thousand seconds.[30] The measured fractional frequency difference between the two clocks was found to be $(-1.8 \pm 0.7) \times 10^{-17}$, consistent with the estimated systematic frequency uncertainty of the $^{27}Al^+/^9Be^+$ clock (2.3×10^{-17}). The systematic uncertainty of the $^{27}Al^+/^{25}Mg^+$ clock was estimated to be 8.6×10^{-18}.

The uncertainties of the best absolute frequency measurements (Figure 10.2), however, are nowhere near as low as this, because they are dominated by the uncertainties of the caesium fountain primary frequency standards used to provide traceability to the SI second. Nevertheless such frequency measurements are important because they currently provide the only indication as to the level of agreement between independent realizations of a particular trapped ion optical atomic clock by different research groups. The best such agreement to date is at the part in 10^{15} level, be-

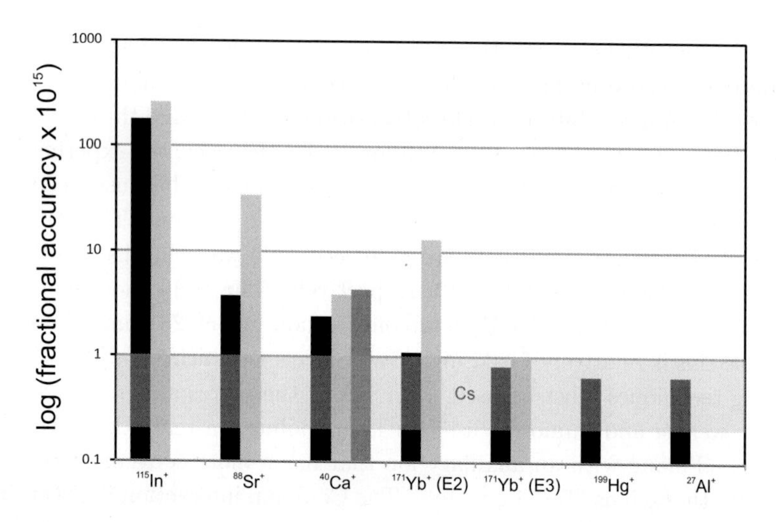

Fig. 10.2. Accuracy of absolute measurements of trapped ion optical clock transition frequencies in singly charged indium,[31,32] strontium,[33,34] calcium,[35–37] ytterbium,[38–41] mercury[42] and aluminium.[25] The uncertainty limit set by the caesium fountain primary frequency standards is also indicated. Where several independent measurements carried out in different laboratories are shown, these are generally in good agreement, except for the case of the indium ion standard, where there is an unresolved discrepancy of 1.3 kHz between the two measurements.

tween two recent absolute optical frequency measurements of the 467 nm electric octupole clock transition frequency in $^{171}Yb^+$ carried out at the National Physical Laboratory (NPL)[41] and the Physikalisch–Technische Bundesanstalt (PTB).[40] However, since direct comparisons between co-located trapped ion optical frequency standards indicate that their reproducibility is much higher than this, methods are required for comparing remotely located optical clocks without the need for an intermediate frequency reference. Current satellite-based techniques are inadequate in terms of the stability and accuracy they can achieve,[43] but transmission over optical fibre networks appears to present a viable alternative, at least for comparisons within Europe.[44]

10.4. Systematic Frequency Shifts

Even though an RF trap provides a relatively low perturbation environment for the ion, the high accuracy level aimed for with an optical atomic clock means that a large number of systematic frequency shifts must be considered to determine the overall uncertainty budget. This includes not only the fundamental effects outlined in this section, but also other shifts that are more technical in nature, one example being potential frequency chirps arising from AOMs.

External magnetic fields in general cause a linear Zeeman shift of atomic transition frequencies. However, for odd isotopes of alkali-like ions such as $^{171}Yb^+$ or $^{199}Hg^+$, which have nuclear spin $I = 1/2$, $m_F = 0 \rightarrow m_F = 0$ clock transitions exist that are field-independent to first order. There will still be a second-order Zeeman shift, but this can be controlled at the 10^{-17}–10^{-18} level by operating in low magnetic fields of around 1 μT. Similar $m_F = 0 \rightarrow m_F = 0$ transitions exist in $^{87}Sr^+$ and $^{43}Ca^+$, but these systems have rather complicated hyperfine structure due to their larger nuclear spin ($I = 9/2$ and $I = 7/2$ respectively). For this reason the even isotopes $^{88}Sr^+$ and $^{40}Ca^+$ are used instead, and the linear Zeeman shift is eliminated by alternately probing two Zeeman components that have equal but opposite frequency shifts from the zero-field centre of the multiplet.[33,34] In this case magnetic shielding is vital to reduce magnetic field drift and noise to a few nT. The advantage of this approach is that a real-time record of the Zeeman splitting, and hence the magnetic field experienced by the trapped ion, is obtained. The $^1S_0$–$^3P_0$ intercombination transitions in $^{27}Al^+$ and $^{115}In^+$ also have a linear Zeeman shift, which is eliminated in the same way.[7] The second-order Zeeman shift for all these systems is negligible at the typical magnetic field strengths used.

The atom flux created during the ion loading process can lead to the formation of contact potentials on the electrode surfaces, which in turn causes residual electric field gradients within the trapping volume. Any uncompensated electric field gradient present at the position of the trapped ion can lead to a frequency shift of the clock transition if the atomic states involved have an electric quadrupole moment. The magnitude of this electric quadrupole shift depends on the ion species. It is zero for $^{115}In^+$ and $^{27}Al^+$ because states with zero angular momentum have no quadrupole moment. Experimental measurements of quadrupole moments have been carried out for the other systems shown in Table 10.1[40,45–48] and with the uncompensated field gradients present in most traps the quadrupole shifts can be several hertz or more. However it is possible to cancel the quadrupole shift to a reasonable level of accuracy by averaging either measurements of several different Zeeman components[34] or measurements made in three mutually orthogonal directions of the applied magnetic field.[49]

The residual thermal motion and micromotion of the ion within the trap lead to both a Stark shift (because the ion experiences a non-zero root mean square value of the electric field) and a second-order Doppler shift. However, with the ion cooled to the Doppler cooling limit and the micromotion properly minimized, it should be possible to reduce both shifts to the parts in 10^{18} level.

The lasers used to cool and probe the trapped ion can also potentially lead to Stark shifts. Apart from the probe laser, these beams can all be switched off during the interrogation of the clock transition, but it is vital to achieve a high extinction ratio. For most trapped ion standards, the Stark shift associated with the probe laser is extremely small, because very low laser intensity is used to drive the clock transition. The exception is the extremely weak electric octupole transition in $^{171}Yb^+$. In this case the transition frequency has to be measured at two or more laser powers with an extrapolation to zero power being used to determine the unshifted value.[40,41] However by performing this extrapolation in real time using an interleaved servo scheme, a fractional uncertainty in the unperturbed transition frequency of 4.2×10^{-17} has been reached.[40] This can potentially be reduced further still by using a probe laser with narrower linewidth or by using a multipulse Ramsey interrogation scheme.[50]

The ion also experiences a blackbody Stark shift due to the temperature of the surrounding apparatus. Since the reference temperature is absolute zero this shift is relatively large, approaching 10^{-15} at 300 K for the worst case ($^{40}Ca^+$). There are two contributions to the uncertainty in the blackbody Stark shift. The first is the uncertainty in the temperature and isotropy of the radiation field experienced by the ion. The other is the uncertainty in the Stark shift coefficients, which may be determined from measurements of the static polarizability of the clock transition.[40] The $^1S_0$–$^3P_0$ transition in $^{27}Al^+$ has the lowest known blackbody Stark shift of $-8(3) \times 10^{-18}$ at a temperature of 300 K.[30]

One final source of systematic frequency uncertainty is common to all frequency standards. This is the gravitational redshift of the clock transition frequency. To compare frequency standards operating in different locations, their frequencies must be corrected to the frequency at which they would run on the geoid (the gravitational equipotential surface closest to mean sea level). Since the gravitational redshift at the surface of the Earth is approximately 1×10^{-16} m^{-1}, the height of the standard above the geoid must be known to an accuracy of 10 cm in order to achieve a fractional accuracy of 10^{-17}.

10.5. Conclusions and Perspectives

Trapped ion optical clock development has advanced rapidly over recent years, with the result that the stability, reproducibility and estimated systematic uncertainty of the most advanced systems now exceeds that of the best caesium fountain primary frequency standards, with further improvements still expected. This progress has led to the concept of secondary representations of the second being introduced by the International Committee for Weights and Measures (CIPM), following advice from the Consultative Committee for Time and Frequency (CCTF).[51] The term "secondary representation" is used because, by definition, the frequency of a standard cannot be determined in SI units with uncertainty lower than that of the best caesium primary standard. Nevertheless, it was considered that this step would help to advance the development of the new generation of highly reproducible frequency standards. Three trapped ion standards, based on the electric quadrupole transitions in $^{199}Hg^{+}$, $^{88}Sr^{+}$ and $^{171}Yb^{+}$, were ratified by the CIPM as secondary representations of the second in 2006 and others will undoubtedly be added to the list over the next few years.

These developments can be viewed as the first step towards a possible redefinition of the SI second in terms of an optical frequency. There are, however, many different species of ion being investigated, in addition to the range of competing optical lattice clocks. At present, none of these stands out clearly from the rest as an obvious candidate for a redefinition, and significant work remains to be carried out to evaluate fully the ultimate limits to reproducibility. The prospects of using a well-controlled ensemble of ions to improve the short-term instability of trapped ion optical clocks is also being investigated.[52,53] Furthermore, although it has been demonstrated that optical fibre networks are suitable for long-distance comparisons of optical clocks,[44] the necessary infrastructure for making routine comparisons between optical clocks located in different countries is not yet in place. For these reasons, it is unlikely that an optical redefinition of the second will occur before 2019.[51]

In the meantime, optical clocks are already being used in a range of experiments designed to test fundamental theories such as general relativity with increasingly high accuracy. For example, the most stringent limit set to date on present-day variation in the fine structure constant α comes from comparisons between $^{199}Hg^{+}$ and $^{27}Al^{+}$ trapped ion standards at NIST over a period of a year,[25] and similar experiments comparing the two optical clock transitions in $^{171}Yb^{+}$ offer the prospect of even higher sensitivity.[54]

Looking further to the future, optical clocks are even being considered for deployment in space missions designed to test the gravitational redshift and other aspects of general relativity at unprecedented levels of precision.[55]

Acknowledgments

The author would like to thank Rachel Godun and Patrick Gill for critical reading of the manuscript. Support is also acknowledged from the UK Department for Business, Innovation and Skills as part of the NMS Electromagnetics and Time Programme.

References

1. R. Li, K. Gibble and K. Szymaniec, Improved accuracy of the NPL-CsF2 primary frequency standard: evaluation of distributed cavity phase and microwave lensing frequency shifts, *Metrologia* **48**, 283–289 (2011).
2. J. Guéna *et al.*, Progress in atomic fountains at LNE-SYRTE, *IEEE Trans. Ultrason. Ferroelectr. Freq. Control* **59**, 391–410 (2012).
3. H. S. Margolis, Optical frequency standards and clocks, *Contemp. Phys.* **51**, 37–58 (2010).
4. H. Katori, Optical lattice clocks and quantum metrology, *Nature Photon.* **5**, 203–210 (2011).
5. W. Paul, Electromagnetic traps for charged and neutral particles, *Rev. Mod. Phys.* **62**, 531–540 (1990).
6. C. Schrama *et al.*, Novel miniature ion traps, *Opt. Commun.* **101**, 32–36 (1993).
7. T. Rosenband *et al.*, Observation of the ${}^1S_0 - {}^3P_0$ clock transition in ${}^{27}Al^+$, *Phys. Rev. Lett.* **98**, 220801 (2007).
8. H. G. Dehmelt, Mono-ion oscillator as potential ultimate laser frequency standard, *IEEE Trans. Instrum. Meas.* **31**, 83–87 (1982).
9. J. C. Bergquist, W. M. Itano and D. J. Wineland, Recoilless optical absorption and Doppler sidebands for a single trapped ion, *Phys. Rev. A* **36**, 428–430 (1987).
10. D. J. Berkeland *et al.*, Minimization of ion micromotion in a Paul trap, *J. Appl. Phys.* **83**, 5025–5033 (1998).
11. W. Nagourney, J. Sandberg and H. G. Dehmelt, Shelved electron amplifier: observation of quantum jumps, *Phys. Rev. Lett.* **56**, 2797–2799 (1986).
12. R. W. P. Drever *et al.*, Laser phase and frequency stabilization using an optical resonator, *Appl. Phys. B* **31**, 97–105 (1983).
13. M. Notcutt *et al.*, Simple and compact 1-Hz laser system via an improved mounting configuration of a reference cavity, *Opt. Lett.* **30**, 1815–1817 (2005).
14. S. A. Webster, M. Oxborrow and P. Gill, Vibration-insensitive optical cavity, *Phys. Rev. A* **75**, 011801(R) (2007).

15. D. R. Leibrandt *et al.*, Spherical reference cavities for frequency stabilization of lasers in non-laboratory environments, *Opt. Express* **19**, 3471–3482 (2011).
16. S. A. Webster and P. Gill, Force-insensitive optical cavity, *Opt. Lett.* **36**, 3572–3574 (2011).
17. K. Numata, A. Kemery and J. Camp, Thermal-noise limit in the frequency stabilization of lasers with rigid cavities, *Phys. Rev. Lett.* **93**, 250602 (2004).
18. T. Kessler, T. Legero and U. Sterr, Thermal noise in optical cavities revisited, *J. Opt. Soc. Am. B* **29**, 178–184 (2012).
19. B. C. Young *et al.*, Visible lasers with subhertz linewidths, *Phys. Rev. Lett.* **82**, 3799–3802 (1999).
20. Y. Y. Jiang *et al.*, Making optical atomic clocks more stable with 10^{-16}-level laser stabilization, *Nature Photon.* **5**, 158–161 (2011).
21. T. Kessler *et al.*, A sub-40-mHz linewidth laser based on a silicon single-crystal optical cavity, *Nature Photon.* **6**, 687–692 (2012).
22. J. L. Hall, Nobel lecture: defining and measuring optical frequencies, *Rev. Mod. Phys.* **78**, 1279–1295 (2006).
23. T. W. Hänsch, Nobel lecture: passion for precision, *Rev. Mod. Phys.* **78**, 1297–1309 (2006).
24. D. J. Jones *et al.*, Carrier-envelope phase control of femtosecond mode-locked lasers and direct optical frequency synthesis, *Science* **288**, 635–639 (2000).
25. T. Rosenband *et al.*, Frequency ratio of Al^+ and Hg^+ single-ion optical clocks; metrology at the 17th decimal place, *Science* **319**, 1808–1812 (2008).
26. M. Roberts *et al.*, Observation of the $^2S_{1/2} - ^2F_{7/2}$ electric octupole transition in a single $^{171}Yb^+$ ion, *Phys. Rev. A* **62**, 020501(R) (2000).
27. E. Peik *et al.*, Sideband cooling of ions in radio-frequency traps, *Phys. Rev. A* **60**, 439–449 (1999).
28. P. O. Schmidt *et al.*, Spectroscopy using quantum logic, *Science* **309**, 749–752 (2005).
29. C. W. Chou *et al.*, Optical clocks and relativity, *Science* **329**, 1630–1633 (2010).
30. C. W. Chou *et al.*, Frequency comparison of two high-accuracy Al^+ optical clocks, *Phys. Rev. Lett.* **104**, 070802 (2010).
31. J. von Zanthier *et al.*, Absolute frequency measurement of the In^+ clock transition with a mode-locked laser, *Opt. Lett.* **25**, 1729–1731 (2000).
32. Y. H. Wang *et al.*, Absolute frequency measurements and high resolution spectroscopy of $^{115}In^+$ $5s^2\,^1S_0 - 5s5p\,^3P_0$ narrowline transition, *Opt. Commun.* **273**, 526–531 (2007).
33. H. S. Margolis *et al.*, Hertz-level measurement of the optical clock frequency in a single $^{88}Sr^+$ ion, *Science* **306**, 1355–1358 (2004).
34. P. Dubé *et al.*, Electric quadrupole shift cancellation in single-ion optical frequency standards, *Phys. Rev. Lett.* **95**, 033001 (2005).
35. M. Chwalla *et al.*, Absolute frequency measurement of the $^{40}Ca^+$ $4s\,^2S_{1/2} - 3d\,^2D_{5/2}$ clock transition, *Phys. Rev. Lett.* **102**, 023002 (2009).
36. Y. Huang *et al.*, Hertz-level measurement of the $^{40}Ca^+$ $4s\,^2S_{1/2}$–$3d\,^2D_{5/2}$ clock transition frequency with respect to the SI second through the Global Positioning System, *Phys. Rev. A* **85**, 030503(R) (2012).

37. K. Matsubara *et al.*, Frequency measurement of the optical clock transition of $^{40}Ca^+$ ions with an uncertainty of 10^{-14} level, *Appl. Phys. Express* **1**, 067011 (2008).
38. C. Tamm *et al.*, Stray-field-induced quadrupole shift and absolute frequency of the 688-THz $^{171}Yb^+$ single-ion optical frequency standard, *Phys. Rev. A* **80**, 043403 (2009).
39. S. A. Webster *et al.*, Frequency measurement of the $^2S_{1/2}$–$^2D_{3/2}$ electric quadrupole transition in a single $^{171}Yb^+$ ion, *IEEE Trans. Ultrason. Ferroelectr. Freq. Control* **57**, 592–599 (2010).
40. N. Huntemann *et al.*, High-accuracy optical clock based on the octupole transition in $^{171}Yb^+$, *Phys. Rev. Lett.* **108**, 090801 (2012).
41. S. A. King *et al.*, Absolute frequency measurement of the $^2S_{1/2}$–$^2F_{7/2}$ electric octupole transition in a single ion of $^{171}Yb^+$ with 10^{-15} fractional uncertainty, *New J. Phys.* **14**, 013045 (2012).
42. J. E. Stalnaker *et al.*, Optical-to-microwave frequency comparison with fractional uncertainty of 10^{-15}, *Appl. Phys. B* **89**, 167–176 (2007).
43. A. Bauch *et al.*, Comparison between frequency standards in Europe and the USA at the 10^{-15} uncertainty level, *Metrologia* **43**, 109–120 (2006).
44. K. Predehl *et al.*, A 920-kilometer optical fiber link for frequency metrology at the 19th decimal place, *Science* **336**, 441–444 (2012).
45. G. P. Barwood *et al.*, Measurement of the electric quadrupole moment of the 4d $^2D_{5/2}$ level in $^{88}Sr^+$, *Phys. Rev. Lett.* **93**, 133001 (2004).
46. W. H. Oskay, W. M. Itano and J. C. Bergquist, Measurement of the $^{199}Hg^+$ $5d^9\,6s^2\,^2D_{5/2}$ electric quadrupole moment and a constraint on the quadrupole shift, *Phys. Rev. Lett.* **94**, 163001 (2005).
47. C. F. Roos *et al.*, 'Designer' atoms for quantum metrology, *Nature* **443**, 316–319 (2006).
48. C. Tamm *et al.*, $^{171}Yb^+$ optical frequency standard at 688 THz, *IEEE Trans. Instrum. Meas.* **56**, 601–604 (2007).
49. W. M. Itano, External field shifts of the $^{199}Hg^+$ optical frequency standard, *J. Res. Natl. Inst. Stand. Technol.* **105**, 829–837 (2000).
50. V. I. Yudin *et al.*, Hyper-Ramsey spectroscopy of optical clock transitions, *Phys. Rev. A.* **82**, 011804(R) (2010).
51. P. Gill, When should we change the definition of the second?, *Phil. Trans. Roy. Soc. A* **369**, 4109–4130 (2011).
52. N. Herschbach *et al.*, Linear Paul trap design for an optical clock with Coulomb crystals, *Appl. Phys. B* **107**, 891–906 (2012).
53. K. Hayasaka, Synthesis of two-species ion chains for a new optical frequency standard with an indium ion, *Appl. Phys. B* **107**, 965–970 (2012).
54. S. N. Lea, Limits to time variation of fundamental constants from comparisons of atomic frequency standards, *Rep. Prog. Phys.* **70**, 1473–1523 (2007).
55. S. Schiller *et al.*, Einstein Gravity Explorer – a medium-class fundamental physics mission, *Experimental Astronomy* **23**, 573–610 (2009).

Chapter 11

Novel Penning Traps

José Verdú
Department of Physics and Astronomy, University of Sussex,
Sussex BN1 9QH, United Kingdom
J.L.Verdu-Galiana@sussex.ac.uk

Penning traps are used in a wide variety of applications, such as mass spectrometry, high precision measurements of atomic and nuclear properties, antihydrogen production and others. The maturity achieved by Penning trap technology has recently led to several theoretical proposals about the construction of a quantum processor using trapped electrons. Novel scalable surface traps have been conceived of and actually built. However, the observation of a single trapped electron and the accurate measurement of its motional frequencies is still an open experimental challenge. At the University of Sussex a new approach has been introduced: the *coplanar-waveguide Penning trap.* We describe this trap in detail and discuss how the compensation of electric anharmonicities might allow for the observation of a single electron.

11.1. Introduction

A single electron in a Penning trap is known as a geonium atom[1] , as named by its inventor, the 1989 Nobel Prize winner Hans Dehmelt. It has been demonstrated as an outstanding system for testing the laws of physics.[1] Examples include the measurement of the g-factor of the free electron[2] and its mass .[3] The energy of an electron in a cryogenic trap can be monitored with great accuracy, at the level of observing quantum jumps between the vibrational Fock states.[4] Additionally, the spin can be coherently manipulated and detected, by means of the continuous Stern–Gerlach effect.[5] Electrons in cryogenic Penning traps have been proposed for implementing a quantum processor.[6–8] Initially motivated by the famous DiVincenzo requirements for quantum computation,[9] a potentially scalable planar Penning trap has been conceived of at the University of Mainz.[10] With that trap the storage of electrons has been demonstrated, both at room temperature and with a cryogenic setup.[11,12] Several optimised versions of the Mainz trap have been proposed at Harvard.[13] The latter designs incorporate the ability of compensating anharmonicities of the electric trapping potential. Such anharmonicities have impeded the observation of a single electron in the first cryogenic planar trap.[12]

In this chapter, the coplanar-waveguide (CPW) Penning trap is discussed in detail.[14] Its design is inspired by the great development of planar microwave technology. Planar superconducting cavities have been built on a chip with very low losses, achieving quality factors in the range of $Q \in [10^5 - 10^6]$.[15] The basis of those cavities is the CPW transmission line, consisting of a central conducting strip and two outer so-called "ground planes".[16] A sketch of it is shown in Figure 11.1 (a). The mentioned high-Q cavities have triggered the field of circuit-quantum electrodynamics (circuit-QED).[17] A prominent example of circuit-QED comes from the University of Yale, where a single microwave photon stored in a CPW cavity at ~ 20 mK has been coherently coupled to an "artificial atom", a superconducting Cooper-pair box.[18] This experiment has achieved the implementation of the Jaynes–Cummings model, which is essential for transferring quantum information between different systems in a chip in a controlled way. As explained in following sections, the CPW Penning trap simply results from the projection of a cylindrical trap onto the surface of a chip. That projection naturally results in a coplanar-waveguide cavity, similar to those used in circuit-QED experiments. Thus, the CPW Penning trap makes it possible for a geonium atom to be integrated within microwave quantum circuits. Besides quantum computation, further possibilities have been envisaged for planar Penning traps, such as the entanglement of electrons for improving the measurement of the g-factor,[19] the quantum simulation of magnetism with electrons in an array of traps[20] and the use of quantum tunneling for interferometry experiments with trapped ions/electrons.[21]

11.2. Penning Traps

In general, a Penning trap consists of a set of electrodes with a static magnetic field $\vec{B} = B \cdot \hat{u}_z$. Its working principle is straightforward: the magnetic field forces the charged particles to follow circular orbits around its axes, thus trapping them in the radial direction $\hat{u}_r \perp \vec{B}$. Trapping along the axial direction $-\hat{u}_z-$ is accomplished by an electrostatic harmonic potential, created by the electrodes. The particle moves around the equilibrium posi-

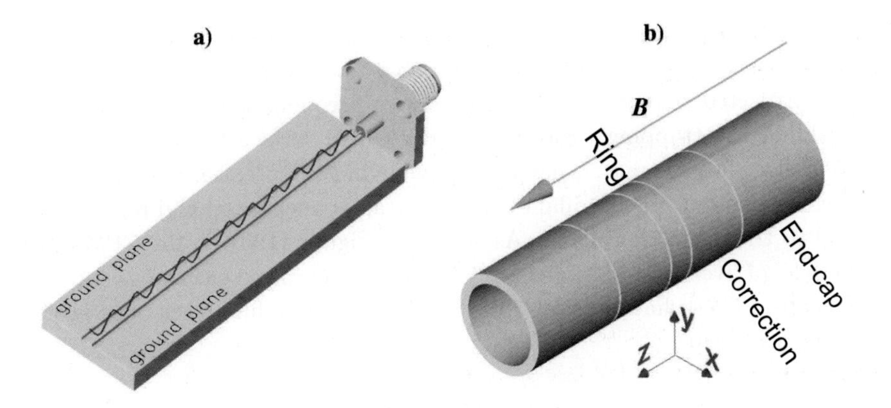

Fig. 11.1. (a) Coplanar-waveguide transmission line. (b) Cylindrical Penning trap.

tion, located at the minimum of the potential energy, in a superposition of three independent periodic motions. These are called cyclotron, magnetron and axial, with the characteristic frequencies $\omega_p = 2\pi\nu_p, \omega_m = 2\pi\nu_m$ and $\omega_z = 2\pi\nu_z$, respectively. Although the properties of a Penning trap can be studied independently of the charge of the particles, electrons will be considered throughout this chapter.

11.2.1. *The cylindrical trap*

The cylindrical Penning trap[22] has demonstrated excellent experimental capabilities[2,23,24] (see sketch in Figure 11.1 (b)). One of its main features is its ability to compensate deviations of the electric potential from the ideal harmonic shape. For that purpose, the trap is provided with compensation (or correction) electrodes. Additionally, it also allows for controlling the spontaneous emission rate of the electron's cyclotron motion.[25] Transferring these important features to the planar domain is one of the key motivations for the novel design of the CPW Penning trap. The origin of this novel, planar trap and its main properties are discussed in the following sections.

11.3. The CPW Penning Trap

The CPW Penning trap is based upon the cylindrical trap. The CPW-trap results from the projection of the latter onto a plane. The transformation is described as follows. An imaginary cut is performed on the surface of a hypothetical cylindrical trap, as shown in Figure 11.2 (a). The originally cylindrical electrodes are then transformed into flat rectangular electrodes and projected onto a surface. Two outer "ground planes" are added for shielding the trapping region above the chip's surface from the underlying dielectric substrate. They also provide an equipotential reference. The magnetic field of the original cylindrical trap is unchanged and results parallel to the projection plane. As shown in Figure 11.2 (b), the CPW-trap consists of five electrodes: the "ring", two compensation electrodes and two endcaps. These have similar roles as the corresponding electrodes in the cylindrical trap.

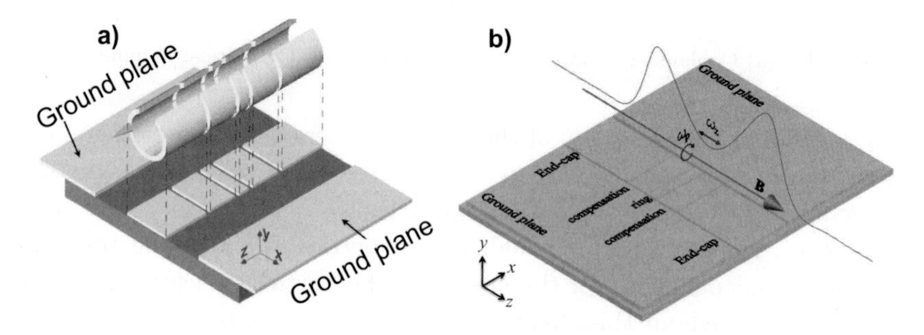

Fig. 11.2. (a) Projection of a standard cylindrical five-pole Penning trap onto a plane. (b) Sketch of the CPW Penning trap, with the cyclotron and axial motions of a trapped particle.

The denomination *coplanar-waveguide* Penning trap is chosen due to the fact that the set of electrodes, together with the ground planes, constitute a section of a coplanar-waveguide transmission line. This allows for integrating the trap into a microwave circuit, as shown in Figure 11.3 (a). For this purpose, the insulating gaps between neighbouring electrodes must be fabricated small enough, so as not to affect the propagation of high-frequency signals along the transmission line. The contacts of the static voltages can be implemented through vias from the rear side of the chip.

11.3.1. *The ideal CPW-trap*

Assuming the chip that houses the trap is an infinite plane, the electrostatic potential created by the electrodes with the voltages applied as in Figure 11.3 (a) can be calculated with standard Green's functions techniques. The

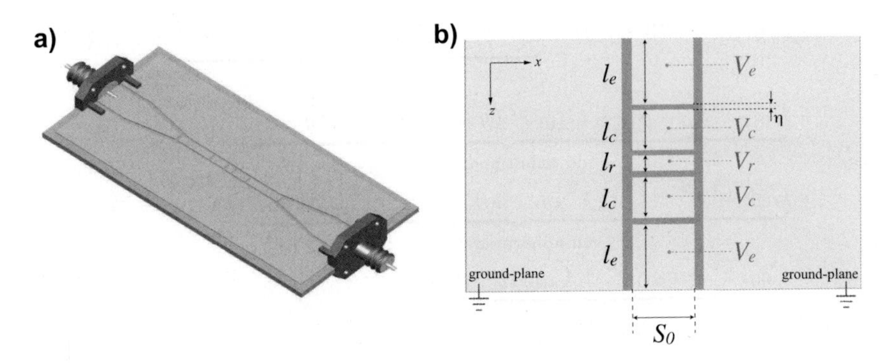

Fig. 11.3. (a) Sketch of the CPW Penning trap integrated within a microwave circuit. (b) Applied voltages and dimensions of the trap.

result is given by a function of the form:

$$\begin{aligned}\phi(x,y,z) &= V_r \cdot f_r(x,y,z) + V_c \cdot f_c(x,y,z) + V_e \cdot f_e(x,y,z) \\ &\quad + f_{gaps}(x,y,z \,|\, V_r, V_c, V_e)\,. \end{aligned} \tag{11.1}$$

In Eq. (11.1), the introduced functions $f_r(x,y,z)$, $f_c(x,y,z)$ and $f_e(x,y,z)$ represent the contributions to the potential by the ring, the compensation electrodes and the endcaps, respectively. They are purely geometric functions, independent of the applied voltages. Exact expressions for these functions can be found in reference[14]. The small insulating gaps between neighbouring electrodes also contribute to the electrostatic potential. This contribution is represented by the function $f_{gaps}(x,y,z \,|\, V_r, V_c, V_e)$, which depends on the geometry and the applied voltages. We now assume that the insulating gaps are vanishing small. With this, we have $f_{gaps}(x,y,z \,|\, V_r, V_c, V_e) \to 0$.[14] The electrostatic potential can be then written as $\phi(x,y,z) = V_r\,(f_r(x,y,z) + T_c \cdot f_c(x,y,z) + T_e \cdot f_e(x,y,z))$. We have introduced the tuning ratio $T_c = V_c/V_r$ and the endcap-to-ring voltage ratio $T_e = V_e/V_r$. The condition for trapping is that the applied voltages should be such that $T_e > T_c \simeq 1$. Now, we can write the series expansion of the potential around the trapping position $(0, y_0, 0)$ as follows:

$$\phi(x,y,z) = \phi(0,y_0,0) + \underbrace{c_{002}\, z^2 + C_{200}\, x^2 + C_{020}\,(y-y_0)^2}_{\phi_{quad}} + \ldots \quad (11.2)$$

$$+ \underbrace{C_{012}\, z^2 (y-y_0) + C_{210}\, x^2\,(y-y_0) + C_{030}\,(y-y_0)^3}_{\text{odd anharmonicities}} + \ldots$$

$$+ \underbrace{C_{202}\, z^2\, x^2 + C_{022}\, z^2\,(y-y_0)^2 + C_{220}\, x^2\,(y-y_0)^2}_{\text{even anharmonicities}} + \ldots$$

$$+ \underbrace{C_{004}\, z^4 + c_{400}\, x^4 + C_{040}\,(y-y_0)^4}_{\text{even anharmonicities}} + \ldots \quad (11.3)$$

The expansion coefficients are given by $C_{ijk} = \frac{1}{i!\,j!\,k!} \cdot \frac{\partial^{i+j+k}\,\phi(x,y,z)}{\partial x^i\,\partial y^j\,\partial z^k}\Big|_{(0,y_0,0)}$. The equilibrium trapping position above the chip's surface, y_0, is the solution to the equation $C_{010}(y_0) = 0$. It is given by the position where the electrostatic field vanishes, as plotted on Figure 11.4 (a). Moreover, Laplace's equation imposes some constraints upon the expansion coefficients, particularly it can be demonstrated that the sum of all second-order coefficients must equal zero: $C_{200} + C_{020} + C_{002} = 0$. With this, the ideal trapping potential, given by the quadrupole contribution ϕ_{quad} alone, can be expressed as:

$$\phi_{quad}(x,y,z) = C_{002} \left\{ \underbrace{\left(z^2 - \frac{x^2 + (y-y_0)^2}{2} \right)}_{\text{standard trap}} + \underbrace{\frac{\epsilon}{2} \cdot (x^2 - (y-y_0)^2)}_{\text{elliptical term}} \right\}. \quad (11.4)$$

The *ellipticity parameter* is given by $\epsilon = \frac{C_{200} - C_{020}}{C_{002}}$. The CPW Penning trap is therefore an *elliptical Penning trap.* Elliptical three-dimensional (3D) Penning traps have been investigated experimentally with argon and fullerene ions.[26] Furthermore, M. Kretzschmar has analytically solved the motion of a single particle in an ideal elliptical Penning trap,[27] i.e. a trap whose electric potential is given by Eq. (11.4), neglecting higher-order terms. The ideal motion is plotted on Figure 11.4 (b). It is the superposition of the three harmonic oscillations: the cyclotron, axial and magnetron motions. Figure 11.4 (b) also shows the ellipticity of the magnetron motion. The orientation of the ellipse is determined by the sign of ϵ[27] and the motion is only stable when $-1 < \epsilon < 1$. For values $|\epsilon| \geq 1$ the motion is unbounded and trapping is not possible.[27]

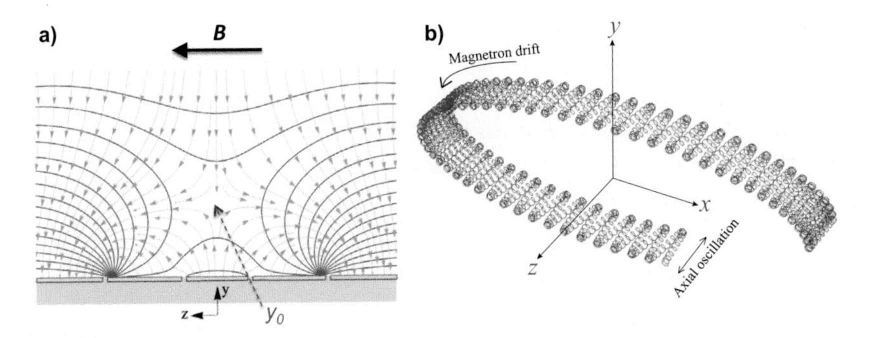

Fig. 11.4. (a) Contour plot of the electrostatic potential. (b) Ideal motion of a single particle in a CPW Penning trap.

11.3.1.1. *Expressions of the frequencies*

The expression for the eigenfrequencies have been calculated analytically by Kretzschmar.[27] For a particle with charge q and mass m these are given by the following expressions:

$$\omega_p = \sqrt{\frac{1}{2}(\omega_c^2 - \omega_z^2) + \frac{1}{2}\sqrt{\omega_c^2\,\omega_1^2 + \epsilon^2\,\omega_z^4}} \tag{11.5}$$

$$\omega_m = \sqrt{\frac{1}{2}(\omega_c^2 - \omega_z^2) - \frac{1}{2}\sqrt{\omega_c^2\,\omega_1^2 + \epsilon^2\,\omega_z^4}}$$

$$\omega_z = \sqrt{2\,C_{002}\,\frac{q}{m}} \quad \text{with:} \quad \omega_c = \frac{q}{m}\cdot B \quad \text{and} \quad \omega_1 = \sqrt{\omega_c^2 - 2\,\omega_z^2}\,.$$

The eigenfrequencies given in Eq. (11.5) are functions of all applied voltages V_r, V_c, V_e. However, it is convenient to express explicitly those dependencies in the following way:

$$y_0 = y_0(T_c, T_e) \;\longrightarrow\; \omega_z = \omega_z(y_0, V_r),\; \omega_{p,m} = \omega_{p,m}(y_0, V_r, B) \tag{11.6}$$

Equation (11.6) reflects that the values of $\omega_{p,m,z}$ can be controlled by the ring voltage V_r (and the magnetic field B) independently of the trapping position y_0 and applied tuning ratio. T_e, and to a small extent also T_c, change the value of y_0, and thus the frequencies might also change with them. However $\omega_{p,m,z}$ can be readjusted to an specified value, at any accessible y_0, by properly changing V_r.

For a single electron, with a magnetic field of $B = 0.5$ T and $V_r = 1$ volt, typical values for the frequencies are $\omega_p \simeq 2\pi \cdot 14$ GHz, $\omega_z \simeq 2\pi \cdot 30 - 90$ MHz and $\omega_m \simeq 2\pi \cdot 1 - 30$ kHz. The exact values depend on the geometry of the trap (see Eq. (11.1)). It is important to observe that the cyclotron frequency is in the microwave domain. This potentially allows for the electron to be coupled with other quantum systems in a microwave network, through the interaction of the cyclotron or spin degrees of freedom with microwave photons. For a single proton, with $B = 1$ T and $V_r = 1$ volt, the frequencies are in the ranges $\omega_p \simeq 2\pi \cdot 15$ MHz, $\omega_z \simeq 2\pi \cdot 0.2 - 1$ MHz and $\omega_m \simeq 2\pi \cdot 1 - 30$ kHz.

11.3.1.2. *The invariance theorem*

From Eq. (11.5) it can be observed that when the ellipticity vanishes, $\epsilon = 0$, the usual expression for the frequencies in a standard, non-elliptical, Penning trap are recovered. Moreover, for any value of the ellipticity, it is easy to demonstrate that the three eigenfrequencies satisfy the well-known Gabrielse-Brown invariance theorem:

$$\sqrt{\omega_p^2 + \omega_z^2 + \omega_m^2} = \frac{q}{m} B \,. \tag{11.7}$$

The Gabrielse–Brown invariance theorem[1] is one of the "holy grails" in Penning trap technology. It enables high precision measurements of masses and other atomic and nuclear properties. Hence, the ellipticity of the CPW Penning trap does not prevent it from being suitable for high precision measurements. However, the deviations of the electrostatic potential from the ideal quadrupole case of Eq. (11.4) still have to be considered.

11.4. The Real CPW Penning trap

Any even C_{ijk} coefficients (i.e. those with all the indices i, j, k even) produce deviations of the ideal trap frequencies which can be calculated by means of first order canonical perturbation theory.[28] The calculation simply demands the averaging of the perturbation Hamiltonian over an oscillation period of the trapped particle. For that purpose, the Hamiltonian must be expressed in terms of appropriate angle-action canonical variables. This method has been used, for instance, for calculating the frequency shifts in a real 3D elliptical trap,[27] where only even anharmonicities are present. On the other hand, the perturbation hamiltonian of any odd C_{ijk} ($= j$ is odd) averages to zero, and thus those coefficients must be treated with second-order perturbation theory methods. Such methods require the solution of the Hamilton–Jacobi equation for the perturbative Hamiltonian. The shifts of the frequencies produced by the particle's energies are then derived from that solution.[28] In particular, all first- and second-order perturbations up to the fourth-order in the expansion of ϕ, $3 \leq i+j+k \leq 4$ (see Eq. (11.2)), produce shifts which scale linearly with the energies of the trapped particle, E_p, E_z, E_m. Thus, they can be expressed in matrix form:

$$\begin{pmatrix} \Delta\nu_p \\ \Delta\nu_z \\ \Delta\nu_m \end{pmatrix} = \underbrace{\begin{pmatrix} M_{1,1} & M_{1,2} & M_{1,3} \\ M_{2,1} & M_{2,2} & M_{2,3} \\ M_{3,1} & M_{3,2} & M_{3,3} \end{pmatrix}}_{M=\text{frequency-shifts matrix}} \cdot \begin{pmatrix} \Delta E_p \\ \Delta E_z \\ \Delta E_m \end{pmatrix}. \tag{11.8}$$

Each anharmonic perturbation in Eq. (11.2) delivers such a *frequency-shifts matrix*. In total the CPW Penning trap requires nine such matrices, denoted by M^{ijk}. Each is generated by the corresponding C_{ijk} perturbative hamiltonian. The expressions for each M^{ijk} for the CPW-cavity trap have been calculated in reference.[14] The overall frequency-shifts matrix is simply the sum of all those:

$$M = M^{004} + M^{220} + M^{202} + M^{022} + M^{400} + M^{040} + M^{012} + M^{210} + M^{030}. \tag{11.9}$$

The term $M_{2,2} = \frac{\Delta\nu_z}{\Delta E_z}$ in the frequency-shifts matrix is the main perturbation which might impede the observation of a single trapped particle and the accurate measurement of its motional frequencies. It can be demonstrated that, in general, all other terms in the frequency-shifts matrix can be ignored.[14] This can be illustrated by an example, calculated with $l_r = 0.9, l_c = 2, l_e = 5, S_0 = 7$ all in mm, for an electron with $B = 0.5$ T. The overall frequency-shifts matrix is:

$$M\Big|_{y_0=1.37\,\text{mm}} = \begin{pmatrix} 5\cdot10^{-6} & 0.5 & -0.9 \\ 1\cdot10^{-3} & 203 & -411 \\ -2\cdot10^{-6} & -0.4 & 2 \end{pmatrix} \text{ Hz/K}. \tag{11.10}$$

The high value $\frac{\Delta\nu_z}{\Delta E_z} = 203$ Hz/K would make the observation of the axial motion of the trapped electron basically impossible. The technique for detecting the motion of the trapped particle is discussed in Chapter 12 of this volume. Here, the actual details of the detection technique are not important. The fluctuations of the axial energy E_z (which follows a Boltzmann distribution around the electron's axial temperature T_z) broaden the electrons detection signal, making its observation basically impossible. For the detection to be successful, the term $M_{2,2}$ must be "compensated", i.e. eliminated. This is achieved by properly choosing the value of the tuning ratio T_c, as explained in the following section.

11.5. Compensation of Electric Anharmonicities

The element $M_{2,2}$ is given by the sum of the two elements $M_{2,2}^{004}$ and $M_{2,2}^{012}$:

$$\left.\frac{\Delta\nu_z}{\Delta E_z}\right|_{M_{2,2}^{004}} = \frac{q}{16\,\pi^4\,m^2}\,\frac{3}{\nu_z^3}\,C_{004} \quad ; \quad \left.\frac{\Delta\nu_z}{\Delta E_z}\right|_{M_{2,2}^{012}} \simeq \frac{q^2}{32\,\pi^6\,m^3}\,\frac{\eta_m^2}{\nu_z^5}\,C_{012}^2\,. \tag{11.11}$$

On the one hand $M_{2,2}^{012}$ is always positive, since it is proportional to the square of C_{012}. On the other hand, $M_{2,2}^{004}$ can be positive or negative, depending on the sign of the coefficient C_{004}. Thus, if an appropriate *optimal tuning ratio* can be found, such that the $M_{2,2}^{004}$ cancels the $M_{2,2}^{012}$, $T_c^{opt} \to M_{2,2}^{012} + M_{2,2}^{004} = 0$, then the linear fluctuations of the axial frequency ν_z with the axial energy can be cancelled. The existence of T_c^{opt} cannot be guaranteed in general, however it is usually the case. This can be seen in the calculated example of Figure 11.5 (a) (assumed trap dimensions as in Eq. (11.10)), where $\frac{\Delta\nu_z}{\Delta E_z}$ is plotted as a function of the applied tuning ratio T_c. One particular value exists, $T_c^{opt} = 1.13440$, which eliminates $M_{2,2}$. For the axial frequency of the trapped particle to be well defined, it is necessary that the tuning ratio is optimised. Modern voltage calibrators achieve resolutions of the order of $1\cdot10^{-6}$ volt at an output voltage around 1 volt, hence for trapping voltages in that range, the optimal tuning ratio can be adjusted within an accuracy of five to six decimal places. This suffices for the axial frequency to be nearly independent of the axial energy.[24]

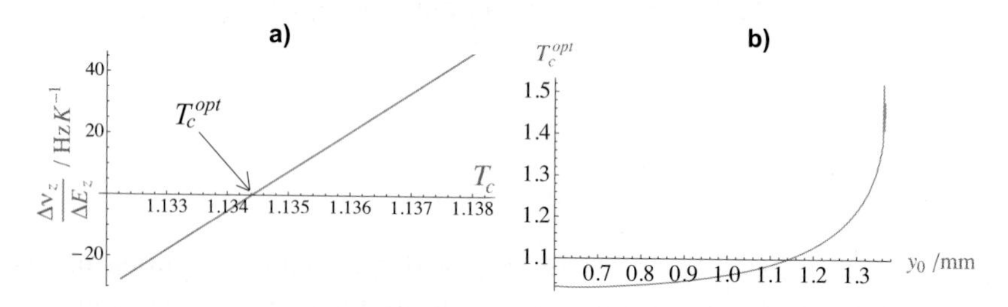

Fig. 11.5. (a) The optimal tuning ratio. (b) The useful trapping interval.

11.5.1. *The useful trapping interval*

The example given in Figure 11.5 (a) has been calculated for a particular value of the trapping position y_0. As mentioned in Section 11.3.1, the trapping position depends on the applied voltages; y_0 particularly is strongly dependent on the endcap-to-ring voltage ratio T_e. Varying T_e allows for changing the value of y_0. For each equilibrium position the optimal tuning ratio which eliminates $\frac{\Delta\nu_z}{\Delta E_z}$ has to be readjusted. This is shown in the calculated example of Figure 11.5 (b) (assumed same trap as in Eq. (11.10)). The figure shows a continuous interval of values of the trapping position within which the optimal tuning ratio, T_c^{opt}, exists. In principle other values of y_0 can be achieved by changing T_e, however only those values within the interval shown in the figure allow for eliminating the dependence of the axial frequency with the axial energy. As already explained, this is a required condition for the trapped particle to be detectable. Therefore, the mentioned interval is denoted the *useful trapping interval*, that is where the trap works properly. For values of y_0 outside that interval the particle is not observable and the trap is therefore useless.

To conclude, it must be mentioned that the compensation process is different from the usual compensation of electrostatic anharmonicities in standard Penning traps. In the latter case, the tuning ratio is chosen to eliminate the coefficient C_{004}.[1] The compensation of the CPW Penning trap presented in Section 11.5 does not in general eliminate the coefficients C_{004} nor C_{012}. However, the CPW-trap can be optimally designed,[14] such that for one particular value of y_0 those coefficients, and even one of a higher-order, all vanish simultaneously: $C_{004} = C_{012} = C_{006} = 0$. The planar trap developed in Harvard also allows for similar optimal design strategies.[13] These should allow for the observation of a single trapped electron (or ion in general) and the accurate measurement of the eigenfrequencies.

11.6. Conclusions

The cylindrical Penning trap has demonstrated outstanding experimental capabilities with trapped electrons. The proposed novel CPW Penning trap design is basically derived from the cylindrical one and keeps its main feature, namely its ability to eliminate the influence upon the particle's motion of the lowest order deviations from harmonicity of the axial trapping potential. The potential compensation guarantees the good behaviour of the trapped electron/ion as an ideal harmonic oscillator; namely as having very sharply defined motional frequencies that can be measured precisely. This in turn permits accurate control over the trapped particle. Besides scalability and ease of fabrication, the essential requirement that a useful planar Penning trap design must fulfil is its ability to be compensated, i.e. the electrodes must provide the means to cancel the strongest electric-potential anharmonicities efficiently, arising from the flatness of the trap. This is an essential condition for manipulating the electrons/ions with high accuracy, stability and repeatability for quantum computation or other applications.

Acknowledgments

The author acknowledges financial support from EPSRC, under grant EP/I012850/1, from the Marie Curie reintegration grant "NGAMIT" and from SEPnet.

References

1. L.S. Brown and G. Gabrielse, Geonium theory: Physics of a single electron or ion in a Penning trap, *Rev. Mod. Phys.* **58**, 233 (1986).
2. D. Hanneke, S. Fogwell and G. Gabrielse, New measurement of the electron magnetic moment and the fine structure constant, *Phys. Rev. Lett.* **100**, 120801 (2008).
3. T. Beier, H. Häffner, N. Hermanspahn, S. G. Karshenboim, H-J. Kluge, W. Quint, S. Stahl, J. Verdú and G. Werth, New determination of the electron's mass, *Phys. Rev. Lett.* **88**, 011603 (2001).
4. S. Peil and G. Gabrielse, Observing the quantum limit of an electron cyclotron: QND measurements of quantum jumps between Fock states, *Phys. Rev. Lett.* **83**, 1287 (1999).
5. H. G. Dehmelt, Continuous Stern-Gerlach effect: principle and idealized apparatus, *Proc. Natl. Acad. Sci. USA* **83**, 2291 (1986).
6. G. Ciaramicoli, I. Marzoli and P. Tombesi, Realization of a quantum algorithm using a trapped electron, *Phys. Rev. A* **63**, 052307 (2001).
7. G. Ciaramicoli, I. Marzoli and P. Tombesi, Scalable quantum processor with trapped electrons, *Phys. Rev. Lett.* **91**, 017901 (2003).
8. G. Ciaramicoli, I. Marzoli and P. Tombesi, Trapped electrons in vacuum for a scalable quantum processor, *Phys. Rev. A* **70**, 032301 (2004).
9. D. P. DiVincenzo, The physical implementation of quantum computation, *Fortschritte der Physik* **48**, 771 (2000).
10. S. Stahl, F. Galve, J. Alonso, S. Djekić, W. Quint, T. Valenzuela, J. Verdú, M. Vogel and G. Werth, A planar Penning trap, *Eur. Phys. J. D* **32**, 139 (2005).
11. F. Galve, P. Fernández and G. Werth, Operation of a planar Penning trap, *Eur. Phys. J. D* **40**, 201 (2006).
12. P. Bushev, S. Stahl, R. Natali, G. Marx, E. Stachowska, G. Werth, M. Hellwig and F. Schmidt-Kaler, Electrons in a cryogenic planar Penning trap and experimental challenges for quantum processing, *Eur. Phys. J. D* **50**, 97 (2008).
13. J. Goldman and G. Gabrielse, Optimized planar Penning traps for quantum-information studies, *Phys. Rev. A* **81**, 052335 (2010).

14. J. Verdú, Theory of the coplanar waveguide Penning trap, *New J. Phys.* **13**, 113029 (2011).
15. P. K. Day, H. G. LeDuc, B. A. Mazin, A. Vayonakis and J. Zmuidzinas, A broadband superconducting detector suitable for use in large arrays, *Nature* **425**, 817 (2003).
16. C.P. Wen, Coplanar waveguide: a surface strip transmission line suitable for nonreciprocal gyromagnetic device applications, *IEEE Trans. Microwave Theory Techn.* **MTT-17**, 1087 (1969).
17. A. Blais, R.-S. Huang, A. Wallraff, S.M. Girvin and R.J. Schoelkopf, Cavity quantum electrodynamics for superconducting electrical circuits: An architecture for quantum computation, *Phys. Rev. A* **69**, 062320 (2004).
18. A. Wallraff, D. I. Schuster, A. Blais, L. Frunzio, R. S. Huang, J. Majer, S. Kumar, S. M. Girvin and R. J. Schoelkopf, Strong coupling of a single photon to a superconducting qubit using circuit quantum electrodynamics, *Nature* **431**, 162 (2004).
19. L. Lamata, D. Porras, J. I. Cirac, J. Goldman and G. Gabrielse, Towards electron-electron entanglement in Penning traps, *Phys. Rev. A* **81**, 022301 (2010).
20. G. Ciaramicoli, I. Marzoli and P. Tombesi, Quantum spin models with electrons in Penning traps, *Phys. Rev. A* **78**, 012338 (2008).
21. A. Retzker, R. C. Thompson, D. M. Segal and M. B. Plenio, Double well potentials and quantum phase transitions in ion traps, *Phys. Rev. Lett.* **101**, 260504 (2008).
22. G. Gabrielse and F. C. MacKintosh, Cylindrical Penning traps with orthogonalized anharmonicity compensation, *Int. J. Mass Spectrom. Ion Processes* **57**, 1 (1984).
23. H. Häffner, T. Beier, N. Hermanspahn, H. -J. Kluge, W. Quint, S. Stahl, J. Verdú and G. Werth, High-accuracy measurement of the magnetic moment anomaly of the electron bound in hydrogenlike carbon, *Phys. Rev. Lett.* **85**, 5308 (2000).
24. J. Verdú, S. Djekic, S. Stahl, T. Valenzuela, M. Vogel, G. Werth, T. Beier, H. -J. Kluge and W. Quint, Electronic g factor of hydrogenlike oxygen $^{16}O^{7+}$, *Phys. Rev. Lett.* **92**, 093002 (2004).
25. G. Gabrielse and H. Dehmelt, Observation of inhibited spontaneous emission, *Phys. Rev. Lett.* **55**, 67 (1985).
26. M. Breitenfeldt, S. Baruah, K. Blaum, A. Herlert, M. Kretzschmar, F. Martinez, G. Marx, L. Schweikhard and N. Walsh, The elliptical Penning trap: Experimental investigations and simulations, *Int. J. Mass Spectrom.* **275**, 34 (2008).
27. M. Kretzschmar, Theory of the elliptical Penning trap, *Int. J. Mass Spectrom.* **275**, 21 (2008).
28. H. Goldstein, *Classical Mechanics* (Addison-Wesley Publishing Company, Cambridge, MA, 1980).

Chapter 12

Trapped Electrons as Electrical (Quantum) Circuits

José Verdú

Department of Physics and Astronomy, University of Sussex, Sussex BN1 9QH, United Kingdom

J.L.Verdu-Galiana@sussex.ac.uk

In this chapter, we present a detailed model of the equivalent electric circuit of a single trapped particle in a coplanar-waveguide (CPW) Penning trap. The CPW-trap, which is essentially a section of coplanar-waveguide transmission-line, is designed to make it compatible with circuit-quantum electrodynamic architectures. This will enable a single trapped electron, or geonium atom, as a potential building block of microwave quantum circuits. The model of the trapped electron as an electric circuit was first introduced by Hans Dehmelt in the 1960s. It is essential for the description of the electronic detection using resonant tank circuits. It is also the basis for the description of the interaction of a geonium atom with other distant quantum systems through electrical (microwave) signals.

12.1. Introduction

The CPW Penning trap[1] has been discussed in detail in the previous chapter. Its design as a section of a surface transmission line allows for its connection to other remote CPW Penning traps or other atomic or solid state systems within a microwave network. This opens potential novel applications in circuit-quantum electrodynamics (circuit-QED), where a trapped electron might be coherently coupled to a superconducting device, such as a Cooper-pair box, or a quantum dot. The spin of the electron is the textbook standard of a quantum two-level system. Due to its magnetic nature, it has potentially very long coherence times. The geonium's cyclotron motion is also a potential candidate for performing quantum operations. At cryogenic temperatures, the electron's cyclotron motion is a pure quantum harmonic oscillator.[2] Moreover, the Purcell effect[3] might help to obtain long coherence times.

The cyclotron motion of a trapped electron can be coupled to propagating microwave photons. Therefore, it can be also coupled to distant solid state devices or other systems with transitions in the microwave domain. A detailed discussion of the possibilities of geonium atoms in circuit-QED is beyond the scope of this chapter. However, it must be mentioned that electrons have already shown applications in this emerging field of research. An example is the recently achieved high-cooperativity coupling of electron-spin ensembles, housed in ruby and diamond coatings, to superconducting CPW microwave cavities.[4] The aim is to provide superconducting circuits with long-coherence quantum memories. Other physical systems have been considered as potential candidates for the storage of quantum information and other circuit-QED applications, such as Bose–Einstein condensates[5] and polar molecules.[6]

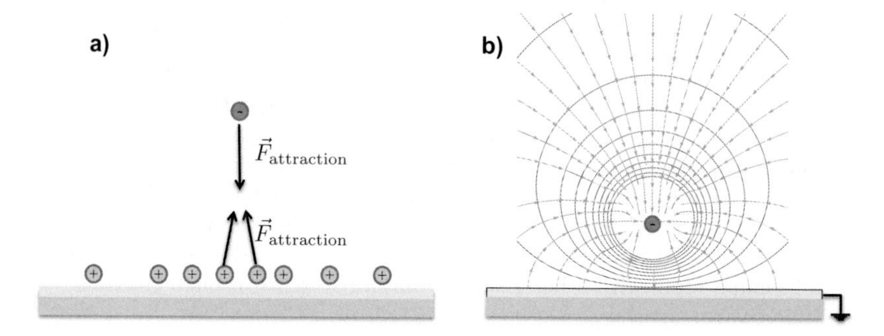

Fig. 12.1. (a) Induced charges and attraction force. (b) The modified electric potential of the trapped particle + induced charges.

In this chapter, we will investigate the electrical signals that might be used to couple electrically a geonium atom stored in a CPW Penning trap to other systems. As described in following sections, the coupling mechanism is intimately linked to the "standard" electronic detection technique of a trapped charged particle. This technique had already been introduced by Dehmelt in the late 1960s.[7] For the sake of clarity, we are going to assume an electron as the trapped particle, however, in a wider context, the discussion might also be applied to other charged particles, such as protons, highly charged ions, etc.

The interaction of the trapped electron with the metallic electrodes of a CPW Penning trap is sketched in Figure 12.1 (a). The trapped particle induces a charge density upon the surface of the electrodes. These induced charges are real and should not be confused with the imaginary "mirror charges". The latter simply represent a mathematical tool used to solve electrostatic problems. The induced charges give rise to an attraction force between the trapped electron and the metallic surface, $\vec{F}_{\text{attraction}}$. In the non-relativistic limit, $\vec{F}_{\text{attraction}}$ is conservative, i.e. it does not dissipate any energy from the electron's motion. It is much weaker than the restoring trapping force, but it does affect the motional frequencies of the trapped particle. Thus, its effect must be considered when performing high precision experiments, such as in references[8,9] and others. Thus, $\vec{F}_{\text{attraction}}$ is not relevant for the cooling or for the detection of the electron. However, with a tuned resonant circuit, the induced charges are the ones permitting the detection of the trapped electron. They also transmit the electrical signal required for coupling a geonium atom to other systems in a circuit. For these reasons, it is essential to describe them and understand their behaviour in detail. For the sake of mathematical simplicity, we centre the discussion on the axial motion of the electron. The axial motion is itself of limited interest for quantum applications, however the theory is easier to develop and can be generalised to the, mathematically more demanding, cyclotron motion. We also assume that the experimental setup, including the CPW Penning trap and the detection system, are within a cryostat, keeping its temperature at 4.2 K (liquid He) or lower. This is required for keeping electrical noise at a sufficiently low level for the detection of a single electron. For applications with superconducting microwave quantum circuits, cryogenic temperatures in the range of 100 mK or lower are required. A dilution refrigerator is employed in these cases.

12.2. The Induced Charge Density

Figure 12.1 (b) shows the electric field and equipotential lines created by the charge of the trapped electron. The electron's electric field is modified by the induced charges sitting on the trap's conducting surface. As is well known from electrostatics, the expression for the modified electron's potential, $\Phi(\vec{r})$, is given by the Green's function for Laplace's equation, fulfilling Dirichlet's boundary conditions[10], $G(\vec{r}\,|\vec{r}\,')$:

$$\Phi(\vec{r}) = \frac{q}{4\pi\epsilon_0} G(\vec{r}\,|\vec{r}\,') \quad ; \quad G(\vec{r}\,|\vec{r}\,') = \frac{1}{|\vec{r} - \vec{r}\,'|} + F(\vec{r}\,|\vec{r}\,'). \tag{12.1}$$

In Eq. (12.1) the symbol q represents the charge of the electron and ϵ_0 is the vacuum's dielectric constant. The function $F(\vec{r}\,|\vec{r}\,')$ is the electric potential created by the imaginary mirror charges. $F(\vec{r}\,|\vec{r}\,')$ is mathematically required for the Green's function to fulfil the boundary conditions on the surface of the trap,[10] that is $G(\vec{r},\vec{r}\,') = 0 \;\; ; \forall\, \vec{r}\,, \vec{r}\,' \in S$ (S is the surface of the chip). We are interested in the real induced surface charge density σ. The density can be obtained from the Green's function. Indeed, in general, the charge density is related to the electric field at the surface of the trap through $\sigma = \epsilon_0\, E_n$.[10] Here E_n represents the electric field along the normal direction of the chip. E_n is not to be confused with the trapping field; rather, it is the field created by the trapped electron and the induced charges. Hence, we have:

$$\Phi(\vec{r}\,) = \tfrac{q}{4\pi\epsilon_0}\, G(\vec{r},\vec{r}\,'\,) \rightarrow E_n = -\frac{\partial\,\Phi(\vec{r}\,)}{\partial\, n}\bigg|_{y=0} \Longrightarrow \sigma(\vec{r},\vec{r}\,') = -\frac{q}{4\pi}\cdot\frac{\partial\, G(\vec{r},\vec{r}\,'\,)}{\partial\, n'}. \tag{12.2}$$

In Eq. (12.2) the vector $\vec{r}$ represents the position of the trapped electron, while $\vec{r}\,'$ is the position on the trap's surface at which the charge density is being computed. Hence, if the Green's function $G(\vec{r},\vec{r}\,'\,)$ is known, then the surface charge density can be easily obtained. Now, assuming the chip which houses the CPW Penning trap is an infinite conducting plane, situated at $y = 0$, we have:[1,10] $G(\vec{r},\vec{r}\,') = 1/\sqrt{(x-x')^2+(y-y')^2+(z-z')^2} - 1/\sqrt{(x-x')^2+(y+y')^2+(z-z')^2}$. Here, we have assumed the usual coordinates system of a CPW Penning trap, where the vertical axis is y (see previous chapter of this volume). With the given Green's function, the detection signal as well as the coupling to external resonators can be computed. This is discussed in detail in the following section.

12.3. Detection of the Electron's Motion

For detecting the trapped particle one particular electrode is chosen to pick up the induced charge. For the axial motion, the most convenient option would be one of the compensation electrodes, while for the cyclotron motion the ring is the best choice.[11] The total charge induced upon the pick-up electrode is simply given by the integral of the charge density over the electrode's surface S: $q_{\text{ind}} = \oint_S \sigma(\vec{r}(t))\, dS'$. The motion of the trapped particle, $\vec{r} = \vec{r}(t)$, generates the induced current: $I_{\text{ind}}(t) = \frac{dq_{\text{ind}}(\vec{r}(t))}{dt}$. This current can be directed to an external resistor, as shown in Figure 12.2 (a).

Thus, the induced current flows to ground, thereby inducing a voltage drop, V_{induced}, across the detection resistor. This voltage can be amplified and then observed in the frequency domain with a Fast Fourier Transform spectrum analyser. The induced voltage can also be expressed as:

$$V_{\text{ind}} = R\,I_{\text{ind}}(t) = R\left(\vec{\nabla}q_{\text{ind}}(\vec{r})\cdot\dot{\vec{r}}(t)\right) = R\left(\frac{\partial q_{\text{ind}}}{\partial x}\dot{x} + \frac{\partial q_{\text{ind}}}{\partial y}\dot{y} + \frac{\partial q_{\text{ind}}}{\partial z}\dot{z}\right). \tag{12.3}$$

Performing the corresponding operations and using the expression of the Green's function, the induced detection signal can be simply expressed as:

$$I_{\text{ind}} = -\frac{q}{D_{\text{eff}}}\cdot\dot{z} \quad ; \quad \frac{1}{D_{\text{eff}}} = \frac{1}{1\text{Volt}}\left|E_z(y_0)\right| . \tag{12.4}$$

In Eq. (12.4) D_{eff} is the so-called *effective coupling distance.*[11] D_{eff} has the dimension of length. Its value depends on the geometry of the CPW Penning trap. It also depends strongly on the position of the trapped electron above the surface of the chip, y_0. The physical meaning of D_{eff} is given in Eq. (12.4). Actually, the effective coupling distance is directly related to the electric field, which the pick-up electrode would create at y_0 when a voltage of 1 volt is applied, while keeping all other electrodes grounded. The strength of the detection signal, and the strength of the coupling of the geonium atom to other electrical components, strongly depends on D_{eff}. Hence, it is essential to know D_{eff} in detail when designing a CPW Penning trap. It must be observed that in Eq. (12.4) the components of the radial motion $(x(t), y(t))$ have been ignored. It can be demonstrated that each component of the particle's motion, (x, y, z), experiences a different effective coupling distance.[11] Hence, the cyclotron and axial motions, and the corresponding induced currents, have different values of D_{eff}. A rigorous mathematical model of the effective coupling distance, with explicit analytic expressions for all motion components, is presented in reference[11] .

The resistor employed for the detection must have a high resistance at the frequencies of the particle's motion (typically 30–90 MHz for the axial motion of one electron). At these frequencies a simple ohmic resistor would be inadequate. The induced current, I_{ind}, would flow to ground through the unavoidable parasitic capacitances (typically several pF) of the CPW Penning trap. The induced current would bypass the resistor R without inducing a detectable voltage signal. To overcome this problem, a tuned detection coil, with inductance L, is used instead of the ohmic resistor. Together with the parasitic capacitances, C, the coil forms a parallel tank circuit. Its resonance resistance is given by $\omega_{\text{LC}} = 1/\sqrt{LC}$. The value of the

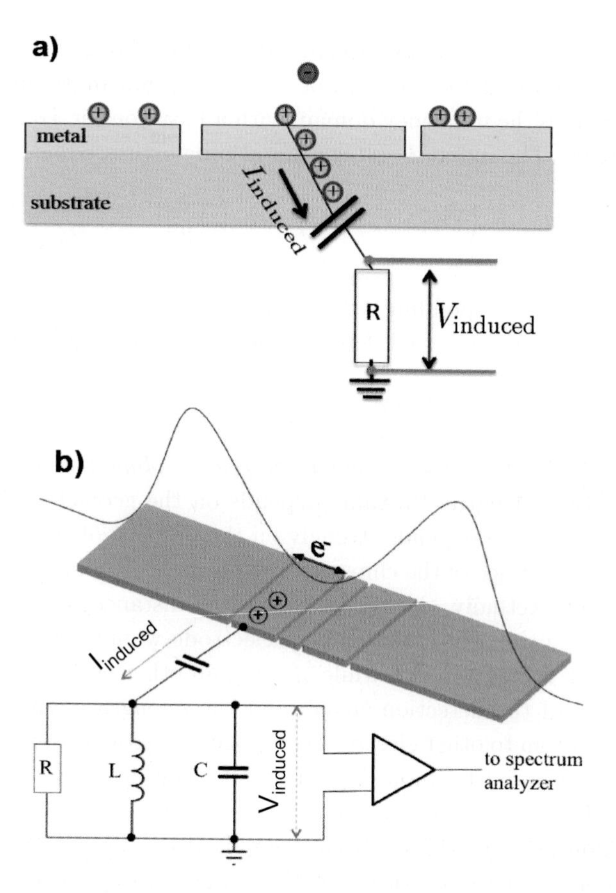

Fig. 12.2. (a) The induced voltage across the capacitively coupled detection resistor. (b) Sketch of the axial detection with a tuned LC-circuit. The resistor R represents the losses of the circuit. It is not a real, explicit ohmic resistor.

coil's inductance must therefore be chosen, such that ω_{LC} equals the axial frequency of the trapped particle ω_z. This maximises the input impedance of the detection circuit, hence also maximising the electron's voltage signal. In general, for the electronic detection to be efficient the coil should have a quality factor, Q, as high as possible. For this reason, it is usually fabricated with superconducting wire.[11] Values of Q above a few thousand can be achieved with radiofrequency coils built with lumped components. Details of the fabrication of these coils can be found in reference[12].

In Figure 12.2 (b) the detection LC-circuit shows a parallel resistor R. This resistor is purely symbolic. It models the effective losses of the

LC-circuit, including all possible loss mechanisms, be it ohmic, radiative, dielectric or diamagnetic. It is important to observe that, the higher the quality factor Q of the detection LC-circuit, the higher the value of the effective resistance R. An ideal LC-circuit, i.e. with no losses, would actually show $R \to \infty$. This can be easily understood from the sketch of the LC-circuit given in Figure 12.2 (b). Indeed, the input impedance of the LC-circuit is given by $\frac{1}{Z_{LC}} = \frac{1}{Z_L} + \frac{1}{Z_C} + \frac{1}{R}$. Hence, in the case $R \to \infty$ the impedance of the circuit is produced exclusively by the impedances of the inductance and the capacitance; it is therefore a purely imaginary impedance corresponding to an electric circuit with no losses. Furthermore, it is important to observe that the impedance is a function of the frequency, $Z_{LC} = Z_{LC}(\omega)$, and, in resonance, its value is a maximum and equals the effective resistance: $Z_{LC}(\omega_{LC}) = R$. This maximises the induced voltage signal, making the trapped electron detectable.

12.4. Equivalent Electrical Circuit of the Trapped Particle

We now want to derive the equivalent electric circuit of the trapped electron. The equivalent circuit gives insight not just into the detection of the particle, but also into the possible coupling mechanism of the geonium atom to other components in an electrical circuit. For simplicity, we will continue the discussion with the electron's axial motion, as in the previous sections. Its generalisation to the cyclotron motion will be discussed at the end of the chapter.

The induced voltage V_{ind}, discussed in Section 12.3, is a real voltage. It is actually acting upon the pick-up electrode. Thus, V_{ind} generates an electric field which, in turn, influences the electron's motion itself. Thus, the total force on the axial motion of the trapped particle is:

$$F_z = m\ddot{z} = -q\frac{V_{\text{total}}}{D_{\text{eff}}} - m\,\omega_z^2 z\,. \tag{12.5}$$

In Eq. (12.5), V_{total} is the total voltage experienced by the trapped electron. The symbol m represents the mass of the particle. Hence, the term $-m\,\omega_z^2 z$ is the restoring force created by the trapping potential. Equation (12.5) can be solved for V_{total}. We have: $-V_{total} = m\,\omega_z^2\,z\frac{D_{\text{eff}}}{q} + m\ddot{z}\frac{D_{\text{eff}}}{q}$. Taking into account the expression for the induced current of Eq. (12.4), we finally obtain the total voltage of the electron:

$$V_{\text{total}} = \underbrace{m\,\frac{D_{\text{eff}}^2}{q^2}\,\dot{I}_{\text{ind}}}_{\text{equivalent inductor}} + \underbrace{m\,\omega_z^2\,\frac{D_{\text{eff}}^2}{q^2}\int dt\cdot I_{\text{ind}}(t)}_{\text{equivalent capacitor}}\,. \tag{12.6}$$

The voltage of Eq. (12.6) is composed of two contributions. The first is the voltage drop across an equivalent inductor, $L_{ion} = \frac{m}{q^2}D_{\text{eff}}^2$. The second is the voltage across an equivalent capacitor C_{ion}, with $\omega_z = 1/\sqrt{L_{ion}C_{ion}}$. Since V_{total} is the sum of two voltages, it is clear that the equivalent inductor and capacitor combine *in series*. Hence, the equivalent electric circuit of the trapped particle is a series LC-circuit. These results where first derived by Dehmelt.[7] It has also been investigated in detail in reference[13]. The circuit representing the trapped particle combines *in parallel* with the external detection tank circuit. This is sketched in Figure 12.3 (a). The impedance resonance curve of the unloaded detection LC-circuit is modified by the electron's axial motion. The latter shunts the input impedance of the detection LC-circuit. The result is that a *dip* appears at its resonance curve. This is plotted in Figure 12.3 (b). The dip allows for the observation of the trapped electron and for the accurate measurement of its axial frequency ω_z. This electronic detection technique is used in many experiments, for instance in references[9,14].

12.4.1. *Resistive cooling time constant*

The induced current dissipates energy in the detection LC-circuit at a rate of $P = I_{\text{ind}}^2 \cdot R$. This is the mechanism responsible for the resistive cooling of the trapped particle.[15,16] The dissipated power is provided by the trapped electron. The latter reduces its energy, until a state of thermal equilibrium with the detection circuit is reached. In this case, the motional energy is no longer well defined; rather, a temperature must be assigned to the trapped particle.[17] The average energy is related to the temperature: $\langle E_z\rangle = \kappa_B \cdot T_z$, where κ_B is Boltzmann's constant. The resistive cooling is characterised by an exponential reduction of the particle's excess energy over time: $E(t) = E_0 e^{-t/\tau}$. Here E_0 is the initial excess energy of the electron. The symbol τ represents the resistive cooling time constant. With the expression for I_{ind} of Eq. (12.4) and with $P = -\frac{dE}{dt}$, it is straightforward to obtain:

$$\tau = \frac{m}{q^2}\cdot\frac{1}{R}\,D_{\text{eff}}^2\,. \tag{12.7}$$

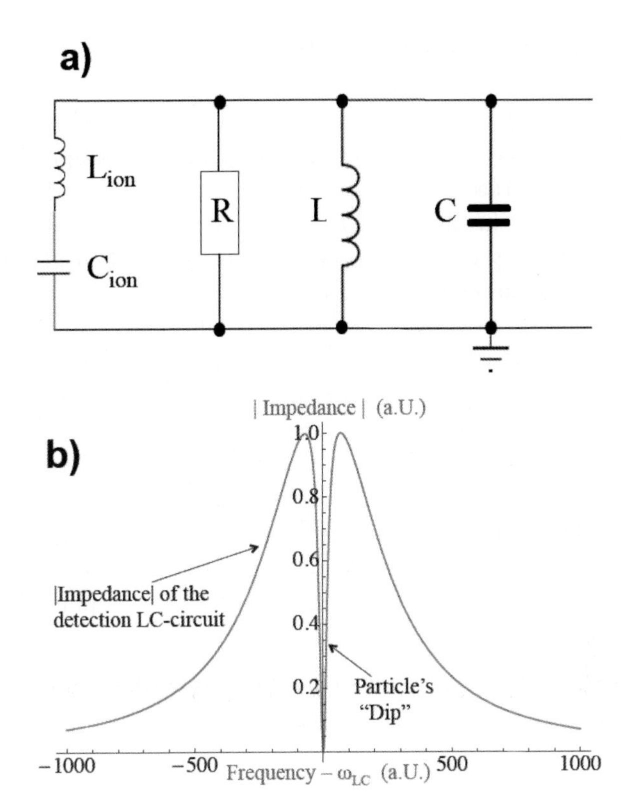

Fig. 12.3. (a) Equivalent electrical circuit of the trapped electron. (b) Resonance curve of the detection LC-circuit modified by the trapped electron. The impedance is plotted in arbitrary units (a.U.).

Values for τ from a few hundred ms to several minutes are typically achieved with radiofrequency detection coils for the axial motion of electrons and the axial and cyclotron motions of heavier ions (see, for instance, reference[17] and references therein).

The cooling time constant τ is also directly related to the detection signal of Figure 12.3 (b). Indeed, the width of the particle's *dip* is given by $\Delta\omega_{dip} = 1/\tau$[13]. Hence, τ also represents the time scale at which energy is exchanged between the trapped particle and the LC-circuit. This time scale, however, must not be confused with the rate at which the LC-circuit itself dissipates energy. As discussed in Section 12.3, a real detection LC-circuit has its losses modelled by the resistance R. The rate at which electromagnetic energy stored in the circuit dissipates is given by the width of its resonance curve (see Figure 12.3 (b)), $\Delta\omega_{LC}$. This width is directly related to the quality factor of the tank circuit, through: $\Delta\omega_{LC} = \omega_{LC}/Q$. For a typical radiofrequency detection circuit, with $Q = 1000$ and resonance frequency 30 MHz (in the range of the axial frequency of a trapped electron), the LC-circuit's loss rate amounts to 30 kHz. On the other hand, the value for $\Delta\omega_{dip}$ is typically in the range of 1 Hz for heavy ions[9] or might achieve a few hundred Hz for the axial motion of an electron.[11] Thus, the detection LC-circuit reaches thermal equilibrium with the environment much faster than the trapped particle reaches thermal equilibrium with the detection circuit. If the coupling of the LC-circuit to the environment is reduced to the level where it is much slower than the coupling rate between the trapped particle and the circuit, $1/\tau$, then the energy of the electron would not be dissipated. In this case the energy would oscillate between the particle and the LC-circuit. A phenomenon similar to the Rabi oscillations of optical cavity-QED. The essential condition for the *coherent coupling* between the trapped electron and an external electrical resonator is therefore the reduction of the losses of the resonator below the coupling rate.

12.5. Coupling the Cyclotron Motion to a Superconducting Cavity

As in the case of the axial motion, the cyclotron motion can also be detected by means of an external resonator. However, for the high frequencies of the (electron's) cyclotron motion, a radiofrequency coil cannot be used for that purpose. Instead, a tuned microwave cavity should play the role of the tank circuit. If the quality factor of the cavity is high enough, then a coherent quantum oscillation of the energy between the trapped electron and the cavity can be achieved. This is the basis for the integration of a geonium atom within a quantum microwave circuit.

The principal quantum number of the axial motion for a trapped electron, at a cryogenic temperature of $T_z = 100$ mK, is of the order of $n \simeq 100$. This is calculated from Planck's law $n = \frac{1}{\exp(\hbar\omega_z/\kappa_B T_z)-1}$. This is a very high quantum number. Hence, the axial motion of the electron is practically always to be considered a classical motion. The situation changes drastically for the cyclotron motion. At a magnetic field of $B = 0.5$ the cyclotron frequency for one electron is $\omega_p = 2\pi \cdot 14$ GHz. The average quantum number for the cyclotron oscillator at that frequency, and at a temperature of 100 mK, amounts to $n \simeq 0.001$. Thus, at that temperature, the cyclotron is in the ground state. It behaves as a pure quantum harmonic oscillator. The choice of 14 GHz to illustrate the orders of magnitude has been also motivated by the fact that this is the range of microwave frequencies where superconducting CPW-cavities are employed in circuit-QED experiments. Superconducting CPW-cavities have been built with very low losses, achieving quality factors in the range of $Q \in [10^5, 10^6]$. A prominent example of the applications of these resonators comes from the University of Yale, where a single microwave photon stored in a CPW-cavity (at 20 mK) has been coherently coupled to a superconducting Cooper-pair box.[18] This experiment has realised a pioneering implementation of the Jaynes–Cummings model in a microwave circuit fabricated in a chip.

A sketch of a CPW-cavity is given in Figure 12.4 (a). The cavity is simply a section of a CPW transmission line. The length of the cavity is chosen to be a multiple integer of $\lambda/4$. The wavelength λ is determined by the resonance frequency, in this case the cyclotron frequency of the electron ω_p. Different types of CPW-resonators can be built. For our purposes, the equivalent electric circuit of the CPW-resonator must be a *parallel* tank circuit, with inductance L_{CPW} and capacitance C_{CPW}, as shown in Figure 12.4 (b). The microwave cavities can be built in different ways, for instance, by terminating a section of CPW transmission line with either an open or a short circuit. Several options are available for obtaining a microwave resonator whose equivalent circuit is a parallel tank circuit, i.e. a circuit with equivalent inductance L and capacitance C connected in parallel.[19] Now, as discussed in Section 12.3, the detection of the trapped particle's axial motion is optimised when the induced voltage, V_{ind}, is maximised. The same argument applies for the cyclotron motion of a trapped electron. The strength of the coupling to an external microwave resonator is therefore optimised when the latter maximises its input impedance at the resonant frequency. This is the case for a parallel LC-circuit.

In Figure 12.4 (a), the length of the CPW-cavity is defined by a small insulating gap at both sides of the resonator. These insulating gaps correspond to the coupling capacitors shown in Figure 12.4 (b). Cavities can be fabricated with low temperature superconductors, such as Niobium, achieving quality factors around 100 000 or better,[20] at cryogenic temperatures of a few hundred mK. The quality factor of these cavities is affected extremely negatively by the presence of a magnetic field.[20] Hence, the microwave cavity must be magnetically isolated from the CPW Penning trap. The magnetic field of the latter would otherwise reduce the value of Q by several orders of magnitude, at a level where the coherent coupling between the cyclotron motion of the electron and the CPW-resonator is no longer possible.

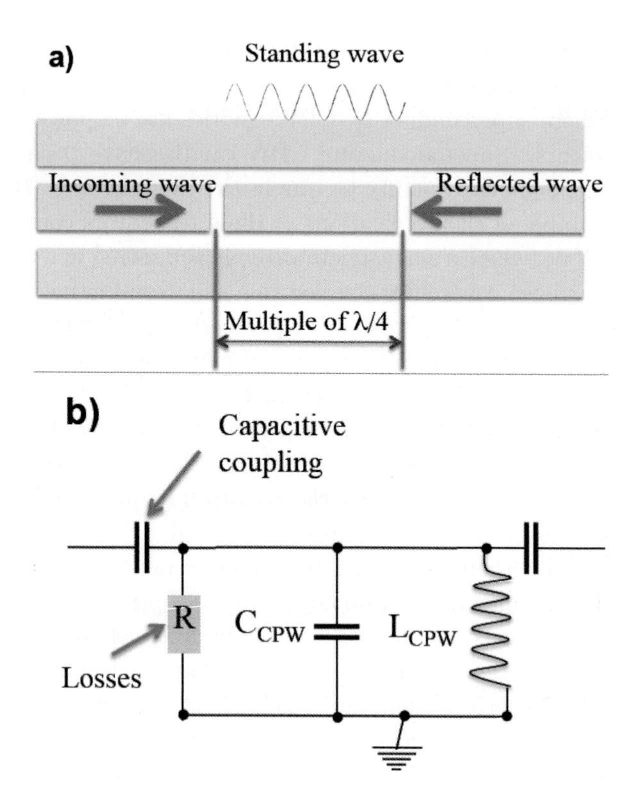

Fig. 12.4. (a) Sketch of a CPW microwave cavity. (b) Equivalent electric circuit of the ($\lambda/4$) CPW-resonator.

The coupling strength between the cyclotron oscillator and the CPW-cavity is given, as for the axial motion, by the inverse of the cooling time constant τ of Eq. (12.7). In this case, the effective coupling distance D_{eff} is also given by Eq. (12.4). However, the axial electric field $E_z(y_0)$ must be substituted by the electric field along the vertical direction y, which couples to the cyclotron motion. Furthermore, the corresponding coupling field, $E_y(y_0)$, cannot be calculated with the electrostatic Green's function given in Section 12.2. Instead, high-frequency techniques must be employed to determine the strength of the electric field of a microwave propagating along a CPW. This calculation is made for example in reference[21] . With this calculation, it can be shown that the coherent coupling of the cyclotron motion of a trapped electron and a distant superconducting CPW microwave cavity is achievable with current fabrication technologies.

12.6. Conclusions

As mentioned in the introduction of this chapter, electrons have already been investigated as potential quantum memories in circuit-QED.[4,22] Recently, it has also been proposed to use electrons trapped on the surface of liquid Helium as potential components in superconducting microwave circuits.[23] Trapped electrons have already demonstrated excellent potential at the quantum level. They have been proposed in the building of a quantum processor.[24] Experimental and theoretical efforts have already started.[25] In this chapter, we have given a brief introduction of how the powerful techniques developed for geonium atoms could be combined with the emerging field of circuit-QED by employing the novel CPW Penning trap.

Acknowledgments

The author acknowledges financial support from EPSRC, under grant EP/I012850/1, from the Marie Curie reintegration grant "NGAMIT"and from SEPnet.

References

1. J. Verdú, Theory of the coplanar waveguide Penning trap, *New J. Phys.* **13**, 113029 (2011).
2. S. Peil and G. Gabrielse, Observing the quantum limit of an electron cyclotron: QND measurements of quantum jumps between Fock states, *Phys. Rev. Lett.* **83**, 1287 (1999).
3. G. Gabrielse and H. Dehmelt, Observation of inhibited spontaneous emission, *Phys. Rev. Lett.* **55**, 67 (1985).
4. D. I. Schuster, A. P. Sears, E.Ginossar, L. DiCarlo, L. Frunzio, J. J. L. Morton, H. Wu, G. A. D. Briggs, B. B. Buckley, D. D. Awschalom and R. J. Schoelkopf, High-cooperativity coupling of electron-spin ensembles to superconducting cavities, *Phys. Rev. Lett.* **105**, 140501 (2010).
5. J. Verdú, H. Zoubi, C. Koller, J. Majer, H. Ritsch and J. Schmiedmayer, Strong magnetic coupling of an ultracold gas to a superconducting waveguide cavity, *Phys. Rev. Lett.* **103**, 043603 (2009).
6. A. André, D. DeMille, J. M. Doyle, M. D. Lukin, P. Rabl, R. J. Schoelkopf and P.Zoller, A coherent all-electrical interface between polar molecules and mesoscopic superconducting resonators, *Nat. Phys.* **2**, 636 (2006).
7. H. G. Dehmelt and F. L. Walls, Bolometric technique for the RF Spectroscopy of Stored Ions, *Phys. Rev. Lett.* **21**, 127 (1968).

8. D. Hanneke, S. Fogwell and G. Gabrielse, New measurement of the electron magnetic moment and the fine structure constant, *Phys. Rev. Lett.* **100**, 120801 (2008).
9. J. Verdú, S. Djekic, S. Stahl, T. Valenzuela, M. Vogel, G. Werth, T. Beier, H. -J. Kluge and W. Quint, Electronic g factor of hydrogenlike oxygen $^{16}O^{7+}$, *Phys. Rev. Lett.* **92**, 093002 (2004).
10. J. D. Jackson, *Classical Electrodynamics* (Wiley and Sons, New York NY, 2005).
11. A. Al-Rjoub and J. Verdú, Electronic detection of a single particle in a coplanar-waveguide Penning trap, *Appl. Phys. B* **107**, 955 (2012).
12. S. Ulmer, H. Kracke, K. Blaum, S. Kreim, A. Mooser, W. Quint, C. C. Rodegheri and J. Walz, The quality factor of a superconducting RF resonator in a magnetic field, *Rev. Sci. Instrum.* **80**, 123302 (2009).
13. X. Feng, M. Charlton, M. Holzscheiter, R. A. Lewis and Y. Yamazaki, Tank circuit model applied to particles in a Penning trap, *J. Appl. Phys.* **79**, 8 (1996).
14. S. Ulmer, C. C. Rodegheri, K. Blaum, H. Kracke, A. Mooser, W. Quint and J. Walz, Observation of spin flips with a single trapped proton, *Phys. Rev. Lett.* **106**, 253001 (2011).
15. D. J. Wineland and H. G. Dehmelt, Principles of the stored ion calorimeter, *J. Appl. Phys.* **46**, 919 (1975).
16. W. M. Itano, J. C. Bergquist, J. J. Bollinger and D. J. Wineland, Cooling methods in ion traps, *Phys. Scripta* **T59**, 106 (1995).
17. S. Djekic, J. Alonso, H. -J. Kluge, W. Quint, S. Stahl, T. Valenzuela, J. Verdú, M. Vogel and G. Werth, Temperature measurement of a single ion in a Penning trap, *Eur. Phys. J. D* **31**, 451 (2004).
18. A. Wallraff, D. I. Schuster, A. Blais, L. Frunzio, R. S. Huang, J. Majer, S. Kumar, S. M. Girvin and R. J. Schoelkopf, Strong coupling of a single photon to a superconducting qubit using circuit quantum electrodynamics, *Nature* **431**, 162 (2004).
19. D. M. Pozar, *Microwave Engineering* (Wiley and Sons, New York NY, 2004).
20. L. Frunzio, A. Wallraff, D. Schuster, J. Majer and R. Schoelkopf, Fabrication and characterization of superconducting Circuit QED devices for quantum computation, *IEEE Trans. Appl. Supercond.* **15**, 860 (2005).
21. R. N. Simons and R. Arora, Coupled slot line field components, *IEEE Trans. Microwave Theory Techn.* **MTT-30**, 1094 (1982).
22. H. Wu, R. E. George, J. H. Wesenberg, K. Mølmer, D. I. Schuster, R. J. Schoelkopf, K. M. Itoh, A. Ardavan, J. J. L. Morton and G. A. D. Briggs, Storage of multiple coherent microwave excitations in an electron spin ensemble, *Phys. Rev. Lett.* **105**, 140503 (2010).
23. D. I. Schuster, A. Fragner, M. I. Dykman, S. A. Lyon and R. J. Schoelkopf, Proposal for manipulating and detecting spin and orbital states of trapped electrons on helium using cavity quantum electrodynamics, *Phys. Rev. Lett.* **105**, 040503 (2010).
24. G. Ciaramicoli, I. Marzoli and P. Tombesi, Scalable quantum processor with trapped electrons, *Phys. Rev. Lett.* **91**, 017901 (2003).

25. I. Marzoli, P. Tombesi, G. Ciaramicoli, G. Werth, P. Bushev, S. Stahl, F. Schmidt-Kaler, M. Hellwig, C. Henkel, G. Marx, I. Jex, E. Stachowska, G. Szawiola and A. Walaszyk, Experimental and theoretical challenges for the trapped electron quantum computer, *J. Phys. B: At. Mol. Opt. Phys.* **42**, 154010 (2009).

Chapter 13

Basics of Charged Particle Beam Dynamics and Application to Electrostatic Storage Rings

Alexander I. Papash

Karlsruhe Institute of Technology,
Hermann-von-Helmholtz-Platz 1,
76344 Eggenstein-Leopoldshafen, Germany
Joint Institute for Nuclear Research,
Joliot-Curie 6, 141980 Dubna, Moscow oblast, Russia
alexander.papash@kit.edu

Carsten P. Welsch

University of Liverpool, Department of Physics,
Liverpool L69 7ZE, United Kingdom
Cockcroft Institute Sci-Tech Daresbury, Keckwick Lane,
Warrington WA4 4AD, United Kingdom
c.p.welsch@liverpool.ac.uk

Electrostatic storage rings (ESRs) are circular accelerators which use electrostatic bending and focusing elements. ESRs are capable of storing a wide variety of ions, from protons and antiprotons to heavy molecular ions. The electric field strength required to bend ions depends on the ion energy and charge, but not on the ion mass, which offers interesting opportunities for storing even very heavy biomolecules. The range of ion energies that can be covered by an ESR is determined by the maximum electric field strength. Typically, ESRs operate at ultralow beam energies in the keV range. In Chapter 13, we first give a basic description of beam motion in storage rings composed of magnetic elements before focusing on the description of particle motion in storage rings that use electrostatic optical elements. This will form the basis for a more detailed analysis of beam motion in electrostatic rings in Chapter 14.

13.1. Introduction

An extensive literature providing a general description of the motion of charged particles in "classical" magnetic storage rings is available, including the books of Livingston,[1] Livingood,[2] Steffen,[3] Bruck,[4] Kolomensky and Lebedev,[5] Wilson,[6] Bryant,[7] Reiser,[8] Lee[9] and many others. The yellow reports of the CERN Accelerator Schools (CAS)[10] and the study materials of the Joint Universities Accelerator School (JUAS)[11] are also an excellent source of detailed information.

Different types of accelerators are used around the world for various purposes, including increasing the energy of a beam of charged particles, decreasing the energy of a beam of particles that were produced at high initial energy, keeping particles circulating on a stable orbit in a storage ring, adding more particles to a stored beam for an increase in beam intensity, or transferring particles from a storage ring to an external experimental area. In this chapter the term "charged particles" is used for all different types of ions, ranging from protons and antiprotons, to light and heavy ions, as well as for electrons and positrons.[10,11]

The basic parameters of some of the most common types of ion accelerators are listed in Table 13.1. The listed parameters are just examples and not all existing types of accelerators are included in this simple table. The individual machine type and range of kinetic energies T_{kin} of particles that can be used are shown in the first and second column. Some important beam parameters are then presented in the following columns: namely the relativistic factor γ which is the ratio of total particle energy to rest energy $\gamma = W / W_0$, the relativistic speed β which is the ratio of particle velocity to the speed of light $\beta = V / c$, the dependence of guiding magnetic field B on time, as well as on radius, and the dependence of orbit radius ρ on particle energy. Finally, the particle rotation frequency F_{rot} as a function of time is shown in the last column. Note that the total ion energy is the sum of rest energy $W_0 = m_0 c^2$ and kinetic energy T_{kin} :

$$W = W_0 + T_{kin} . \tag{13.1}$$

Table 13.1. Examples of different types of ion accelerators.

Machine type	Ion energy	Relativistic factor	Ion velocity	Bending field	Orbit radius	Rotation frequency
	W_{kin}	$\gamma = W / W_0$	$\beta = V / c$	$\boldsymbol{B}$	ρ	$\boldsymbol{F}$
Electrostatic Van de Graaff and tandems	2-30 MeV	1.002-1.03	$6 \cdot 10^{-2}$-$3 \cdot 10^{-1}$	No magnetic field	Linear	--
Isochronous cyclotrons	10-600 MeV	1.01-1.64	0.14-0.79	$B(t) = const$ $dB / dR > 0$ $B(r) \propto \gamma \cdot B(0)$	Spiral $\rho \propto p$	$F = const$
Synchro-cyclotrons	100-1000 MeV	1.1-2.06	0.43-0.87	$B(t) = const$ $dB / dR < 0$	Spiral $\rho \propto p$	Reduced $F(t) \propto B / \gamma W_0$
Linear accelerators	50-800 MeV	1.05-1.85	0.31-0.84	--	Linear	$F = const$
Proton/ion Synchrotrons	0.2-7000 GeV	1.2-7400	0.56-0.9999	$B(t) \propto \beta \cdot \gamma$	Closed orbit $\rho = const$	$F \propto \beta$
Electrostatic Storage Rings	1-100 keV	~1	10^{-3}-10^{-2} $<< 1$	$U_{ESD} \propto W_{kin}$	Closed orbit $R = const$	$F \propto \beta$

In linear electrostatic accelerators, such as the Cockcroft–Walton generator, or the Van de Graaff and tandem accelerators, the energy of particles is increased by applying a static direct current (DC) voltage to a sequence of electrodes. The maximum reachable energy in such accelerators is limited by discharge problems between the electrodes and the ground, as well as sparks between the electrodes themselves.

In linear accelerators, so-called "linacs", particles are accelerated by applying a radiofrequency (RF) voltage to a sequence of accelerating gaps or RF cavities. The central beam trajectory is essentially a straight line for both types of accelerators, and particles cross each acceleration gap only once. No guiding magnetic field is needed in linear accelerators. Focusing particles along the central trajectory is achieved by applying a quadrupolar magnetic field in the direction perpendicular to the beam trajectory. Usually, magnetic quadrupole lenses are used for beam focusing.

The principle of particle energy increase by repeated crossing of acceleration gaps is exploited in circular accelerators. In this case particles move on either circular or spiral reference orbits where they circulate many times. Beam rotation in circular accelerators is usually realized by applying a magnetic dipole field to the charged particle beam. Commonly used types of circular accelerators are discussed below.

The ion beam trajectory in a *cyclotron* is a spiral and the ion's bending radius increases each turn as its energy increases, see Figure 13.1(a).[11] The accelerating RF field shown by the arrow in this figure is created by an RF voltage applied between a pair of D-shaped electrodes. The applied magnetic dipole field is constant in time and has a special configuration in three-dimensional (3D) space in order to provide simultaneous beam focusing. Ions which are crossing the accelerating gaps in phase with the RF voltage increase their energy while particles crossing the gaps out of phase are lost. The relativistic growth of ion mass with an increase in energy is compensated in isochronous cyclotrons by a special shaping of the guiding magnetic field $B(r) \propto \gamma \cdot B(0)$, where γ is the relativistic factor.

The maximum achievable energy in cyclotrons is limited by relativistic effects and a reasonable size and strength of the bending magnetic field to around 600 MeV. In contrast, the RF frequency is kept constant in *isochronous cyclotrons* where ions can be accelerated in continuous wave (CW) mode. At very high energies, relativistic growth of ion mass with increasing particle energy leads to a reduction in ion velocity and causes ions to lose synchronicity with the RF. This is why the RF frequency is decreased in *synchrocyclotrons* by a factor $F(t) \propto B / \gamma(t) W_0$, in order to allow further acceleration to take place. The maximum achievable beam energy in proton synchrocyclotrons is around 1000 MeV.

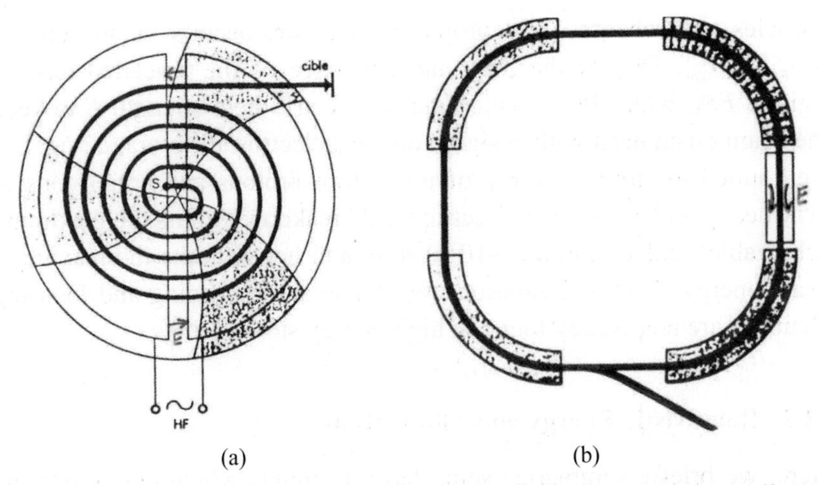

(a) (b)

Fig. 13.1. Schematic view of two commonly used types of circular accelerators.[11] a) In cyclotrons the magnetic field $\boldsymbol{B}$ is constant in time and the orbit radius $\boldsymbol{\rho}$ is increased while the beam is accelerated; b) in synchrotrons, the orbit radius is fixed and the magnetic field grows as the beam energy is increased.

Finally, the *synchrotron* is a circular accelerator where the energy gain of the so-called synchronous particle is matched to the increase of the magnetic field with time. In such an accelerator particles move on a closed reference orbit,[11] see Figure 13.1(b). RF cavities, injection and extraction elements, beam diagnostics, in-ring experiments, etc. are accommodated in straight sections between bending magnets.

The magnetic field needs to be increased during acceleration by a factor of $B(t) \propto \beta \cdot \gamma \cdot B(0)$ in order to follow the gain in beam energy. In addition, the RF frequency needs to increase in proportion to $F(t) \propto \beta \cdot F(0)$ in order to keep particles rotating along the closed orbit at a constant radius. The beam energy in an ion synchrotron is limited by the size of the machine and the maximum magnetic field strength in the bending magnets.

In contrast to magnetic storage rings, ESRs rely on the use of electric fields to bend and focus charged particles. Beam motion in this type of accelerator is fundamentally different from motion in magnetic rings. All existing ESRs operate at ultralow energies in the keV range. Voltages U_{ESD} that need to be applied to the bending electrodes to store a beam of particles with charge q are proportional to the beam's kinetic energy $qU_{ESD} \propto W_{kin}$. Due to the mass independence of the so-called electric rigidity ER_0, which defines the required electric field strength E to keep the beam on an orbit with design radius R_0, electrostatic storage rings are well suited to store a variety of ions, from protons and antiprotons to very heavy molecular ions. Electric field breakdown limits the maximum achievable field strength to ~10 kV/mm and hence limits the maximum beam energy. This is the reason why electrostatic bending and focusing elements are not usually found in high-energy storage rings.

13.2. Relativistic Energy and Momentum

Here, we briefly summarize some basic formulae which define particle energy and momentum. The total energy of a particle $W = mc^2$ is related to the particle's rest energy $W_0 = m_0c^2$ by

$$W = \gamma W_0 , \tag{13.2}$$

where the relativistic factor γ is related to the relativistic velocity $\beta = V/c$ as

$$\gamma^2 = \frac{1}{1-\beta^2} . \tag{13.3}$$

The momentum of a relativistic particle is defined by

$$p = mv = m_0\gamma\beta c . \tag{13.4}$$

The combination of Eqs (13.2), (13.3) and (13.4) results in the well-known equation for total energy

$$W^2 = W_0^2 + p^2c^2 . \tag{13.5}$$

In the ultra-relativistic case, where the velocity of the beam is very close to the velocity of light, the total particle energy W is much greater than the rest energy W_0, and hence the relativistic factor $\gamma >> 1$ and $\beta \approx 1$. The total ion energy can then be approximated by:

$$W \approx T_k \approx pc . \tag{13.6}$$

In ESRs the kinetic energy of the ions is much smaller than the rest energy $T_{kin} << W_0$. Thus, the relativistic factor $\gamma \approx 1$ and $\beta << 1$. The ion momentum is then given by $p = m_0 V$ and the kinetic energy of ions is described by the well-known classical formula:

$$T_k = \frac{m_0 V^2}{2}. \tag{13.7}$$

13.3. Basic Features of Magnetic and Electrostatic Bends

Any circular accelerator (decelerator) should be able to bend, focus, store and accelerate (or decelerate) ions. An accelerator increases (a decelerator reduces) the kinetic energy of the stored particles hence increasing (or decreasing) their momentum. This is done by applying an electric field $\vec{E}$ in the direction of particle motion either parallel to the ion momentum $\vec{p}$ (acceleration) or opposite to the ion momentum $\bar{p}$ (deceleration). The change of the ion momentum due to the action of this electric field is given by

$$\frac{d\vec{p}}{dt} = q\vec{E}, \tag{13.8}$$

where q is the charge of the particle. The movement of ions in a bending magnet is defined by the Lorenz force $\vec{F}_L$ acting on the charged particle moving through the magnetic field $\vec{B}$ (see Figure 13.2),

$$\vec{F}_L = \frac{d\vec{p}}{dt} = \frac{d(m\vec{V})}{dt} = \frac{q}{c}\left[\vec{B} \times \vec{V}\right]. \tag{13.9}$$

A centripetal force is produced by the magnetic field $\vec{B}$ and is perpendicular to the plane defined by the magnetic field and the ion velocity $\vec{V}$. Rather than using a magnetic field, ion deflection can also be achieved by an electric field $\vec{E}_\perp$ in the plane of the ion's motion, but perpendicular to its velocity $\vec{V}$ (see Figure 13.3). Such a radial electric field can be generated in an electrostatic bend composed of two curved electrodes (see Figure 13.4).

There, a voltage $\pm U$ is applied to opposite electrodes separated by a distance *d*. Ions are moving between the electrodes in the horizontal plane. The electrodes in such electrostatic deflectors (ESDs) are identical to curved capacitor plates.

Neglecting fringe fields at the exit and entrance of the electrostatic bend, the central ion trajectory inside the ESD is perpendicular to the electric field $\vec{V} \perp \vec{E}$. The angular and vertical components of the ESD's electric field at the position of the central orbit are equal to zero, $E_\theta = E_Z = 0$, and the radial component is $E_R = E_0 = 2U/d$. The electric potential φ is equal to zero in the center and the electric field acts on the ions as a centripetal force, i.e. it deflects charged particles (see Figure 13.3) but does not change their energy along the central orbit. Ions initially displaced in coordinate or angle with respect to the central trajectory will experience stable or unstable oscillations around the orbit where the voltage is equal to zero.

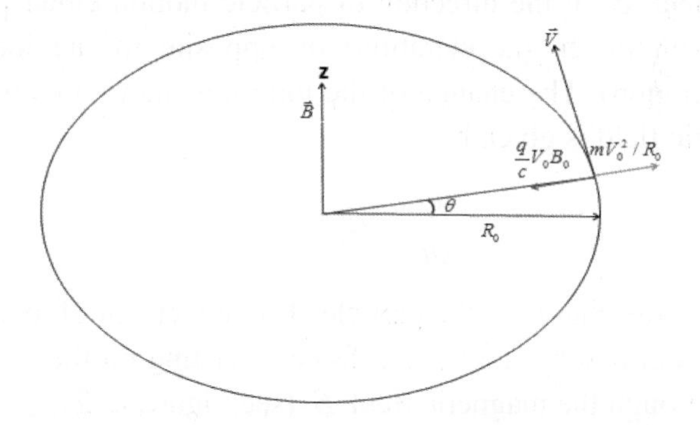

Fig. 13.2. Centripetal Lorentz force and centrifugal force acting on a particle with charge ***q*** moving in a magnetic field. The field ***B*** is perpendicular to the plane of ion motion and to the ion velocity ***V***.

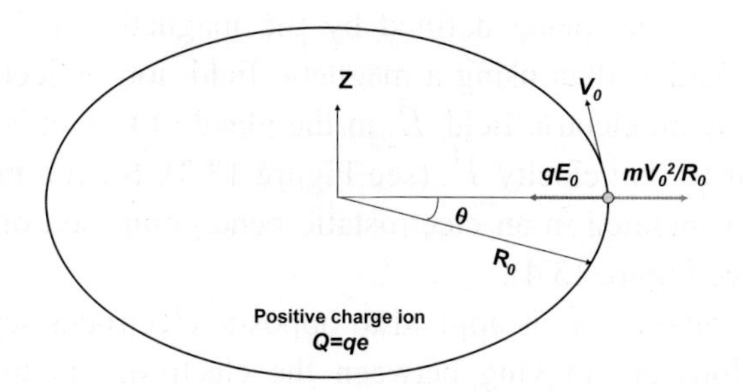

Fig. 13.3. Force acting on a charged particle ***q*** moving in an electric field ***E*** located in the plane of ion motion, but perpendicular to the ion velocity ***V***.

The energy of ions inside an ESD changes because of the tangential component of the electric field. This will accelerate or decelerate the charged particles. For now, we will neglect this effect caused by the ESD fringe fields.

The central ion trajectory within a magnetic or electric bend is defined by the equilibrium between centrifugal and centripetal forces acting on a charged particle q moving either in a magnetic field $\vec{B}$ perpendicular to the plane of ion motion or in a radial electric field $\vec{E}_R$. A particle with velocity V_0 will move under the action of a magnetic field B_0 on a circle of radius R_0 which satisfies the relation

$$\frac{mV_0^2}{R_0} = -\frac{q}{c}V_0B_0. \tag{13.10}$$

The orbit where the Lorentz force is equal to the centripetal force is called the equilibrium orbit (EO). A similar expression to Eq. (13.10) defines the forces acting on a charge particle moving in a radial electric field E_0:

$$\frac{mV_0^2}{R_0} = -qE_0. \tag{13.11}$$

Note the negative sign in Eqs (13.10) and (13.11) which indicates the opposite direction of centripetal and centrifugal forces, seen in Figures 13.2 and 13.3. Here, we focus on a discussion of the magnitude of the forces on the stored particles and will neglect the negative sign in the following discussion.

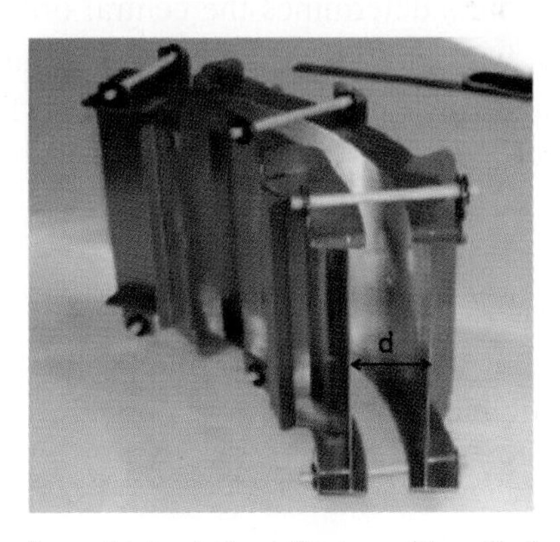

Fig. 13.4. Photograph of an electrostatic deflector with cylinder electrodes which are separated by a gap ***d***.[13]

13.3.1. Magnetic rigidity

The efficiency of a magnet to bend a charged particle can be described by the magnetic rigidity. It is defined as the product of the magnetic field B_0 and the bending radius R_0. One may rewrite Eq. (13.10) as

$$qB_0R_0 = mV_0c = p_0c\,. \tag{13.12}$$

Thus, the product of magnetic field and bending radius is proportional to particle momentum. Assuming relativistic energies one may write Eq. (13.12) in the following form

$$B_0R_0 = \frac{1}{q}\beta\gamma m_0c^2 = \frac{A}{q}\beta\gamma W_0^{proton}, \tag{13.13}$$

where A is the atomic mass number and W_0^{proton} is the rest energy of the proton. At low energies when the ion kinetic energy T_{kin} is much lower than the rest energy W_0 and the relativistic speed $\beta \ll 1$, the magnetic rigidity is approximately proportional to the square root of ion mass to charge ratio $\sqrt{A/q}$:

$$B_0R_0 = \frac{1}{q}p_0c \approx \frac{1}{q}\sqrt{2T_kW_0} = \sqrt{\frac{A}{q}}\cdot\sqrt{2\left(\frac{T_k}{q}\right)W_0^{proton}}\,. \tag{13.14}$$

13.3.2. Electric rigidity

The electric rigidity E_0R_0 determines the central orbit and ion energy in ESRs in the same way in which magnetic rigidity defines the ion momentum in storage rings using magnet dipoles. The electric rigidity can be derived from Eq. (13.11) using the relativistic parameters β and γ:

$$\left|E_0R_0\right| = \frac{1}{q}mV_0^2 = \frac{1}{q}m\beta^2c^2 = \frac{1}{q}mc^2\left(1-\frac{1}{\gamma^2}\right). \tag{13.15}$$

From Eq. (13.1) one may express the kinetic energy as product of rest energy and relativistic factor $(\gamma-1)$:

$$T_k = m_0c^2(\gamma-1)\,. \tag{13.16}$$

One can then rewrite Eq. (13.1) in the following form:

$$mc^2 = \frac{T_k}{\gamma - 1} + T_k = T_k \cdot \frac{\gamma}{\gamma - 1}. \qquad (13.17)$$

Substituting Eq. (13.17) into Eq. (13.15) gives a relativistic expression for the electric rigidity:

$$\left|E_0 R_0\right| = \frac{1}{q} mc^2 \left(1 - \frac{1}{\gamma^2}\right) = \frac{1}{q}\left(\frac{\gamma+1}{\gamma}\right) T_k. \qquad (13.18)$$

At low energies, where $\gamma \approx 1$, the electric rigidity is twice the kinetic energy divided by the ion charge:

$$\left|E_0 R_0\right| \approx \frac{2T_k}{q}. \qquad (13.19)$$

Therefore, the ability of the electric field to deflect charged particles depends on the ratio of kinetic energy and ion charge, regardless of the mass of the particle. The corresponding voltage $\pm U$ between two electrodes separated by a distance d needed to bend a beam of charged particles along a radius R_0 is then given by:

$$U_{\pm} = \frac{1}{q} T_k \frac{d}{R_o}. \qquad (13.20)$$

Light and heavy ions of the same electric charge which are injected into an ESR at identical kinetic energy will hence require the same voltages on both electrodes, independent of their mass. This is why ESRs are capable of storing a very wide range of ions ranging from protons and light ions to heavy molecules with atomic masses of up to $A \approx 1000$.

The main limitation of electrostatic deflectors is that the maximum possible beam energy is limited by the achievable electric field strength. This can be estimated by:

$$\frac{U_{\pm}}{d}\left[kV / cm\right] = 300\beta \cdot B_0\left[kGs\right]. \qquad (13.21)$$

At low energies, where $\beta << 1$, the required voltages are of the order of a few kV, but quickly reach unpractical limits at higher beam energies. Therefore, the use of electrostatic deflectors is limited to beam energies below a few hundreds of keV.

13.4. Betatron Oscillations

Next, we will consider the motion of particles which are displaced from the EO (see Figure 13.5). These trajectories describe the motion of many particles and these make up a real beam. We will then identify conditions where this motion is stable, i.e. where the trajectories are confined around a reference orbit. The motion is *stable* if the amplitude of oscillations is limited and *unstable* if it is growing steadily.

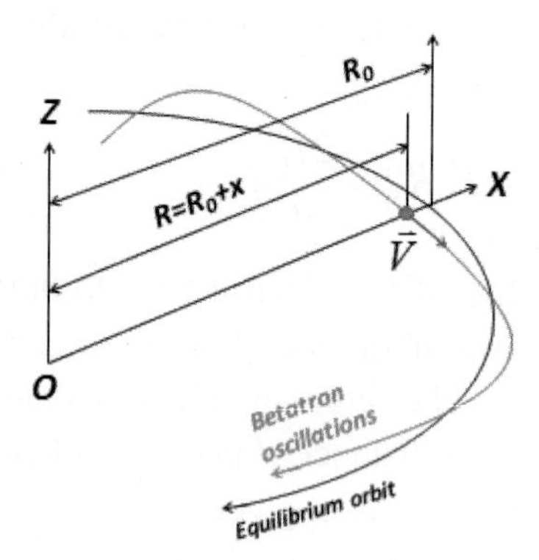

Fig. 13.5. Particle oscillations around the so-called EO. It is assumed that the amplitude of oscillations is much less than the radial position of the ion $x \ll R_0$.

The EO is defined as the trajectory where the centripetal force is equal to the centrifugal force, see Eqs (13.10) and (13.11). Here, we first derive linear equations of ion oscillations around the EO in a magnetic dipole (see Figure 13.6), and then nonlinear equations that describe motion within an electrostatic bend.

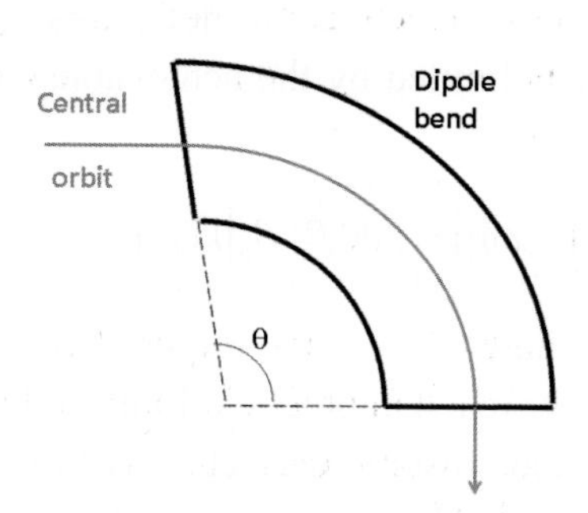

Fig. 13.6. Schematic view of a dipole magnet with bending angle θ.

13.4.1. Radial motion

The radial position of a particle displaced from the reference orbit at distance x is $R = R_0 + x$, where R_0 defines the position of the EO (see Figure 13.5). It is assumed that the displacement x is much smaller than the radial coordinate of the ion, i.e. $x << R_0$. A force will then act on the displaced particle in the radial direction because the centripetal force is not equal to the centrifugal force at any position $R \neq R_0$:

$$F_R = m\frac{d^2R}{dt^2} - m\frac{V_0^2}{R} = \frac{e}{c}V_0 B_z . \tag{13.22}$$

By introducing a local coordinate system with a center at the EO,

$$s = R_0\theta \qquad x = R - R_0 \qquad z = z, \tag{13.23}$$

one can transform Eq. (13.22) to the (s,x,z) coordinate system, substituting the time derivative by:

$$\frac{d}{dt} \equiv V_0 \frac{d}{ds} . \tag{13.24}$$

From Eq. (13.10) one may substitute $\frac{e}{mc} = \frac{V_0}{B_0 R_0}$ into Eq. (13.22) and get the following expression:

$$\frac{d^2x}{ds^2} - \frac{1}{R_0 + x} = \frac{e}{mcV_0} B_z . \tag{13.25}$$

At small displacements $x << R_0$ one may use the approximation

$$\frac{1}{R_0 + x} \approx \frac{1}{R_0}\left(1 - \frac{x}{R_0}\right). \tag{13.26}$$

The vertical component of the magnetic field $B_z(R)$ can then be expanded into

$$B_z(x) = B_0 + \left(\frac{\partial B_z}{\partial x}\right)_0 x + ... \tag{13.27}$$

By introducing the normalized gradient of the magnetic field in radial direction k_x

$$k_x = -\frac{1}{B_0 R_0}\left(\frac{\partial B_z}{\partial x}\right)_0 , \tag{13.28}$$

one can derive an equation describing the horizontal oscillations around the EO in linear approximation:

$$\frac{d^2x}{ds^2}+\left(\frac{1}{R_0^2}-k_x\right)x=0\,. \tag{13.29}$$

For off-momentum particles, where $p=p_0+\Delta p$, one may express the force acting on the particle as

$$F_R=\frac{d}{dt}\left[(m+\Delta m)\frac{d}{dt}R\right]-(m+\Delta m)\frac{(V_0+\Delta V)^2}{R}=\frac{q}{c}(V_0+\Delta V)B_z\,. \tag{13.30}$$

Transforming time to distance yields

$$\frac{d}{dt}\equiv(V_0+\Delta V)\frac{d}{ds}\,. \tag{13.31}$$

Keeping only first-order terms then yields a linear equation for the off-momentum particles:

$$\frac{d^2x}{ds^2}+\left(\frac{1}{R_0^2}-k_x\right)x=\frac{1}{R_0}\frac{\Delta p}{p_0}\,. \tag{13.32}$$

These will oscillate around a new EO which is displaced by

$$R_{\Delta p}=R_0+D\frac{\Delta p}{p_0}, \tag{13.33}$$

with respect to the nominal EO. The so-called linear dispersion function D is a feature of an accelerator indicating how much particles with a momentum offset of $\Delta p=p_0$ will be displaced from the nominal orbit R_0. We would like to remind the reader that linear motion is limited to small oscillations and in real accelerators momentum offset cannot exceed a few percent of the nominal momentum. Generally speaking, most storage rings allow the operation of beams with a momentum spread of no larger than $\Delta p/p_0=10^{-3}-10^{-2}$.

13.4.2. Vertical motion

In the vertical direction, the magnetic field in a dipole is symmetric around $z=0$ and the plane of symmetry is called the median plane. The radial component of the magnetic field $B_x(z=0)=0$ in this plane. A Taylor expansion of the radial component of the magnetic field to first order in vertical coordinate z yields:

$$B_x(z) = \left(\frac{\partial B_x}{\partial z} \right)_0 z + ... \tag{13.34}$$

By defining the normalized vertical gradient of the magnetic field in the vertical direction k_z as

$$k_z = -\frac{1}{B_0 R_0} \left(\frac{\partial B_x}{\partial z} \right)_0 , \tag{13.35}$$

one may derive a linear equation describing vertical oscillations in a similar way as in the previous section:

$$\frac{d^2 z}{ds^2} + k_z z = 0 \,. \tag{13.36}$$

Provided there is no dispersion in the vertical direction, Eq. (13.36) is valid in linear approximation for on-momentum, as well as for off-momentum particles.

13.4.3. Solution of the equation of motion

The equation of motion is:

$$y'' + Ky = 0 , \tag{13.37}$$

where y stands for either the radial x or the vertical z coordinate and the parameter K is piecewise constant and defined as $K_x = \left(1/R_0^2 - k_x\right)$ in the radial direction and $K_z = k_z$ in the vertical direction. The general solution is a linear combination of two independent solutions

$$y(s) = a_1 \cdot \cos(\omega \cdot s) + a_2 \cdot \sin(\omega \cdot s) \,. \tag{13.38}$$

By taking the first and second derivatives of the general solution (13.38) and combining with Eq. (13.37) one obtains $\omega = \sqrt{K}$. Thus, a general solution can be written as

$$y(s) = a_1 \cdot \cos(\sqrt{K} \cdot s) + a_2 \cdot \sin(\sqrt{K} \cdot s) \,. \tag{13.39}$$

The derivative of $y(s)$ is

$$y'(s) = -a_1 \sqrt{K} \cdot \sin(\sqrt{K} \cdot s) + a_2 \sqrt{K} \cdot \cos(\sqrt{K} \cdot s) \,. \tag{13.40}$$

At distance $s=0$ the initial coordinate is $y(0)=y_0$ and the initial velocity is $y'(0)=y'_0$. Therefore, the constants in Eq. (13.39) are given by $a_1=y_0$ and $a_2=y'_0/\sqrt{K}$.

The linear motion of a particle from point s_0 to a point s_1 can be described by the following transfer matrix:

$$\begin{pmatrix} y \\ y' \end{pmatrix}_{s_1} = M \cdot \begin{pmatrix} y \\ y' \end{pmatrix}_{s_0} = \begin{Bmatrix} m_{11} & m_{12} \\ m_{21} & m_{22} \end{Bmatrix} \begin{pmatrix} y \\ y' \end{pmatrix}_{s_0} . \tag{13.41}$$

M can either be a focusing matrix where *K>0*

$$M_{foc} = \begin{Bmatrix} \cos\left(\sqrt{|K|}s\right) & \frac{1}{\sqrt{K}}\sin\left(\sqrt{|K|}s\right) \\ -\sqrt{|K|}\sin\left(\sqrt{|K|}s\right) & \cos\left(\sqrt{|K|}s\right) \end{Bmatrix}_{s_0} , \tag{13.42}$$

or a defocusing matrix where *K<0*

$$M_{defoc} = \begin{Bmatrix} \cosh\left(\sqrt{|K|}s\right) & \frac{1}{\sqrt{K}}\sinh\left(\sqrt{|K|}s\right) \\ -\sqrt{|K|}\sinh\left(\sqrt{|K|}s\right) & \cosh\left(\sqrt{|K|}s\right) \end{Bmatrix}_{s_0} . \tag{13.43}$$

It is possible to provide simultaneous focusing in both transverse planes if the horizontal $K_x=\left(1/R_0^2-k_x\right)>0$ and the vertical $K_z=k_z>0$. Motion is stable in both planes if the normalized magnetic gradient satisfies $\left(1/R_0^2\right)>k>0$. This condition is referred to as "weak focusing" and was realized in the bending magnets of synchrotrons and cyclotrons that were built in the 1950s. It was achieved by providing a slight increase of the size of the magnet gap towards larger radii.

13.5. Quadrupole Magnets

A magnetic field with quadrupole characteristics is commonly used to focus (or defocus) a beam of charged particles. A sketch of a quadrupole magnet is shown in Figure 13.7. Electrostatic quadrupoles are used for the same purpose and there is no fundamental difference in the focusing features between magnetic and electrostatic quadrupoles. The four poles in a magnetic quadrupole generate a field which is zero on the central axis, $B(R=0)=0$. The orbit of the beam is perpendicular to the poles and exactly in the middle between the poles. The magnetic field increases linearly with displacement from the center of the quadrupole:

$$B_z = \left(\frac{\partial B_z}{\partial x}\right) \cdot x = g \cdot x\,. \tag{13.44}$$

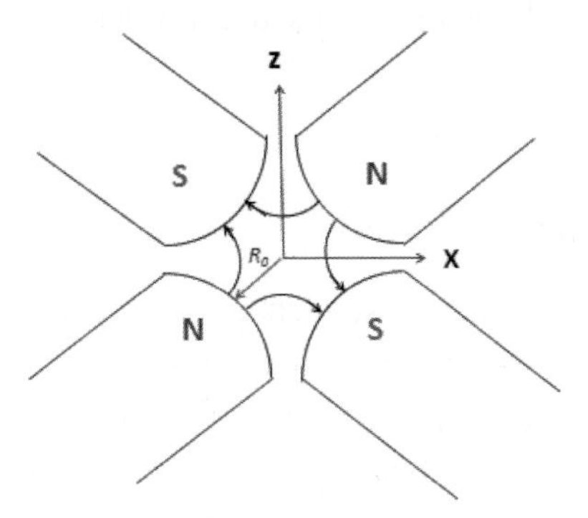

Fig. 13.7. Illustration of a quadrupole magnet. The radius of the inscribed circle is $\boldsymbol{R_0}$.

From Maxwell's equation $\vec{\nabla} \times \vec{B} = 0$, it follows that the field gradient is $g = (\partial B_z / \partial x) = (\partial B_x / \partial z)$. The horizontal component of the magnetic field is given by:

$$B_x = -\left(\frac{\partial B_x}{\partial z}\right) \cdot z = -g \cdot z \tag{13.45}$$

A quadrupole focuses a beam in the horizontal plane while it defocuses it in the vertical plane. Therefore, a combination of at least two quadrupoles is required to provide an overall focusing effect in both transverse planes. The Lorentz force acting on a particle is proportional to its displacement from the axis:

$$F = -k \cdot x\,, \tag{13.46}$$

where k is given by

$$k = k_x = -k_z = \frac{1}{B_0 R_0}\left(\frac{\partial B_x}{\partial z}\right). \tag{13.47}$$

The magnetic field gradient is proportional to the magnetic permeability μ_0 and the number of ampere-turns $N \times I$ of the quadrupole coils and inversely proportional to the radius of an inscribed circle between the poles R_0^{-2} (see Figure 13.7). This yields:

$$g = \frac{2\mu_0 N \cdot I}{R_0^2}. \tag{13.48}$$

Similar to a bending magnet, linear motion in a quadrupole is described by differential equations of second order:

$$x'' + kx = 0, \qquad z'' - kz = 0. \tag{13.49}$$

13.6. Strong Focusing

Storage rings consist of a sequence of bending elements and focusing/defocusing quadrupoles which form the ring's lattice. By placing focusing and defocusing quadrupoles appropriately, stability of motion can be achieved and a beam of charged particles can be stored for long times. The repeating sequence of main ion optical elements is referred to as the "cell". As an example the layout of the ultralow-energy electrostatic storage ring (USR) is shown in Figure 13.8.[12]

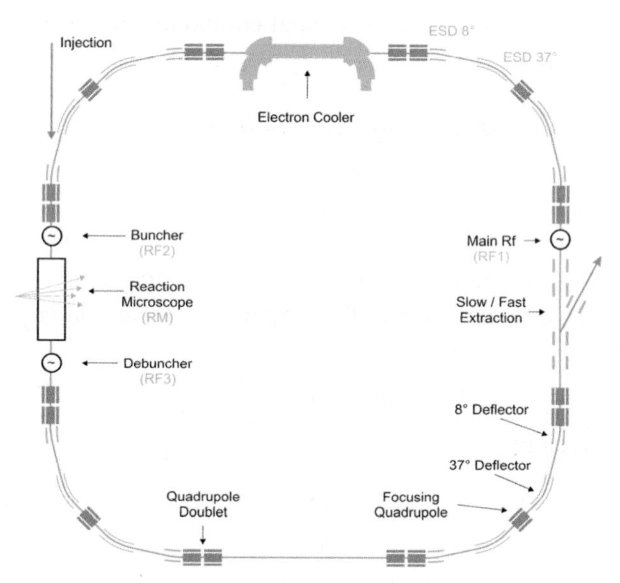

Fig. 13.8. Layout of the ultralow-energy storage ring with its four-fold symmetry lattice.[12]

In this case each cell consists of five quadrupoles (Q), four bending elements (B) and drifts between these elements (–). The cell can then be written as $-QQ-B-B-Q-B-B-QQ-$. The USR has a four-fold symmetry lattice with four identical cells. Generally speaking, the number of cells in a storage ring is not limited to any particular value. With a length of each cell L_c, N cells form the ring circumference $C = N \cdot L_c$.

The motion in any storage ring can be described by a differential equation with periodic focusing properties:

$$y''(s) + k(s) \cdot y(s) = 0. \tag{13.50}$$

This is referred to as "Hill's equation". The restoring force $k(s)$ depends on the position around the machine circumference and is a periodic function of longitudinal position s:

$$k(s + L_c) = k(s). \tag{13.51}$$

The solutions of Eq. (13.50) are quasi-harmonic oscillations where amplitude and phase are periodic functions of longitudinal position

$$y(s) = \sqrt{\varepsilon} \cdot \sqrt{\beta(s)} \cdot \cos(\psi(s) + \phi). \tag{13.52}$$

The parameters ε and ϕ are integration constants defined by the initial conditions. The so-called "betatron function" $\beta(s)$ describes the periodic amplitude of the betatron oscillations and is given by the focusing features of the accelerator lattice:

$$\beta(s + L_c) = \beta(s). \tag{13.53}$$

By substituting Eq. (13.52) into Eq. (13.50) one can find the phase advance of the oscillations at distance s:

$$\psi(s) = \int_0^s \frac{ds'}{\beta(s')}. \tag{13.54}$$

Over one revolution in a storage ring the particles complete $\psi(C)/2\pi$ oscillations. The fractional part of the number of betatron oscillations per turn is called the "betatron tune" of the machine:

$$Q_y = \frac{N}{2\pi}\psi(L_c) = \frac{1}{2\pi}\oint \frac{ds}{\beta(s)}. \tag{13.55}$$

A general solution of Hill's equation is provided by Eq. (13.52), while a solution for the particle velocity in the transverse direction, i.e. for angular deviation $y'(s)$, is given by:

$$y'(s) = -\frac{\sqrt{\varepsilon}}{\sqrt{\beta(s)}} \cdot \{\alpha(s) \cdot \cos(\psi(s) + \phi) + \sin(\psi(s) + \varphi)\}. \tag{13.56}$$

The integration constant ε is called the "emittance" and the parameters α, β, γ are given by:

$$\alpha(s) = -\frac{\beta'(s)}{2}, \quad \gamma(s) = \frac{1 + \alpha^2(s)}{\beta(s)}, \quad \alpha = \sqrt{\beta\gamma - 1}. \tag{13.57}$$

Combining solutions of Hill's equation for coordinate (13.52) and velocity (13.56) and solving them for a constant value of ε yields a parametric representation of particle motion in phase space (y,y'):

$$\varepsilon = \gamma \cdot y^2 + 2\alpha \cdot y \cdot y' + \beta \cdot y'^2. \tag{13.58}$$

The emittance ε is a constant of motion and does not depend on the particle coordinate. Equation (13.58) is an equation of an ellipse. The meaning of the α, β and γ parameters, describing the ellipse envelope in (y,y') phase space, is illustrated in Figure 13.9.

In a slightly simplified description each particle in a beam corresponds to a single point in (y,y') phase space and all particles together are contained within the emittance ellipse. The shape and orientation of the ellipse are given by the so-called Twiss parameters α, β and γ which are properties of the accelerator lattice. The area covered by particles in phase space $A = \pi\varepsilon$ remains constant if no dissipative forces, such as scattering on the residual gas, are present. Here, we consider the motion of charged particles in magnetic and electric fields, i.e. a situation where the beam is under the influence of conservative forces only.

Liouville's theorem, which follows from the energy conservation law, requires that the phase space area occupied by a beam remains constant under the action of conservative force or, in other words, under canonical transformations. The beam emittance is hence a constant beam parameter and cannot be changed by focusing or bending elements. During acceleration, however, i.e. when the beam energy is increased, the emittance is reduced and vice versa. This is because the angular component y' is reduced with increasing ion velocity β:

$$y' = \frac{dx}{ds} = \frac{dy}{dt}\frac{dt}{ds} = \frac{\beta_y}{\beta}. \tag{13.59}$$

Here, β_y is the transverse component of the particle velocity. It can then be seen that the emittance is inversely proportional to the relativistic parameters β and γ:

$$\varepsilon = \frac{1}{\pi}\int y'dy \propto \frac{1}{\beta\gamma}. \tag{13.60}$$

In contrast the normalized emittance $\varepsilon_{norm} = \beta \cdot \gamma \cdot \varepsilon$ is a constant of motion and is conserved even if the beam energy changes.

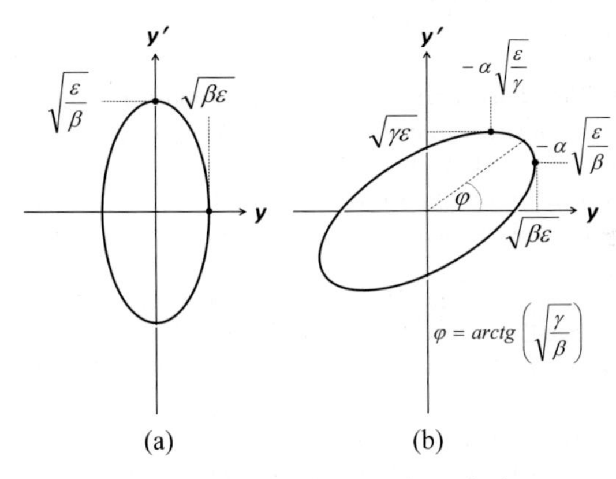

Fig. 13.9. Phase space ellipse: a) A focused beam at its waist is represented by an upright ellipse; b) a tilted ellipse with an angle $0 < \varphi < 90°$ represents a divergent beam, while a tilted ellipse with an angle $90° < \varphi < 180°$ represents a convergent beam.

13.7. Summary

In this chapter, we introduced some basic concepts regarding the description of beam motion in a circular accelerator with a focus on motion in electrostatic storage rings. For a more detailed discussion we would like to refer the reader to the proceedings of the CERN and Joint Universities Accelerator Schools,[10,11] where these concepts are covered in great detail.

The next chapter will concentrate on the specifics of beam motion in electrostatic storage rings, including nonlinear effects, long-term beam dynamics and ion kinetics.

Acknowledgments

The generous support of HGF and GSI under contract VH-NG-328 and STFC under the Cockcroft Institute grant is acknowledged. We would like to thank Yu. Senichev for the provision of material used in the lecture and S. P. Moeller, A. Dolinsky and H. Knudsen for valuable comments and discussions.

References

1. M. S. Livingston and J. B. Blewett, *Particle Accelerators* (Mc Graw Hill Book Comp.Inc., New York, 1962).
2. J. J. Livingood, *Principles of Cyclic Particle Accelerators* (D.Van Nostrand Co Ltd., New York, 1961).
3. K. G. Steffen, *High Energy Optics* (Interscience Publishers, New York, 1965).
4. H. Bruck, *Accelerateurs Circulaires de Particules* (PUF, Paris, 1966).
5. A. A. Kolomensky and A. W. Lebedev, *Theory of Cyclic Accelerators* (North Holland Publishers Company, Amsterdam, 1966).
6. E. J. N. Wilson, *An Introduction to Particle Accelerators* (Oxford University Press, Oxford, 2001).
7. P. J. Bryant and K. Johnsen, *The Principles of Circular Accelerators and Storage Rings* (Cambridge University Press. Cambridge, 1993).
8. M. Reiser, *Theory and Design of Charged Particles Beams* (J. Wiley & Sons, New York, 1994).
9. S. Y. Lee, *Accelerator Physics* (World Scientific, Singapore, 1999).
10. http://cas.web.cern.ch/cas/CAS_Proceedings.html
11. https://espace.cern.ch/juas/SitePages/Home.aspx
12. A. I. Papash, C. P. Welsch, An Update of the USR Lattice: Towards a True Multi-User Experimental Facility, *Proc. Part. Accel. Conf.* 4335–4337 (2009).
13. K. E. Stiebing, FLSR – The Frankfurt low energy storage ring, *Nucl. Instr. Meth. Phys. Res.* **A614**, 10–16 (2010).

Chapter 14

Electrostatic Storage Rings – An Ideal Tool for Experiments at Ultralow Energies

Alexander I. Papash

Karlsruhe Institute of Technology,
Hermann-von-Helmholtz-Platz 1,
76344 Eggenstein-Leopoldshafen, Germany
Joint Institute for Nuclear Research,
Joliot-Curie 6, 141980 Dubna, Moscow oblast, Russia
alexander.papash@kit.edu

Alexander V. Smirnov

Joint Institute for Nuclear Research,
Joliot-Curie 6, 141980 Dubna, Moscow oblast, Russia
smirnov@jinr.ru

Carsten P. Welsch

University of Liverpool, Department of Physics,
Liverpool L69 7ZE, UK
Cockcroft Institute, Sci-Tech Daresbury, Keckwick Lane,
Warrington WA4 4AD,UK
c.p.welsch@liverpool.ac.uk

Electrostatic storage rings (ESRs) have proven to be invaluable tools for atomic and molecular physics at the ultralow-energy range from 1 to 100 keV/q. Due to the mass independence of the electrostatic rigidity, these machines are able to store a wide range of different particles, including antiprotons, light ions and heavy singly charged bio-molecules. However, earlier measurements at the ELISA facility in Aarhus, Denmark, showed strong limitations on beam intensity and lifetime.

Here, we first discuss common features of existing and planned ultralow-energy storage rings, before introducing the basic equations describing the motion of a beam of charged particles through electrostatic elements. Finally, we describe the effects from nonlinear fields on long-term beam dynamics and stability.

14.1. Introduction

Storage rings with magnetic bending and focusing elements were initially developed for the accumulation of high-energy particles. During the last three decades, however, such machines have also been operated in the intermediate and low-energy range. In particular, the LEAR ring at CERN was the first machine to store, cool and decelerate antiprotons down to 5 MeV.[1] Negative ions of $^{4}He^{-}$ and carbon molecule ions $^{12}C_{70}^{-}$ have been stored at energies of 5 and 25 keV, respectively, in the ASTRID storage ring.[2]

The world's first electrostatic storage ring was built in 1953 as an electron analogue of the Brookhaven alternating gradient synchrotron.[3] More commonly known, however, is the use of electrostatic traps and storage rings for operation with ion beams in the keV energy range.[4] Ion traps are designed to store ions for as long as possible and to localize the stored particles in space.[5] Another type of storage device, developed in response to the needs of the astro-, atomic and molecular physics communities, are ESRs, which rely on electrostatic bending and focusing elements.[6] As opposed to magnetic storage rings, ESRs provide access to much lower beam energies. Due to the mass independence of the electric rigidity, there is also no intrinsic limit on the mass of particles that can be stored in these facilities, ranging from light ions to clusters and very heavy bio-molecules. ESRs have been successfully used to study the following problems:[4,6,7]

- collision phenomena and plasma properties of astrophysical objects, e.g. molecular clouds or quasars;
- electron impact rotational and vibrational excitation of cold molecular ions;
- quantum reaction dynamics of cold molecular ions;
- gas-phase spectroscopy of biomolecular ions;
- rotational effects in the process of dissociative recombination of molecular ions with low temperature electrons (<10 K);

- molecular dynamics - to achieve Coulomb crystallization for fast stored ions and study the phase transition to a crystalline beam;
- fundamental few-body Coulomb problem for single and multiple ionization;
- measurements of single and multiple ionization cross-sections (total and differential) of antiprotons colliding with atoms of a crossed supersonic gas jet target;
- ion-impact ionization to benchmark theoretical predictions;
- antihydrogen studies by merging antiprotons and positrons;
- study of the lifetime of metastable atomic states;
- investigations into single-component plasmas.

14.2. Common Features of Electrostatic Storage Rings

ESRs are in some ways complementary to ion traps and allow relaxing the ion's internal energy to almost its ground state. In ESRs, ions circulate in one direction while in ion traps there is no designated direction of motion. One can outline the following common features of all ESRs:

- ability to store ions at keV beam energies;
- no transitions between hyperfine levels of the circulating ions are induced, due to the absence of magnetic fields;
- no remanence or hysteresis effects;
- possibility of fast acceleration/deceleration due to absence of eddy currents;
- mass independence of electrostatic rigidity, allowing storage of a wide range of particles, from light ions, protons and antiprotons to heavy molecular ions, with positive and negative charge;
- comparably compact and cheap as compared to their magnetic counterparts.

ESRs mostly operate at keV beam energies because the maximum ion energy in these rings is limited by the electric field strength between the deflecting electrodes. A clear advantage of ESRs is that in-ring experiments with circulating ions can be done over many turns, thus multiplying the number of interactions. This is unlike single pass experiments, for example those realized behind the RFQ-D at the AD facility at CERN.[8]

In combination with a so-called reaction microscope incorporated into the ring lattice, ESRs are a powerful tool for atomic physics experiments to study collisions between a crossed gas jet target and the stored ion beam. This promises access to kinematically complete measurements of the collision dynamics of fundamental few-body quantum systems on the level of differential cross-sections.

The parameters of some existing and future electrostatic storage rings are shown in Table 14.1.

Table 14.1. Overview of ESRs worldwide.

Ring	**ELISA**[9,10]	**ESR**[11]	**FLSR**[13]	**DESIREE**[14]	**CSR**[15,16,17]	**USR**[18,19,20]
Location	Aarhus University Denmark	KEK Tsukuba Japan	Frankfurt University Germ	Stockholm University Sweden	MPI Heidelberg Germany	FAIR-GSI Darmstadt Germany
Ions	$A \leq 100$	$A \leq 100$	$A \leq 100$	$A \leq 100$	$A \leq 100$	Antiprotons
Energy, keV	$(5–25)\cdot Q$	$20\cdot Q$	50	$(25\text{-}100)\cdot Q$	$(300\text{-}20)\cdot Q$	300-20
Type	Race track	Race track	Race track	2 x Race tracks	four-fold quadratic	two / four-fold, achromatic
Symmetry	2	2	2	2 x 2	4	2 / 4
Circum-ference, m	7.62	8.14	14.23	9.2 x 9.2	35.2	46
Revolution time, μs	3.5 (p) 93 (C_{80})	4 (p) 22 (N_2^+)	4.5 (p)	4–60	4–180	5.67–22
Deflector Angles	SPH/CYL 160°+10°	CYL 160°+10°	CYL 75°+15°	CYL 160°+10°	CYL 39°+6°	CYL 37°+8°
Bending radius, mm	250	250	250	250	2000 + 1000	2000 + 1000
E-cool, eV	N/A	N/A	N/A	N/A	162–11	162–11
Lifetime, s	10-30	12-20	--	--	10–100	~10
Operation modes	Storage	Storage	Int. target	Colliding beams	Cooling, storage, deceleration.	Cooling, storage, deceleration, short bunches, slow extraction
Vac. mbar	10^{-11}	$5\cdot 10^{-11}$	10^{-12}	10^{-12} (10°K)	10^{-15} (2°K)	10^{-11}
Status	Running	Running	Commis-sioning	Assembly	Assembly	Project

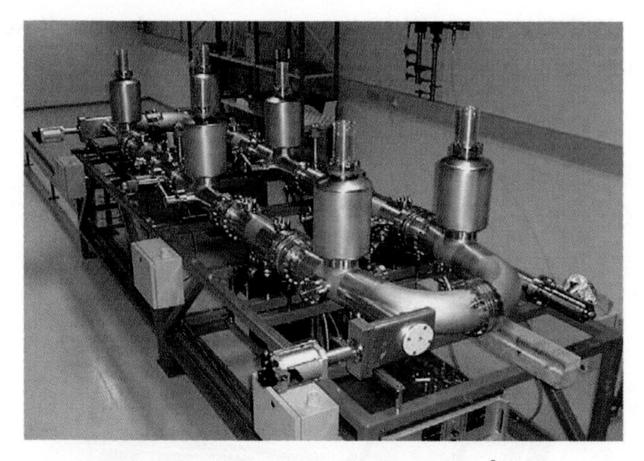

Fig. 14.1. Photograph of the electrostatic storage ring ELISA.[9]

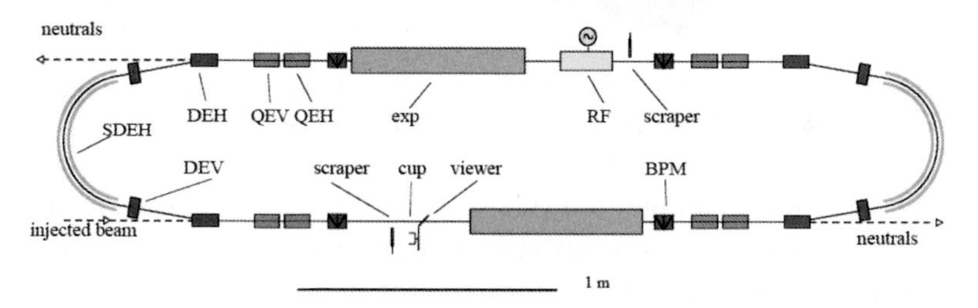

Fig. 14.2. Layout of the ELISA storage ring. Neutrals can be detected behind the 10° parallel plate deflectors.[10] DEH – DEflector Horizontal; DEV – DEflector Vertical; QEH – Quadrupole Electric Horizontal; QEV – Quadrupole Electric Vertical; SDEH – Spherical Deflector Electric Horizontal; BPM – Beam Position Monitor; RF – Radio Frequency device.

The first ESR dedicated to atomic physics experiments was built in Aarhus,[9] see Figure 14.1. In ELISA, two 160° cylinder deflectors and two 10° parallel plate deflectors, together with four sets of electrostatic quadrupoles, form a simple race track structure,[10] as shown in Figure 14.2. The split of the 180° bend into two parallel plate 10° deflectors and one 160° deflector of cylindrical shape allows for detection of neutral particles at the end of the straight sections and for simple injection into the machine. Rings with similar lattices were built at KEK[11] and Tokyo University.[12] Based on the experience gained during the ELISA operation, the Frankfurt low-energy storage ring (FLSR) was designed, built and tested.[13] In the DESIREE double ring project, two ESRs of the same race track type as ELISA are overlapped, to allow for ion–molecular head-on collisions,[14] see Figure 14.3.

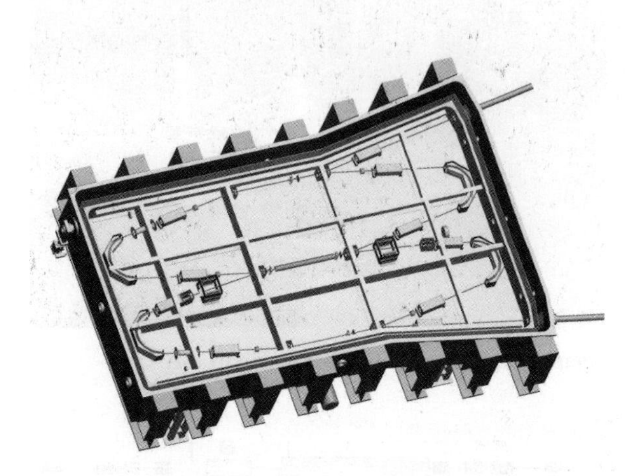

Fig. 14.3. Illustration of the DESIREE ring. Two ESRs of race track shape are overlapped to allow for ion–molecular head-on collision studies.[14]

The Cryogenic storage ring (CSR) at the MPI for Nuclear Physics in Heidelberg, Germany, is a next-generation low-energy storage ring for essentially all ion species – from hydrogen ions to molecular ions, macro- and biomolecules, clusters, and atomic ions at extreme charge states,[15] see Figure 14.4. The kinetic energy of the stored ions is between 20 and 300 keV. All storage ring components can be cooled to a few kelvin and a vacuum better than 10^{-15} mbar is anticipated.[16]

The low temperatures are expected to yield a unique collision and blackbody radiation-free environment for radiative relaxation of molecular species and for lengthy storage of keV beams (q·keV kinetic energy) of ions in charge state q. In this ring, collision velocities near the atomic velocity will become available for use with various supersonic gas jet targets and an in-ring reaction microscope, as well as for electron collisions at a few K in a merged, velocity matched, i.e. eV kinetic energy, electron beam. The four-fold symmetry ring has a split achromatic lattice with two short (6°) and two long (39°) electrostatic deflectors (ESDs) in each 90° bend. A reaction microscope, electron cooling, injection, and an RF (radiofrequency) drift tube are accommodated in its 2.8 m long straight sections.[17]

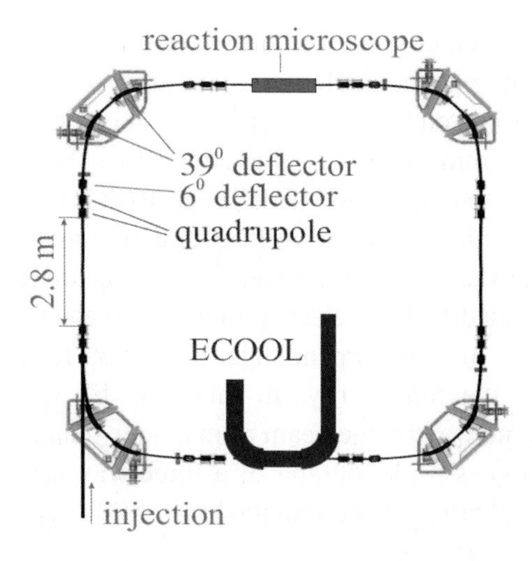

Fig. 14.4. Layout of the CSR. The ring has a four-fold symmetric lattice, each cell consists of two 39°, two 6° deflectors and two quadrupole doublets.[17]

Finally, the ultralow-energy storage ring (USR) will provide cooled beams of antiprotons down to energies of 20 keV in the future Facility for Low-energy Antiproton and Ion Research (FLAIR) at FAIR.[18] A four-fold symmetric, split-achromat geometry was chosen for the USR lattice, see Figure 13.8 in Chapter 13.[19] A field-free drift between the 8° and 37° electrostatic bending deflectors is introduced for the detection of neutral particles leaving the ring after the reaction microscope.[20,21] As a potential test facility for the USR a small recycling ring has been designed to house a re-circulating reaction microscope setup at the CERN-AD.[22] This very compact facility could be used as a bridge in the energy gap between 3 and 30 keV and as a complementary component to the magnetic ring ELENA currently being built at the AD.[23]

14.3. Electrostatic Deflectors of Different Shapes

The central trajectory of particles moving inside electrostatic deflectors has been considered in the previous chapter and all further discussion will make use of the principles developed there. The theory of ion motion in electrostatic bending and focusing elements is well developed[24-28] and we will provide some basic formulae assuming that the particle energy is low, i.e. $\beta \ll 1$.

Note that the velocity of an ion varies as it passes through an electrostatic bending element. Here, we will treat transverse motion only and will refer to some conditions described in more detail by P. Bryant in his JUAS lecture from 2008.[29] It is assumed that the angular deviation and transverse excursion are both small, i.e. $y/R << 1$, where y is the ion coordinate in either radial or vertical direction and R is the central bending radius in the electrostatic element. In our discussion so far we have always assumed that the electric field in the electrostatic deflectors and quadrupoles is always perpendicular to the ion trajectory and hence does not change the ion energy. In this case ES quadrupoles act in essentially the same way on the beam as magnetic quadrupoles.

Figure 14.5 shows a photograph of a three-way electrostatic bend.[29] Its electrodes of spherical shape provide beam focusing in both planes.

Fig. 14.5. Photograph of a three-way electrostatic bend: the beam can go left, right and straight through. The spherical electrodes provide focusing in both horizontal planes.[29]

Similar to the field index in a magnetic dipole bend, the electric field index n_E is defined as the relative change in electric field strength E_R with radius R in the bending plane. The electrodes can be curved in both horizontal and vertical planes and we define ρ as a radius of electrode surface in the vertical plane. The field index n_E can then be expressed as

$$n_E = -\left(\frac{R}{E_R}\right)\frac{dE_R}{dR} \cong 1 + \frac{R}{\rho}. \tag{14.1}$$

The focusing strength of an ESD in the horizontal plane k_X is defined as

$$k_X = \frac{(3 - n_E - \beta^2)}{R^2}, \tag{14.2}$$

while the focusing strength in the vertical plane k_Y is given by

$$k_Y = \frac{(n_E - 1)}{R^2}. \tag{14.3}$$

At low energies the relativistic term β^2 in Eq. (14.2) can be neglected. The focusing features of ESDs with electrodes of different shapes are summarized in Table 14.2.

A spherical deflector is composed of two concentric spheres of radius R_1 and R_2. The horizontal and vertical radii of this bend are identical, i.e. $R = \rho$, and its electric field index is $n_E^{sph} = 2$. A deflector with spherical electrodes provides equal focusing in both horizontal and vertical planes: $k_X^{sph} = k_Y^{sph} = k^{sph} = 1/R^2$ and its focal length is $f_Y^{sph} = f_Y^{sph} = \left(k^{sph}\right)^{-1/2} = R$.

The surface radius of a cylindrical electrode in the vertical plane is infinity $\rho = \infty$ and the electric field index is $n_E^{cyl} = 1$. A cylinder deflector focuses a beam with double the strength of a spherical deflector in the horizontal plane, $k_X^{cyl} = 2/R^2$, but acts as a drift with no focusing at all in the vertical plane, $k_Y^{cyl} = 0$. The focal length of a cylinder deflector in the horizontal plane is $f_X^{cyl} = R/\sqrt{2}$ and infinity in the vertical plane, $f_Y^{cyl} = \infty$. Finally, also electrodes of anti-spherical $R = -\rho$ or hyperbolic shape $R = -2\rho$ could be considered for beam bending. These electrodes would simultaneously provide strong focusing in the bending plane and strong defocusing in the vertical plane.

However, the nonlinear fields in these deflectors are rather strong and limit their use in a storage ring considerably. Cylinder deflectors are most commonly used in ESRs.

Table 14.2. Parameters of ESDs of different shapes.

Electrode shape	Cylindrical	Spherical	Hyperbolic	Anti-spherical
ρ	$\rho=\infty$	$\rho=R$	$\rho=-R/2$	$\rho=-R$
n_E	1	2	−1	0
k_X	$2/R^2$	$1/R^2$	$4/R^2$	$3/R^2$
k_Y	0	$1/R^2$	$-2/R^2$	$-1/R^2$
	focus in x drift in y	identical focus $fx=fy$	focus in x defocus in y	focus in x defocus in y

14.4. Electric Field Distribution in Electrostatic Deflectors

The equations of betatron oscillations in linear approximation have been described in Chapter 13. There, we have used the example of a magnetic bend. On this basis, one can also derive the linear equations of motion in an electrostatic bend. However, to understand the importance of nonlinear effects one should simulate the ion orbits using equations of motion with high-order terms derived from a Fourier harmonic analysis of the electric field distribution in the ESDs. For a detailed discussion of the field distribution and derivation of the equations of ion motion in ESDs we refer to references[24,25,27]. Here, we will use the example of a spherical ESD.

Assume two concentric spheres charged with the same, but opposite potential, $\pm U_0$. Gauss's law yields:

$$E_R \cdot 4\pi R^2 = \frac{q}{\varepsilon}. \tag{14.4}$$

The potential difference between the two spheres is then given by

$$\Delta\varphi = \int_{R_1}^{R_2} \frac{q}{4\pi\varepsilon \cdot R^2} dr = \frac{q}{4\pi\varepsilon}\left(\frac{1}{R_1} - \frac{1}{R_2}\right), \tag{14.5}$$

where the radius of the inner sphere is R_1 and the outer one R_2.

The electric field is then given by the gradient of the potential distribution between the plates. The angular and vertical components of the electric field are $E_\theta = E_Z = 0$ along the central trajectory. The radial component is $E_R(R) = -d\varphi / dR$. Provided the boundary conditions are $\varphi(R_1) = -U_0$ and $\varphi(R_2) = U_0$, the potential between the spherical electrodes is given by:

$$\varphi^{sph}(R) = -\frac{2U_0}{R_2 - R_1} R_2 R_1 \cdot \frac{1}{R} + \frac{U_0}{R_2 - R_1}(R_2 + R_1). \tag{14.6}$$

The radial distribution of the electric field between two spheres can then be expressed by:

$$E_R^{sph}(R) = 2U_0 \frac{R_1 R_2}{R_2 - R_1} \cdot \frac{1}{R^2}. \tag{14.7}$$

It can be seen that the radial component of the electric field in a spherical deflector decreases with radius as $E_R^{sph} \propto 1/R^2$.

A cylinder deflector is identical to a spherical one with infinite electrode radius in the vertical plane. The potential distribution between the two electrodes is then given by:

$$\varphi^{cyl}(R) = -U_0 + \frac{2U_0}{\ln \frac{R_2}{R_1}} \cdot \ln \frac{R}{R_1}. \tag{14.8}$$

The radial component of the electric field in a cylinder deflector is hence inversely proportional to the radial position between the electrodes $E_R^{cyl} \propto 1/R$ and given by:

$$E_R^{cyl}(R) = \frac{2U_0}{\ln(R_2/R_1)} \cdot \frac{1}{R}. \tag{14.9}$$

From Eqs (14.6) and (14.8) one can find the equilibrium radius of the central trajectory $R_{\varphi=0}$ where the potential $\varphi = 0$. The radius of the equilibrium orbit inside the cylinder deflector is:

$$R_{\varphi=0}^{cyl} = R_1 \cdot \sqrt{(R_2/R_1)} = R_1 \sqrt{1 + \frac{d}{R_1}}, \tag{14.10}$$

where the distance between the electrodes is $d = R_2 - R_1$, see Figure 13.4 in the previous chapter. Note that $R^{cyl}_{\varphi=0}$ is slightly lower than the radius of the geometric center between the electrodes $R_0 = R_1 + d/2$. For ESDs with large bending radii, the gap between the electrodes is $d << R_1$ and $R^{cyl}_{\varphi=0} \approx R_1 + d/2$.

In a spherical deflector the radial position of the equilibrium orbit $R^{sph}_{\varphi=0}$ where the potential is zero is given by

$$R^{sph}_{\varphi=0} = \frac{2R_1 \cdot R_2}{R_1 + R_2}, \tag{14.11}$$

and is smaller than the radius of the geometric center between the electrodes. Note that the difference between the two is always positive, i.e. the zero potential orbit is slightly closer to the inner electrode, provided the voltage on both electrodes is $\pm U_0$,

$$R_0 - R^{sph}_{\varphi=0} = \frac{(R_2 - R_1)^2}{2(R_1 + R_2)} > 0. \tag{14.12}$$

In real ESRs the voltage on the inner electrode is slightly lower than the voltage applied to the outer electrode.[10,12] It is determined such that the central trajectory lies in the middle between both electrodes.

14.4.1. Electric field in the vicinity of the equilibrium orbit

In order to find the electric field distribution close to the central trajectory one may use a local coordinate system originating on the reference orbit, as illustrated in Figure 14.6:

$$s = R_0 \vartheta, \qquad x = r - R_0, \qquad z = z. \tag{14.13}$$

Here, s is the longitudinal coordinate along the trajectory, r is the projection of the particle's radius-vector $\vec{R}$ on the horizontal plane, x is the horizontal deviation from the equilibrium orbit of radius R_0 and z is the vertical deviation of the particle from the median plane. The gap between the electrodes is $d = R_2 - R_1$.

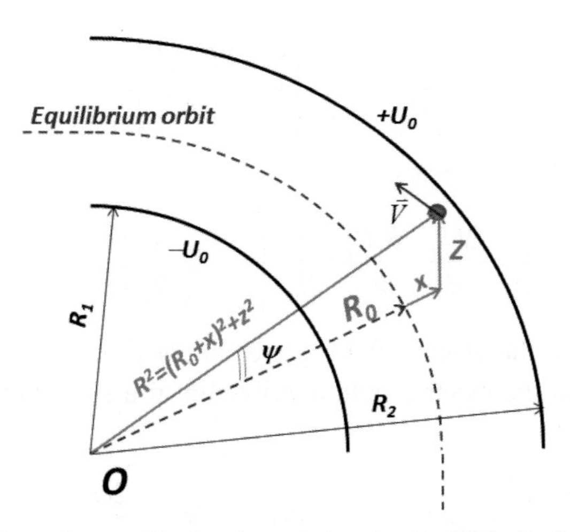

Fig. 14.6. Motion of a positively charged ion in an ESD. Radial and vertical displacements are much smaller than the radius of the equilibrium orbit.

In the following we assume that any deviation from the equilibrium orbit is small in both planes, i.e. $(x/R_0) << 1$ and $(z/R_0) << 1$. Also, we neglect the effect of fringe fields and therefore the longitudinal component of the electric field is $E_S = 0$.
The horizontal component of the electric field in a spherical deflector is $E_x = E_R \cdot \cos\psi$ and can derived from Eq. (14.7):

$$E_x(x,z,s) = \frac{2U_0}{d} \cdot \frac{R_1 R_2}{r^2 + z^2} \cdot \frac{r}{\left[r^2 + z^2\right]^{1/2}} = \frac{2U_0}{d} \cdot \frac{R_1 R_2}{r^2} \cdot \frac{1}{\left(1 + \frac{z^2}{r^2}\right)^{3/2}} \cdot \quad (14.14)$$

The vertical component of the electric field in a spherical deflector is $E_z = E_R \cdot \sin\psi$ and can be written as:

$$E_z(x,z,s) = \frac{2U_0}{d} \cdot \frac{R_1 R_2}{r^2 + z^2} \cdot \frac{z}{\left[r^2 + z^2\right]^{1/2}} = \frac{2U_0}{d} \cdot \frac{R_1 R_2}{r^2} \cdot \frac{z}{r} \cdot \frac{1}{\left(1 + \frac{z^2}{r^2}\right)^{3/2}} \cdot \quad (14.15)$$

Taking into account that $r = R_0 + x$ and truncating both field components to terms not higher than second order of trajectory deviation from the reference orbit yields an expression for the horizontal,

$$E_x = \frac{2U_0}{d} \cdot \frac{R_1 R_2}{r^2} \left[1 - \frac{3}{2} \left(\frac{z}{r} \right)^2 \right], \tag{14.16}$$

and vertical,

$$E_z = \frac{2U_0}{d} \cdot \frac{R_1 R_2}{r^2} \cdot \frac{z}{R_0} \cdot \left[1 - 2 \left(\frac{x}{R_0} \right) \right], \tag{14.17}$$

components of the electric field in the spherical deflector.[19,25] In linear approximation the electric field distribution of a spherical deflector can be simplified to

$$E_x \approx \frac{2U_0}{d} \cdot \left[1 - 2\frac{x}{R_0} \right] \quad \text{and} \quad E_z \approx \frac{2U_0}{d} \cdot \frac{z}{R_0}. \tag{14.18}$$

From the above it can be seen that sextupole components $(x / R_0)^2$, $(z / R_0)^2$, as well as correlation terms $(x \cdot z / R_0^2)$, which are responsible for a mixture of horizontal and vertical oscillations, are all present in the electric field distribution of a spherical deflector.

14.5. Equations of Motion in an Electrostatic Deflector

In Chapter 13 we derived a set of linear equations that described the betatron oscillations around the equilibrium orbit. In order to find nonlinear equations one needs to apply the Lagrangian formalism to ion motion in an ESD. The Lagrangian is defined as the difference of kinetic and potential energy, that is $L = T - V$, where the potential energy is $V(R) = q \cdot \varphi(R)$. In a cylindrical coordinate system (r, ϑ, z) the Lagrangian can be written as:

$$L = \frac{m}{2}\left(\dot{r}^2 + r^2\dot{\vartheta}^2 + \dot{z}^2\right) + \frac{2qU_0}{d} R_2 R_1 \cdot \frac{1}{\sqrt{r^2 + z^2}} - \frac{qU_0}{d}(R_2 + R_1). \tag{14.19}$$

The potential $\varphi^{sph}(R)$ is defined by Eq. (14.6). Using canonical equations for conjugate variables,

$$\frac{d}{dt}\left(\frac{\partial L}{\partial \dot{r}}\right)=\frac{\partial L}{\partial r} \qquad \frac{d}{dt}\left(\frac{\partial L}{\partial \dot{\vartheta}}\right)=\frac{\partial L}{\partial \vartheta} \qquad \frac{d}{dt}\left(\frac{\partial L}{\partial \dot{z}}\right)=\frac{\partial L}{\partial z}, \tag{14.20}$$

and applying the condition of spherical symmetry $(\partial L / \partial \vartheta = 0)$ and consequently $(\partial L / \partial \dot{\vartheta} = const)$ one may derive a system of three differential equations:

$$\begin{aligned} m\ddot{r} &= mr\dot{\vartheta}^2 - qE_x \\ mr^2\dot{\vartheta} &= M = const \\ m\ddot{z} &= -qE_z . \end{aligned} \tag{14.21}$$

The second equation is the conservation law for angular momentum $M = mv_\vartheta r$ where the azimuthal velocity is $v_\vartheta = r\dot{\vartheta}$. The expression $mr\dot{\vartheta}^2$ is just a centrifugal force:

$$mr\dot{\vartheta}^2 = \frac{mv_\vartheta^2}{r}. \tag{14.22}$$

Replacing the second order time derivative by the second order derivative in distance s,

$$r'' = \frac{d^2 r}{ds^2} = \frac{1}{v_\vartheta^2}\frac{d^2 r}{dt^2} \equiv \frac{1}{v_\vartheta^2}\ddot{r}, \tag{14.23}$$

in the first equation of Eq. (14.21) and using Eq. (14.14) for the electric field $E_x(x,z,s)$, one can derive a second-order differential equation of ion motion in a spherical deflector:

$$r'' = \frac{1}{r} - \frac{qE_x}{mv_\vartheta^2} = \frac{1}{r} - \frac{2qmU_0}{M^2}\cdot\frac{R_1R_2}{d}\cdot\frac{1}{\left(1+\frac{z^2}{r^2}\right)^{3/2}}, \tag{14.24}$$

where the angular velocity is expressed by the angular momentum $mv_\vartheta^2 = \frac{M^2}{mr^2}$. The angular momentum M is an invariant of motion, see Eq. (14.21).

The equilibrium radius of the ion trajectory can be found by using $r'' = 0$ and $z = 0$ in Eq. (14.24):

$$R_{eq} \equiv R_0 = M^2 \frac{1}{2qmU_0} \cdot \frac{d}{R_1 R_2}. \tag{14.25}$$

After expanding Eq. (14.24) around the equilibrium orbit $r = R_0 (1 + x / R_0)$ and limiting the resulting Taylor series to second order in $(x / R_0)^2$ and $(z / R_0)^2$, one gets two differential equations that describe betatron oscillations in a spherical ESD:

$$\begin{aligned} & x'' + \frac{1}{R_0^2} x - \frac{1}{R_0^3} x^2 - \frac{3}{2 \cdot R_0^3} z^2 = 0 \\ & z'' + \frac{1}{R_0^2} z - \frac{3}{R_0^3} xz = 0 \end{aligned} \quad . \tag{14.26}$$

These nonlinear equations include sextupole components, as well as coupling elements between the radial and vertical planes. In linear approximation the spherical deflector provides equal focusing in both horizontal and vertical planes:

$$\begin{aligned} & x'' + \frac{1}{R_0^2} x = 0 \\ & z'' + \frac{1}{R_0^2} z = 0 \end{aligned} \quad . \tag{14.27}$$

The solutions of Eq. (14.27) are harmonic oscillations with an identical oscillation period in both planes. In a similar way, motion in a cylindrical deflector can be described by:

$$\begin{aligned} & x'' + \frac{2}{R_0^2} x - \frac{1}{R_0^3} x^2 = 0 \\ & z'' = 0 \end{aligned} \quad . \tag{14.28}$$

Here, the sextupole component is smaller as compared to the spherical deflector and affects only the horizontal plane. Also, there is no coupling between both planes, up to the second order at least. The linear equations of motion under the assumption of small oscillations where $y / R << 1$ are similar to those for a sector magnet:

$$x'' + \frac{2}{R_0^2}x = 0 \quad . \qquad z'' = 0 \tag{14.29}$$

The solutions of Eq. (14.29) are harmonic oscillations in the horizontal plane. The focusing strength of a cylinder deflector in the horizontal (bending) direction is twice as strong as that of a spherical deflector. In the vertical direction a cylinder deflector can be considered as a drift.

At low beam energies the linear transfer matrix $\Re$ of an electrostatic deflector with bending angle θ, bending radius R and effective length $L=\theta{\cdot}R$ might be written in a non-relativistic form,[19] where the electric field index is n=1 for a cylinder deflector and n=2 for a spherical deflector:

$$\Re = \begin{Bmatrix} \cos\sqrt{(3-n)}\theta & \frac{R}{\sqrt{3-n}}\sin\sqrt{(3-n)}\theta & 0 & 0 & 0 & \frac{2R}{3-n}\left[1-\cos\sqrt{(3-n)}\theta\right] \\ -\frac{\sqrt{3-n}}{R}\sin\sqrt{(3-n)}\theta & \cos\sqrt{(3-n)}\theta & 0 & 0 & 0 & \frac{2}{\sqrt{3-n}}\sin\sqrt{(3-n)}\theta \\ 0 & 0 & \cos\sqrt{(n-1)}\theta & \frac{R}{\sqrt{n-1}}\sin\sqrt{(n-1)}\theta & 0 & 0 \\ 0 & 0 & -\frac{\sqrt{n-1}}{R}\sin\sqrt{(n-1)}\theta & \cos\sqrt{(n-1)}\theta & 0 & 0 \\ -\frac{1}{\sqrt{3-n}}\sin\sqrt{(3-n)}\theta & \frac{R}{3-n}\left[\cos\sqrt{(3-n)}\theta-1\right] & 0 & 0 & 1 & L-\frac{2R}{3-n}\left[\theta-\frac{\sin\sqrt{3-n}\theta}{\sqrt{3-n}}\right] \\ 0 & 0 & 0 & 0 & 0 & 1 \end{Bmatrix} \tag{14.30}$$

14.6. Nonlinear Effects in ESRs

In order to study the effects from nonlinear fields on the stored ion beam one needs to solve the equations of motion with high-order terms as derived from a Fourier analysis of the electric field distribution in the ESDs. Alternatively, ions can be tracked in a computer model of a ring using the relaxation electric field maps of all bending and focusing elements. This will allow also for an investigation of transition processes, equilibrium conditions and long-term beam dynamics. As an example we select the ELISA ring to explain these advanced studies.[9] ELISA is the first ESR that was dedicated to atomic physics experiments and has been successfully operated in Aarhus, Denmark, since the late 1990s.[10] Initially deflectors of spherical shape had been used to provide identical focusing in both horizontal and vertical planes, but have been substituted by cylinder deflectors later on since systematic studies showed strong limitations on beam current and reduced lifetime for spherical deflectors.

It was found in experiments[30] and proven by multi-turn tracking simulations[31-33] that the nonlinear terms of the electric field distribution reduce the area of stable oscillations in the ring and that the sextuple component is the dominant one. The so-called dynamic aperture which presents the stability limits for beam storage is significantly smaller when spherical deflectors are used as compared to cylinder deflectors, see Figure 14.7.

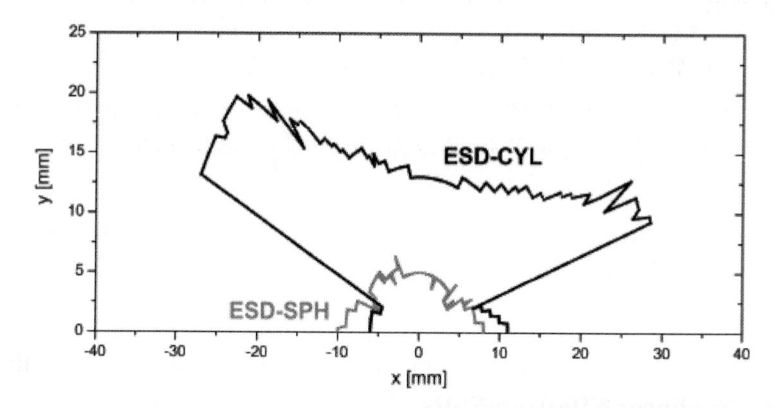

Fig. 14.7. Dynamic aperture of the electrostatic ring ELISA.[31] ESD-CYL denotes the area of stable oscillations in the ring when using cylinder deflectors. ESD-SPH denotes the area of stable beam motion when using spherical deflectors.

In the following we will briefly describe the main forces that drive low-energy ions to the periphery of the reduced ring acceptance.

14.7. Ion Kinetics and Long-term Beam Dynamics in Electrostatic Storage Rings

The transition processes and equilibrium conditions in ultralow-energy storage rings need to be studied to find the driving forces responsible for pushing ions out of the aperture of the ring, as well as for understanding the reason for reduced beam lifetime or intensity losses. This requires simulation of the evolution of the ion distribution function with time.[34,35]

Long-term beam dynamics processes are slow in comparison with the ion revolution period in a ring. Heating effects, such as the interaction of the circulating ions with the residual gas leading to multiple scattering of ions, interaction with different kinds of internal targets, multiple scattering of ions between each other, so-called intra-beam scattering (IBS), external heating of the ion beam, or different kinds of particle losses all lead to a variation of the ion distribution function in six-dimensional phase space. The beam area in phase space is increased due to these heating effects and the ions will eventually reach the limiting aperture and are lost. Heating processes might be counteracted by frictional forces, such as stochastic, electron or laser cooling.

The betatron motion of the ions is assumed to be stable and is usually treated in linear approximation. Nonlinear effects strongly restrict the amplitude of stable betatron oscillations in storage rings and are taken into account in ion kinetics studies by the proper choice of the ring acceptance.[33,34]

14.7.1. *Kinetic equations*

Studies of ion kinetics in storage rings are based on the application of stochastic (kinetic) differential equations to the periodic motion with an assumption that the incidence of diffusion, i.e. heating processes, as well as friction, i.e. cooling processes, act on the beam. It is assumed that the forces resulting from these processes are small as compared to the forces caused by external bending and focusing ion optical elements.[34,35]

Instead of describing each particle, the probability or distribution function of the beam is used. The evolution of the ion distribution function is described by the Fokker–Plank equation:[36]

$$\frac{\partial f}{\partial t} = -\sum_{i=1}^{N} \frac{\partial}{\partial x_i}\left[D_i^1(x_1,\ldots,x_N)f\right] + \sum_{i=1}^{N}\sum_{j=1}^{N} \frac{\partial^2}{\partial x_i \partial x_j}\left[D_{ij}^2(x_1,\ldots,x_N)f\right], \quad (14.31)$$

where f is the probability density of the ion distribution function in the ring, D^1 is the drift vector and D^2 is the diffusion tensor (the latter is a result of the presence of stochastic forces). The different friction and diffusion terms depend on the actual distribution function. The main methods employed in solving the stochastic differential equations are based on the transformation of the initial kinetic equation to the form of the Fokker–Planck equation and finding a solution for the density distribution probability function. In some cases, when the effects acting on the distribution function do not lead to a change in its shape, the Fokker–Plank equation can be reduced to an equation for the second-order moments of the distribution function or a Langevin equation in invariant or momentum space.

The second momentum of the ion coordinate $y = R - R_0$ is a mean value defined by the equation:

$$\langle y^2 \rangle = \iint y^2 f(y, y') dy dy', \tag{14.32}$$

and the quantity $y_{rms} = \sqrt{\langle y^2 \rangle}$ is called the root mean square (RMS) ion coordinate.

Assuming a Gaussian distribution in both position (y) and angle (y') coordinates, where $(y),(y')$ can describe either the horizontal or vertical position/angle, the two-dimensional probability distribution function $\rho(y, y')$ describing a projection of six-dimensional phase space on the two-dimensional phase plane (y, y') , can be written as

$$\rho(y, y') = \frac{1}{2\pi\varepsilon_{rms}} \exp\left(-\frac{\varepsilon}{2\varepsilon_{rms}} \right) . \tag{14.33}$$

If the initial beam distribution is Gaussian then the shape of the distribution function does not change with time. The RMS beam distribution can be approximated by an equivalent ellipse described by the so-called Twiss parameters α , β and γ. The equivalent ellipse with an area $\pi\varepsilon$ as defined by the Courant–Snyder invariant,

$$0.5I = \varepsilon = \gamma y^2 + 2\alpha y y' + \beta y'^2 , \tag{14.34}$$

and is a curve of constant density $\rho(y, y')$ in phase space enclosing a certain percentage of the ions in a beam. The RMS emittance ε_{rms} is given by mean square values (second moments) of a normal (Gaussian) beam distribution:

$$\varepsilon_{rms} = \sqrt{\langle y^2\rangle\langle y'^2\rangle - \langle yy'\rangle^2} = \sigma_y \sigma'_y \sqrt{1-r^2}\ , \tag{14.35}$$

where $\sigma_y = \sqrt{\langle y^2\rangle}$ is the RMS beam width and $\sigma_{y'} = \sqrt{\langle y'^2\rangle}$ is the RMS beam angular width. The correlation term $r = \sigma_{yy'} = \langle yy'\rangle$ vanishes in a focal plane where the phase space ellipse is "upright" and the tilt angle is zero. The RMS emittance can be estimated as:

$$0\,\varepsilon_{rms} = \sigma_y^2 / \beta = \sigma_{y'}^2 / \gamma\ . \tag{14.36}$$

The normalized emittance ε_y^{norm} is proportional to the percentage of beam enclosed by the phase space ellipse and approximated by:

$$\varepsilon_y^{norm} = \beta\gamma n_\sigma \varepsilon_y^{rms}\ , \tag{14.37}$$

where $n_\sigma = -2\ln(1-\alpha/100)$ is the number of standard deviations σ calculated via a given percentage of the beam intensity α[%] for a Gaussian distribution. $n_\sigma = 1$, $n_\sigma = 4$ and $n_\sigma = 5.991$ correspond to 39.3%, 86% and 95% of the total beam intensity respectively.

RMS dynamics calculate the evolution of the RMS parameters of the ion distribution function during motion in a storage ring under the action of heating and cooling effects. It is based on the solution of a system of four differential equations:

$$\frac{d\varepsilon_x}{dt} = \frac{\varepsilon_x}{\tau_{hor}};\quad \frac{d\varepsilon_y}{dt} = \frac{\varepsilon_y}{\tau_{vert}};\quad \frac{d\varepsilon_{long}}{dt} = \frac{\varepsilon_{long}}{\tau_{long}};\quad \frac{dN}{dt} = \frac{N}{\tau_{life}}, \tag{14.38}$$

where ε_x and ε_y are $\varepsilon_{long} = \sigma_{long}^2 = \left(\delta p^{rms}/p\right)^2$ are the RMS emittances in the horizontal, vertical and longitudinal phase spaces and N is the number of particles. Characteristic times τ_i of the evolution of beam parameters are usually estimated assuming a Gaussian shape of the distribution function:

$$\frac{1}{\tau_i} = \frac{\varepsilon_{fin} - \varepsilon_{in}}{\varepsilon_{in} T_{rev}}, \tag{14.39}$$

where T_{rev} is the revolution time. The rates $1/\tau_i$ are positive for heating processes and negative for beam cooling. Beam lifetime is inversely proportional to the average value of ion loss probability $1/\tau_{life} = -\langle P_{loss}\rangle / T_{rev}$.

14.7.2. *Multiple scattering of ions*

In ESRs, beam emittance growth is caused mainly by small angle multiple Coulomb scattering of the circulating ions on the atoms and molecules of the residual gas, as well as by IBS if the beam intensity is high. If an internal target is installed in a ring then circulating ions are often lost since the projectile beam is scattered on the atoms of the target.

14.7.2.1. *Rate of RMS emittance growth*

Here we briefly outline the effects of an internal thin target, based on the formalism proposed in reference[37] in order to simulate the interaction of the beam with the target as well as vacuum losses. Assuming a Gaussian distribution in both position (y) and angle (y') coordinates, the two-dimensional probability distribution function $\rho(y, y')$ is described by Eq. (14.33). The gas jet target is treated as the equivalent of a thin lens in ion optics.[37] After crossing the target, the ion does not alter the coordinate but experiences angular scattering, i.e. all three components of particle momentum are changed and it is most likely the particle will be lost. A change in transverse momentum is mainly related to a multiple Coulomb scattering of ions on the nuclei of target atoms. The longitudinal momentum of the ion is reduced as a result of the ionization and the excitation energy losses which occur during the interaction of the ion with the electron shell of the target atom. Fluctuation of ionization energy losses leads to the growth of the beam momentum spread.

A collision of the incident ion with the thin target atom produces a small angle kick $\Delta y'$ which is of a stochastic nature. Introducing the mean square scattering angle per target traversal, θ_{rms}^2, one can describe the scattering heating rate in the horizontal or vertical transverse plane by the formula:

$$\frac{1}{\tau_{h,v}} = \frac{1}{2}\frac{\beta_{h,v}}{\varepsilon_{h,v}}\frac{\theta_{rms}^2}{T_{rev}}. \tag{14.40}$$

In the presence of dispersion D and the derivative of dispersion D', the emittance growth after N target traversals depends not only on the RMS scattering angle. Due to the presence of dispersion in a ring the emittance growth also depends on the momentum spread δ_{rms}^2 introduced by fluctuations of the ionization and excitation energy losses:

$$\Delta\varepsilon_{h,v} = \varepsilon - \varepsilon_0 = N\frac{\beta_{h,v}\theta_{rms}^2}{2} + \frac{N}{2}\left(\frac{(1+\alpha_{h,v})^2}{\beta_{h,v}}D_{h,v}^2 + 2\alpha_{h,v}D_{h,v}D'_{h,v} + \beta_{h,v}D'^{2}_{h,v}\right)\delta_{rms}^2, \quad (14.41)$$

where α, β and γ are the Twiss parameters. The mean square relative momentum deviation per target traversal is $\delta_{rms}^2 = (\delta p/p)^2$ for a coasting beam.

14.7.2.2. *Mean square scattering angle*

Small angle Coulomb scattering is well studied at intermediate and high energies and can be understood on the basis of Rutherford scattering.[38] An expression for the mean square scattering angle of one scattering event is based on Molière's formula for the screened ion-atom potential[39] of the Thomas–Fermi type,[40] with Fano's contribution of ion inelastic collisions with atomic electrons:[41]

$$\theta_{str}^2 = 2\pi\rho x\left(\frac{Z_T Z_i r_p}{A_i\beta^2\gamma}\right)^2\left[\ln\left(\frac{\alpha_2^2}{\chi^2}\right) - 1 + \Delta b\right]. \quad (14.42)$$

Here ρ and x are the target density and thickness respectively, whereas $\rho\cdot x$ is the linear target thickness, i.e. the number of target atoms per unit area. A_T, A_i, Z_T and Z_i are the mass number and charge of the target and projectile ion, β and γ are relativistic factors and r_p is the classical proton radius. The parameter α_2 is the upper integration constant of the RMS scattering angle. The screening angle χ takes into account deviation from the Born approximation.[42] The inelastic collision factor with atomic electrons Δb was introduced by Fano[43] and includes a constant determined by the electron configuration of the target atom which is estimated from the Thomas–Fermi model.

Assuming a Gaussian distribution of the processes one can calculate a particle's longitudinal and transverse momentum variations after a single crossing of the target:

$$x'_f = x'_0 + \sqrt{\frac{\theta_{str}^2}{2}}\times\xi_1 \qquad y'_f = y'_0 + \sqrt{\frac{\theta_{str}^2}{2}}\times\xi_2, \quad (14.43)$$

where ξ_1 and ξ_2 are independent random values of the scattering angle with a Gaussian distribution at unit standard deviation.

14.7.2.3. *Ionization energy losses*

When a beam of ions crosses a target, it loses energy due to excitation and ionization of the target electrons. The expected mean energy loss ΔE_{BB} for one target traversal can be approximated using a simplified version of the Bethe–Bloch equation:[44]

$$\Delta E_{BB} = 2\xi\left[\ln\frac{E_{\max}}{I} - \beta^2\right]. \tag{14.44}$$

The parameter ξ is proportional to the linear target thickness $\rho \cdot x$ and inversely proportional to the beam energy β^{-2}:

$$\xi = 0.1535\left[\frac{MeV \cdot cm^2}{g}\right]\frac{Z_P^2}{\beta^2}\frac{Z_T}{A_T}\rho x. \tag{14.45}$$

Here, Z_P and Z_T are the charge number of projectile and target atoms and A_T is the target atomic number. The mean ionization energy I is 13.6 eV for hydrogen and $I \sim 16 \cdot Z_T^{0.9}$ for heavier atoms.

The maximum transferable energy during impact $E_{\max}$ is determined by kinematic parameters:

$$E_{\max} = \frac{2m_e c^2 \beta^2 \gamma^2}{1 + 2\gamma\dfrac{m_e}{M_i} + \left(\dfrac{m_e}{M_i}\right)^2}, \tag{14.46}$$

where m_e and M_i are electron and projectile ion mass, respectively and β and γ are the relativistic factors. Fluctuations in energy loss can be estimated by a square of the standard deviation of the ion distribution function in energy space. The mean energy loss leads to a deceleration of the beam while fluctuations cause a growth in ion momentum spread.

14.7.2.4. *Multiple scattering on residual gas*

Growth rates due to scattering of ions on the residual gas are calculated in the same way as for an internal gas target. The residual gas model is then composed of gas cell targets which are distributed along the entire circumference of the ring. Residual gas heating rates are integrated over the entire lattice structure using the lattice functions of each ion optic element. The effective partial densities of the residual gas components are calculated on the basis of pressure and gas composition. Energy losses are estimated with the Bethe–Bloch formula. Characteristic growth rates $1/\tau_i$ are calculated on the basis of the actual residual gas composition.

14.7.3. *Particle loss probability*

The ion loss probability after crossing the target is proportional to the linear target density $P_{loss} = \sigma_{total} \cdot \rho \cdot x$. The total cross-section of ion losses is the sum of the cross-sections of the different processes leading to particle losses. Amongst the most important ones are large angles single scattering σ_{ss}, electron capture or charge-exchange σ_{ec}, and nuclear reactions in the target at high energies σ_{nr}.

The cross-section of single scattering at angles larger than the acceptance angle θ_{acc} is calculated by the formula:

$$\sigma_{ss} = 4\pi \left(\frac{Z_T Z_p r_p}{A \gamma \beta^2} \right)^2 \frac{1}{\theta_{acc}^2}, \tag{14.47}$$

where r_p is the classic proton radius. Similar expressions are used to estimate the beam losses due to the multiple scattering of ions on the residual gas.

14.7.4. *Intra-beam scattering*

IBS refers to small angle Coulomb scattering of the ions in a beam. Here, we are interested in IBS effects in circular accelerators. IBS is essentially a diffusion process and leads to the relaxation of ions at thermal equilibrium and to diffusion growth of beam volume in six-dimensional phase space due to a variation of the lattice parameters along the ring circumference, as well as the presence of dispersion in the ring and coupling between longitudinal and transverse motion. One can find the algorithm used to describe IBS in a storage ring in the BETACOOL physics guide[35] and here we will just discuss some basic principles.

Analytical calculations of IBS growth rates are based on an estimation of the particle momentum variation caused by Coulomb interaction with other ions of a beam. IBS growth rates are averaged over all elements of a ring. There are several different approaches to describe IBS, e.g. the Piwinski,[45] Martini,[46] Bjorken–Mtingwa,[47] and the Jie–Wei[48] models.

Calculations of IBS growth rates are typically based on an analytical model of the collisions first proposed by A. Piwinski.[49] The relative change of particle momentum after the collision of two ions leads to a corresponding change of the Courant–Snyder invariant $I = 2\varepsilon$, see Eq. (14.34). The variation of the invariant δI is calculated under the assumption that the position of an ion is not affected during the interaction of the ions. Also, the variation of the longitudinal component of particle momentum $\delta p_{//}/p$ leads to a change of the transverse motion invariant δI_x in the horizontal plane in parts of a storage ring where the dispersion function D and/or the first derivative of the dispersion function D' are not equal to zero.

The resulting variation of beam phase space volume can be calculated by averaging the particle invariant change δI over all collisions in accordance with the particle density distribution function $\overline{P}$. The latter is a product of the independent probability functions of two interacting particles. Note that the bar indicates values in the center-of-mass (CM) reference frame.

The time derivative of the average radial emittance for all particles, denoted $\langle\varepsilon\rangle$, can be expressed in the CM reference system by the formula:

$$\left\langle \frac{d}{d\overline{t}} \frac{\langle\varepsilon\rangle}{\beta_x} \right\rangle = \int 2c\overline{\beta} \cdot \overline{P} \int_{\Psi_m}^{\pi} d\overline{\Psi} \int_{0}^{2\pi} d\overline{\phi} \, \frac{d\overline{\sigma}}{d\overline{\Omega}} \frac{\delta I_1}{\beta_x} \sin \overline{\Psi} d\overline{\tau}\,, \tag{14.48}$$

where the outer brackets $\langle ... \rangle$ indicate a value averaged around the whole ring circumference. The first integral extends over all phase space betatron coordinates, momentum spread values and the azimuth location of the two interacting particles. The $d\overline{\sigma}/d\overline{\Omega}$ is the Rutherford cross-section in the CM reference frame for the scattering into the element of the solid angle at the given axial $\overline{\Psi}$ and azimuth $\overline{\phi}$ scattering angles.

$\bar{\beta}c$ is the particle velocity in the CM coordinate system under the assumption $\bar{\beta} << 1$. Finally, the integration variable $d\bar{\tau}$ is the infinitesimal element of phase space volume and $\bar{\Psi}_m$ is the smallest scattering angle defined by the impact parameter $\bar{b}$:

$$tg\bar{\Psi}_m = \frac{r_i}{2\bar{\beta}^2\bar{b}}, \tag{14.49}$$

where r_i is the classic ion radius. Formula (14.48), which describes the rate of emittance growth due to IBS is derived without any assumption about the particle distribution.

14.8. Benchmarking of Experiments

Systematic studies in the ELISA storage ring showed strong limitations in maximum beam current and reduced beam lifetime at higher beam intensities. A fast intensity decay was measured for negative oxygen ions stored at 22 keV and experimental data is shown in Figure 14.8(a).[30] It can be seen that the ion losses in this ESR depend on the beam intensity, in particular for the ring configuration using spherical deflectors. The reason for these losses had not been fully understood for a long time and required detailed numerical studies.

Nonlinear and long-term beam dynamics and ion kinetics in ELISA have been benchmarked[31-33] against experimental data[10,30] using the kinetic theory described in references[34,35]. Transition processes, such as growth rates of beam emittance and momentum spread, beam intensity decay, and equilibrium conditions have been estimated by simulation of RMS parameters of ion distribution function over the period of time. Only heating processes were considered in these simulations.

The measured rates of beam intensity decay of a 22 keV beam of O^- ions as indicated in Figure 14.8(a) have been reproduced in simulations with BETACOOL with very good accuracy, see Figure 14.8(b). The main reasons for beam size growth were identified as multiple scattering of ions on the atoms and molecules of the residual gas, i.e. vacuum losses, as well as Coulomb repulsion of the ions from each other at higher intensities, i.e. IBS. As a consequence of fast beam growth, the ions are lost on the ring aperture because of a rather small ring acceptance.[31] The rate of beam losses increases at higher intensities because IBS adds to vacuum losses. The initial shape of the decay curve depends on the emittance, the momentum spread and the intensity of the injected beam.

The kinetics of negative oxygen ions has been studied for different initial intensities, ranging from $5{\cdot}10^5$ up to $1.6{\cdot}10^7$ particles.

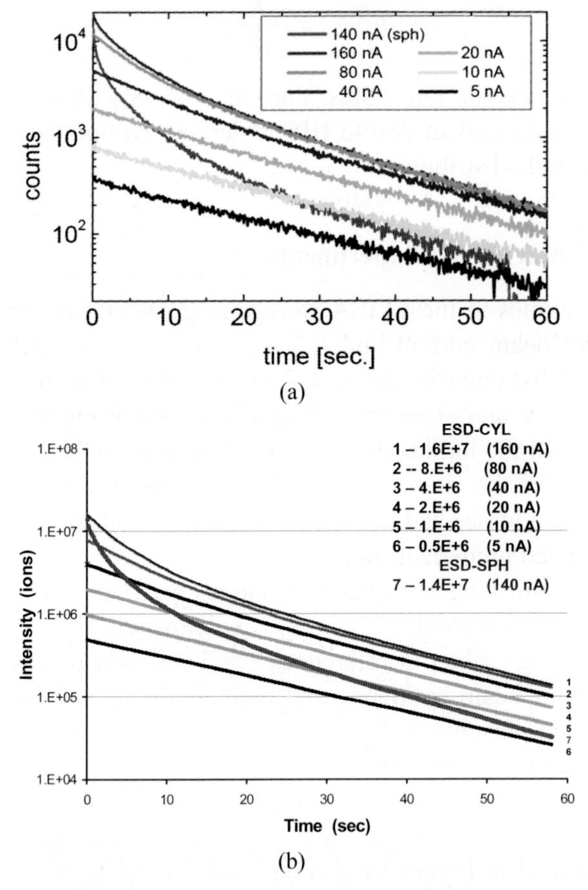

Fig. 14.8. Decay of stored negative oxygen O^- ions in the ELISA at 22 keV beam energy: a) Experimental data taken from reference [30], beam lifetime τ =12 s; (b) BETACOOL simulations.[32,33] The pink curve shows a significant drop in beam intensity when using spherical electrodes. All other curves relate to measurements with cylinder electrodes.

The IBS effect can clearly be seen in Figures 14.8(a) and 14.8(b) to be an excessive drop of beam current during the first few seconds when the beam intensity is still high. The long-term slope of this loss curve is determined by the ring acceptance and by the rate of multiple scattering on the residual gas, i.e. the pressure level in the ring. The slope of the decay curve also depends on the lifetime due to electron detachment for negative ions or electron stripping for positive ions. At very low beam intensities, when IBS is negligible and only multiple scattering on the residual gas is present, the loss rate does not depend any more on beam current or details of the ring lattice. It is for this reason that all graphs in Figure 14.8 are almost parallel towards the end of the cycle. The acceptance of an ELISA using cylinder deflectors was estimated to be $A_{cyl} \approx 10\pi$ mm·mrad, see Figure 14.8(b).

An exception can be seen in the pink curve, which shows a significantly faster decay. This has been measured in a ring configuration using spherical deflectors. In this case significantly higher loss rates are caused by a reduced ring acceptance. The ELISA ring acceptance is mainly restricted by the sextupole component of the electric field of its ESDs. The long range slope of the decay curve depends on the ring acceptance and supports the assumption that the ring acceptance is much smaller using spherical deflectors. In this case the acceptance was estimated to be $A_{sph} \approx 6\pi$ mm·mrad, see Figure 14.8(b).

14.9. Conclusion and Outlook

In this chapter we first discussed the field distribution in different electrostatic ion optic elements. We then derived the equations of motion of charged particles travelling through these fields. We then explained the processes that contribute to beam heating and cooling and how a model can be developed that allows us to understand beam stability, lifetime and losses. Finally, we showed how numerical simulations that take into account all these processes, as well as the ring lattice, can reproduce experimental data with excellent accuracy. Such models can then be used to make predictions about beam lifetime and maximum intensities in future storage rings and are a very powerful tool to optimize future facilities.

Acknowledgments

The generous support of HGF and GSI under Contract VH-NG-328 and STFC under the Cockcroft Institute core grant is acknowledged. We would like to thank Yu. Senichev for providing material used in this chapter, as well as S. P. Møller, A. Dolinsky and H. Knudsen for their valuable comments and discussions.

References

1. P. Lefèvre and D. Möhl, A Low Energy Accumulation Ring of Ions for LHC, *CERN/PS* **93–62**(DI), 259 (1993).
2. S. P. Moller, The Aarhus Storage Ring for Ions and Electrons ASTRID, *Proc. Part. Accel. Conf.*, Washington, USA, 1741–1743 (1993).
3. M. Plotkin, The Brookhaven Electron analogue, *BNL Report* **45058** 1953–1957 (1991).
4. D. Zajfman *et al.*, Physics with electrostatic rings and traps, *J. Phys. B: At. Mol. Opt. Phys.* **37.** 57–88 (2004).
5. P. Reinhed, *Ions in Cold Electrostatic Storage Devices,.* PhD Dissertation, Stockholm University (2008).
6. C.P. Welsch and J. Ullrich, FLAIR – A Facility for Low-Energy Antiproton and Ion Research at GSI, *Hyperfine Interactions*, **172** (1–3) 71–80 (2007).
7. D. Zaifman and J. Ullrich, Physics with colder molecular ions: The Heidelberg Cryogenic Storage Ring CSR, *J. Phys..Sixth Int. Conf. on Dissoc. Recombination* **4** 206–215 (2005).
8. R. Hayano *et al.*, ASACUSA Status Report-2010, *CERN-SPSC*-2010-005/*SPSC-SR*-056 (2010).
9. S. P. Moller, Design and first operation of the electro-static storage ring ELISA, *Proc. Europ. Part. Accel. Conf.*, Stockholm, Sweden, 73–77 (1998).
10. S. P. Moller and U. Pedersen, Operational experience with the electrostatic Storage ring ELISA, *Proc. Part. Accel. Conf.*, New York, USA, 2295–2297 (1999).
11. T. Tanabe and K. Noda, Storage of bio-molecular ions in the electrostatic storage ring, *Nucl. Instr. Meth. Phys. Res.* **A496,** 105–110 (2003).
12. T. Azuma, TMU electrostatic ion storage ring designed for operation at liquid nitrogen temperature, *Nucl. Instr. Meth. Phys. Res.* **A532**, 477–482 (2004).
13. K. E. Stiebing, FLSR – The Frankfurt low energy storage ring, *Nucl. Instr. Meth. Phys. Res.* **A614**, 10–16 (2010).
14. P. Löfgren *et al.*, Status of the Electrostatic and Cryogenic Double ring DESIREE, *Proc. Europ. Part. Accel. Conf.* Genoa, Italy, 331–333 (2008).
15. J. Ullrich *et al.*, Next-Generation Low-energy Storage Rings for Antiprotons, Molecules, and Atomic Ions in Extreme Charge States, *Proposal to the Max Planck Society* (2004).
16. R. von Hahn *et al.*, CSR – a Cryogenic Storage Ring at MPI-K, *Proc. Europ. Part. Accel. Conf.* Lucerne, Switzerland, 1237–1239 (2004).
17. H. Fatil *et al.*, Finite elements Calculations of the Lattice and Ring Acceptance of the Heidelberg CSR, *Proc. Europ. Part. Accel. Conf.* Edinburgh, Scotland, 1960–1962 (2006).
18. C. P. Welsch *et al.*, Ultra-low energy storage ring at FLAIR, *Hyperfine Interact.* DOI 10.1007/s10751-011-0460-z, 11–16 (2011).

19. A. I. Papash and C. P. Welsch, An update of the USR lattice: towards a true multi-user experimental facility, *Proc. Part. Accel. Conf.* Canada, 4335–4337 (2009).
20. A. I. Papash and C. P. Welsch, Realization of nanosecond antiproton pulses in the ultra-low energy storage ring, *Nucl. Inst. Meth. Phys. Res.* **A620,** 128–141 (2010).
21. C. P. Welsch *et al.*, An ultra-low-energy storage ring at FLAIR, *Nucl. Inst. Meth. Phys. Res.* **A546** 405–417 (2005).
22. M. R. F. Siggel-King *et al.*, Electrostatic Low-EnergyAntiproton Recycling Ring, *Hyperfine Interactions* **199** (1), 311–319 (2011).
23. ELENA Proposal to the CERN SPS, http://cdsweb.cern.ch/record/1206242/files/SPSC-P-338.pdf.
24. Yu. Senichev, Beam Dynamics in Electrostatic Rings, *Proc.Europ. Part. Accel. Conf.* Vienna, Austria, 794–796 (2000).
25. Yu. Senichev, The features of beam dynamics in the electrostatic storage rings, *FZJ IKP report,* Juelich, 23 (2000).
26. http://cosyinfinity.org
27. S. R. Mane, Orbital dynamics in a storage ring with electrostatic bending, *Nucl. Instr. Meth. Phys. Res.* **A596,** 288–294 (2008).
28. F. Hinterberger, Ion optics with electrostatic lenses, *CERN School on Small Accelerators* Zeegse, The Netherlands, 27–44 (2005).
29. P. Bryant, Transverse motion and electrostatic elements, *Joint Univ. Accel. School* Archamps, France, 22 (2008).
30. S. P. Moller *et al.*, Intensity Limitations of the Electrostatic Storage Ring ELISA, *Proc. Europ. Part. Accel. Conf.* Vienna, Austria, 788–790 (2000).
31. A. I. Papash and C. P. Welsch, Simulations of space charge effects in low energy electro-static storage rings, *Proc. Inter. Part. Accel. Conf.* Kyoto, Japan, 1952–1954 (2010).
32. A. I. Papash, A. V. Smirnov and C. P. Welsch, Long term beam dynamics in ultra-low energy storage rings, *Proc. Inter. Part. Accel. Conf. San-Sebastian* Spain, 2166– 2168 (2011).
33. A. V. Smirnov, A. I. Papash and C. P. Welsch, Nonlinear and long-term beam dynamics in low energy storage rings, *Phys. Rev. ST Accel. Beams* **16**, 060101 (2013)
34. A. O. Sidorin, I. N. Meshkov *et al.*, BETACOOL program for simulation of beam dynamics in storage rings, *Nucl. Instr. Meth. Phys. Res.* **A558,** 325–328 (2006).
35. A. V. Smirnov *et al.*, BETACOOL Physics Guide, http://betacool.jinr.ru (2007).
36. H. Risken, *The Fokker–Planck Equation. Methods of Solution and Applications*, Springer series edition, 472 (1996).
37. F. Hinterberger and D. Prasuhn, Analysis of Internal Target Effects in Light Ion Storage Rings, *Nucl. Instr. Meth. Phys. Res.* **A279**, 413–422 (1989).
38. E. Rutherford, The Scattering of α and β Particles by Matter and the Structure of the Atom, *Philos. Mag.* Series 6, Vol. 21, Issue 125, 669–688 (1911).
39. G. Molière, Theory of Scattering of Fast Charged Particles. I. Single Scattering in a Screened Coulomb Field, *Zeitschrift Naturforschung*, **2a**, 133–149 (1947).
40. L. H. Thomas, The calculation of atomic fields, *Math. Proc. Cambridge Phil. Soc.* **23** (5): 542–548 (1927).

41. U. Fano, Inelastic collisions and the Molière theory of multiple scattering, *Phys. Rev.* **93**, 117–120 (1954).
42. G. Molière, Theorie der Streuung schneller geladener Teilchen II. Mehrfach- und Vielfachstreuung, *Zeitschrift Naturforschung*, **3a**, 78–96 (1948).
43. U. Fano, Penetration of protons, alpha particles and Mesons, *Ann. Rev. Nucl. Sci.* **13**, 1–66 (1963).
44. H. A. Bethe, Molière's theory of multiple scattering, *Phys. Rev.* **89**(6), 1256–1266 (1953).
45. A. Piwinski, Intrabeam scattering, *Proc. 9th Int. Conf. on High Energy Accelerators* SLAC, USA, 405–410 (1974).
46. M. Martini and M. Conte, Intra-beam scattering in the CERN Antiproton Accumulator, *Proc. Part. Accel. Conf.* **17**, 1–10 (1985).
47. D. Bjorken and S. K. Mtingwa, Intrabeam scattering, *Proc. Part. Accel. Conf.* **13**, 115–120 (1983).
48. J. Wei, Evolution of Hadron Beams Under Intra-beam Scattering, *Proc. Part. Accel. Conf.* Washington, USA, 3651–3654 (1993).
49. A. Piwinski, Beam losses and lifetime, *Proc. CERN Accelerator School* Gifsur-Yvette, Paris, 432–462 (1984).

Index